Computational Methods
for Process Simulation

Computational Methods for Process Simulation

Second edition

W. Fred Ramirez
Professor of Chemical Engineering
University of Colorado
Boulder, Colorado

Butterworth-Heinemann
Linacre House, Jordan Hill, Oxford OX2 8DP
A division of Reed Educational and Professional Publishing Ltd

A member of the Reed Elsevier plc group

OXFORD BOSTON JOHANNESBURG
MELBOURNE NEW DELHI SINGAPORE

First published 1989
Second edition 1997

© Reed Educational and Professional Publishing Ltd 1989, 1997

All rights reserved. No part of this publication
may be reproduced in any material form (including
photocopying or storing in any medium by electronic
means and whether or not transiently or incidentally
to some other use of this publication) without the
written permission of the copyright holder except
in accordance with the provisions of the Copyright,
Designs and Patents Act 1988 or under the terms of a
licence issued by the Copyright Licensing Agency Ltd,
90 Tottenham Court Road, London, England W1P 9HE.
Applications for the copyright holder's written permission
to reproduce any part of this publication should be addressed
to the publishers

British Library Cataloguing in Publication Data
A catalogue record for this book is available from the British Library

ISBN 0 7506 3541 X

Library of Congress Cataloguing in Publication Data
A catalogue record for this book is available from the Library of Congress

Printed and bound by Antony Rowe Ltd, Eastbourne

CONTENTS

Preface 1
Acknowledgments 3
Introduction 5
   Definition of the Problem 6
   Mathematical Modeling of the Process 7
   Equation Organization 7
   Computation 7
   Interpretation of Results 8
   Limitations of Process Simulation 8
   Usefulness of Process Simulation 8
   Reference... 9

Chapter 1: Development of Macroscopic Mass, Energy, and Momentum Balances 11
   1.1 Conservation of Total Mass 12
         1.1.1 Tapered Tube Geometry 13
   1.2 Conservation of Component i 14
   1.3 Method of Working Problems 14
   1.4 Conservation of Total Energy 19
         1.4.1 Tapered Tube Geometry 22
   1.5 Method of Working Problems 23
   1.6 Mechanical Energy Balance 27
         1.6.1 Tapered Tube Geometry 29
   1.7 Conservation of Momentum 32
         1.7.1 Tapered Tube Geometry 34
         1.7.2 Comparison Between Mechanical Energy and
               Momentum Balances 34
   Problems 38
   References 43

Chapter 2: Steady–State Lumped Systems 45
   2.1 Methods 46
         2.1.1 Partitioning Equations 46
         2.1.2 Tearing Equations..................................... 47
         2.1.3 Simultaneous Solution................................. 47
   2.2 Simultaneous Solution of Linear Equations...................... 47
         2.2.1 MATLAB Software 53
         2.2.2 Linear Algebra Routines in MATLAB.................... 54
         2.2.3 Other Matrix Capabilities in MATLAB 64
               2.2.3.1 Singular Value Decomposition........ 64
               2.2.3.2 The Pseudo–Inverse.................. 65
               2.2.3.3 Sparse Matrices..................... 66
   2.3 Solution of Nonlinear Equations 67
         2.3.1 Solving a Single Nonlinear Equation in One
               Unknown 67
               2.3.1.1 Half–Interval (Bisection) 68

     2.3.1.2 Linear Inverse Interpolation
        (Regula Falsi) 70
     2.3.1.3 Direct Substitution 73
     2.3.1.4 Wegstein Method 75
     2.3.1.5 Newton Method 80
  2.3.2 Simultaneous Solution of Nonlinear Algebraic
     Equations 82
2.4 Structural Analysis and Solution of Systems of
  Algebraic Equations 87
  2.4.1 The Functionality Matrix 88
  2.4.2 An Optimal Solution Strategy 92
  2.4.3 Simple Example of Structural Analysis 96
  2.4.4 Computer Implementation of Structural Analysis . 99
  2.4.5 Mixer–Exchanger–Mixer Design 101
     2.4.5.1 Nomenclature for Mixer–Exchanger–
        Mixer Design 106
Problems 111
References 122

Chapter 3: Unsteady–State Lumped Systems 125
3.1 Single Step Algorithms for Numerical Integration 125
  3.1.1 Euler Method 125
  3.1.2 Runge–Kutta Methods 129
  3.1.3 MATLAB Runge–Kutta Routines 132
3.2 Basic Stirred Tank Modeling 135
3.3 Multistep Methods 141
3.4 Stirred Tanks with Flow Rates a Function of Level 144
3.5 Enclosed Tank Vessel 152
3.6 Stirred Tank with Heating Jacket 156
3.7 Energy Balances with Variable Properties 158
3.8 Tanks with Multicomponent Feeds 160
3.9 Stiff Differential Equations 162
3.10 Catalytic Fluidized Beds 164
Problems 167
References 176

Chapter 4: Reaction–Kinetic Systems 177
4.1 Chlorination of Benzene 177
  4.1.1 Order of Magnitude Analysis for Chlorination
     of Benzene................................... 179
4.2 Autocatalytic Reactions 182
4.3 Temperature Effects in Stirred Tank Reactors 184
  4.3.1 Mathematical Modeling of a Laboratory
     Stirred Tank Reactor 189
     4.3.1.1 Experimental 189
     4.3.1.2 Modeling 194
  4.3.2 Dynamics of Batch Fermentation 198
Problems 209

Contents   vii

References 216
Chapter 5: Vapor–Liquid Equilibrium Operations 217
  5.1 Boiling in an Open Vessel 217
  5.2 Boiling in a Jacketed Vessel (Boiler) 218
  5.3 Multicomponent Boiling—Vapor–Liquid Equilibrium 227
  5.4 Batch Distillation 229
  5.5 Binary Distillation Columns 231
        5.5.1 A Tray 233
        5.5.2 The Reboiler 234
        5.5.3 The Condenser 235
  5.6 Multicomponent Distillation Columns 235
  Problems 255
  References 256
Chapter 6: Microscopic Balances 257
  6.1 Conservation of Total Mass (Equation of Continuity) 257
  6.2 Conservation of Component i 259
  6.3 Dispersion Description 260
  6.4 Method of Working Problems 263
  6.5 Stagnant Film Diffusion 263
  6.6 Conservation of Momentum (Equation of Motion) 264
  6.7 Dispersion Description 266
  6.8 Pipe Flow of a Newtonian Fluid 266
  6.9 Development of Microscopic Mechanical Energy
       Equation and Its Application 272
  6.10 Pipeline Gas Flow 274
  6.11 Development of Microscopic Thermal Energy Balance
       and Its Application 275
  6.12 Heat Conduction Through Composite Cylindrical Walls ... 277
  6.13 Heat Conduction with Chemical Heat Source 282
  6.14 Mathematical Modeling for a Styrene Monomer
       Tubular Reactor 283
        6.14.1 Gas Phase Energy Balance 286
        6.14.2 Catalyst Bed Energy Balance 290
        6.14.3 Equation of Motion 291
        6.14.4 Material Balances 292
        6.14.5 Steady-State Model Solution 292
  Problems 297
  References 303
Chapter 7: Solution of Split Boundary–Value Problems 305
  7.1 Digital Implementation of Shooting Techniques:
      Tubular Reactor with Dispersion 305
  7.2 A Generalized Shooting Technique 311
  7.3 Superposition Principle and Linear Boundary–Value
      Problems 316
  7.4 Superposition Principle: Radial Temperature Gradients
      in an Annular Chemical Reactor 320

7.5 Quasilinearization 322
7.6 Nonlinear Tubular Reactor with Dispersion:
    Quasilinearization Solution 327
7.7 The Method of Adjoints 330
7.8 Modeling of Packed Bed Superheaters 334
    7.8.1 Single–Phase Fluid Flow Energy Balance 336
    7.8.2 Two–Phase Fluid Flow Energy Balance 339
    7.8.3 Superheater Wall Energy Balance 340
    7.8.4 Endcap Model 341
    7.8.5 Boundary Conditions 343
    7.8.6 Solution Method 344
    7.8.7 Results 346
Problem 349
References 351

Chapter 8: Solution of Partial Differential Equations 353
8.1 Techniques for Convection Problems 353
8.2 Unsteady–State Steam Heat Exchanger: Explicit
    Centered–Difference Problem 355
8.3 Unsteady–State Countercurrent Heat Exchanger:
    Implicit Centered–Difference Problem 359
8.4 Techniques for Diffusive Problems 370
8.5 Unsteady–State Heat Conduction in a Rod 372
8.6 Techniques for Problems with Both Convective and
    Diffusion Effects: The State–Variable Formulation 374
8.7 Modeling of Miscible Flow of Surfactant in Porous Media . 378
8.8 Unsteady–State Response of a Nonlinear Tubular Reactor 382
8.9 Two–Phase Flow Through Porous Media 392
8.10 Two–Dimensional Flow Through Porous Media 400
8.11 Weighted Residuals 408
    8.11.1 One–Dimensional Heat Conduction 409
    8.11.2 Two–Dimensional Heat Conduction 412
    8.11.3 Finite Elements 413
8.12 Orthogonal Collocation 414
    8.12.1 Shifted Legendre Polynomials 414
    8.12.2 Heat Conduction in an Insulated Bar 416
    8.12.3 Jacobi Polynomials 418
    8.12.4 Diffusion in Spherical Coordinates 420
    8.12.5 Summary 423
Problems 423
References 430

Nomenclature 431

Appendix A: Analytical Solutions to Ordinary Differential
Equations 435
A.1 First–Order Equations 435
A.2 N^{th} Order Linear Differential Equations with
    Constant Coefficients 440

Reference 444
Appendix B: MATLAB Reference Tables 445
Index 455

PREFACE

The purpose of this book is to present a time domain approach to modern process control. The time domain approach has several advantages including the fact that process models are naturally developed through conservation laws and mechanistic phenomena in the time domain. This approach also allows for the formulation of precise performance objectives that can be extremized. There is a definite need in the process industries for improved control. New hardware and software tools now allow the control engineer to consider the implementation of more sophisticated control strategies that address critical and difficult process control problems. In general, it is necessary to incorporate process knowledge into the control design in order to improve process operation. Advanced control designs require more engineering analysis but can lead to significant improvements in process behavior and profitability.

The reader will notice that I have tried to include practical examples throughout the book in order to illustrate theoretical concepts. This approach also allows the reader to be aware of computational issues of implementation as well as the interpretation of the results of process testing.

Chapter 1 presents basic time domain system concepts that are needed to mathematically describe an advanced process control problem. The important concepts of observability and controllability are introduced. Observability is used in the design of the measurement system, and controllability is important for the specification of the control variables of the system. The software package MATLAB is introduced. It simplifies many of the control design calculations.

Chapter 2 treats the topic of steady-state optimization. Necessary conditions for extrema of functions are derived using variational principles. These steady-state optimization techniques are used for the determination of optimal setpoints for regulators used in supervisory computer control.

Chapter 3 gives the fundamental mathematical principles of the calculus of variations used for the optimization of dynamic systems. Classical results of the Euler equation for functional extrema and those of constrained optimization given by the Euler–Lagrange equation are developed.

Chapter 4 applies variational calculus to problems that include control variables as well as state variables. Optimal control strategies are developed that extremize precise performance criteria. Necessary conditions for optimization are shown to be conveniently expressed in terms of a mathematical function called the Hamiltonian. Pontryagin's maximum principle is developed for systems that have control constraints. Process applications of optimal control are presented.

Chapter 5 considers optimal regulator control problems. The Kalman linear quadratic regulator (LQR) problem is developed, and this optimal multivariable proportional controller is shown to be easily computable using the Riccati matrix differential equation. The regulator problem with unmeasurable load disturbances is shown to lead to an optimal multivariable proportional-integral feedback structure.

Chapter 6 develops model predictive control concepts. This structure allows for the inclusion of predictive feed-forward control into the optimal control problem. We consider design strategies for completely measurable disturbances as well as systems with both measurable and unmeasurable disturbances.

Chapter 7 discusses robust control. This allows for the inclusion of uncertainty of process parameters in the control design. The concept of robustness refers to the preservation of closed-loop stability under allowable variations in system parameters. General stability results and integrity results are given for the LQR problem.

Chapter 8 considers optimal control problems for systems that are either linear or nonlinear in the state variables but are linear in the controls. The solution of this class of problems leads to bang-bang control strategies. The existence of singular or intermediate control must also be investigated. Both time-optimal control and minimum integral square error problems are discussed.

Chapter 9 develops necessary conditions for optimality of discrete time problems. In implementing optimal control problems using digital computers, the control is usually kept constant over a period of time. Problems that were originally described by differential equations defined over a continuous time domain are transformed to problems that are described by a set of discrete algebraic equations. Necessary conditions for optimality are derived for this class of problems and are applied to several process control situations.

Chapter 10 discusses state and parameter identification. Using uncertainty concepts, an optimal estimate of the state for a linear system is obtained based upon available measurements. The result is the Kalman filter. The Kalman filter is extended for nonlinear systems and discrete-time models. Kalman filtering is also shown to be effective for the estimation of model parameters.

Chapter 11 presents the use of sequential least squares techniques for the recursive estimation of uncertain model parameters. There is a statistical advantage in taking this approach to model parameter identification over that of incorporating model parameter estimation directly into Kalman filtering.

Chapter 12 considers the combination of optimal control with state and parameter estimation. The separation principle is developed, which states that the design of a control problem with measurement and model uncertainty can be treated by first performing a Kalman filter estimate of the states and then developing the optimal control law based upon the estimated states. For linear regulator problems, the problem is known as the linear quadratic Gaussian (LQG) problem. The inclusion of model parameter identification results in adaptive control algorithms.

ACKNOWLEDGMENTS

This book is a result of the author's research and teaching career in the area of optimal process control and identification. I gratefully acknowledge the contributions of my research students to the development of many of the ideas contained in this book. I have been fortunate to have had a group of research students who have stimulated new and creative insights into process control. They deserve credit for many of the novel and important ideas found in this work.

Special thanks are due to Ellen Romig who did the technical word processing and layout of this book. Her talents and personal concern for this project are truly appreciated.

I also thank the University of Colorado for awarding me a faculty fellowship, which provided the time needed to prepare the final manuscript in the excellent academic environment of Cambridge University.

Finally, I want to thank my wife, Marion, who has been my personal inspiration for many years.

INTRODUCTION

Process modeling and computer simulation have proved to be extremely successful engineering tools for the design and optimization of physical, chemical, and biological processes. The use of simulation has expanded rapidly during the past three decades because of the availability of high–speed computers and computer workstations. In the chemical process industry, large, realistic nonlinear problems are now routinely being solved via computer simulation. Also, the recent trends toward personal computing and specialized, industrial software allow for the expanded use of computers in engineering practice. This means that virtually all engineering computations will shortly be computerized and engineers need to understand the principles behind available software and how to effectively use software to solve pertinent process engineering problems.

The increasing use of computer simulation techniques has broadened the usefulness of the scientific approach to engineering. Developing competency in process simulation requires that the engineer develop the following skills:

1. A sound understanding of engineering fundamentals: The engineer must be familiar with the physical system and its mechanisms in order to be able to intelligently simulate a real process and evaluate that simulation. The process cannot be viewed as a black box.

2. Modeling skills: The engineer has to be able to develop a set of mathematical relations which *adequately* describes the significant process behavior.

3. Computational skills: Rapid and inexpensive solutions to simulation problems must be obtained. The engineer must be capable of choosing and using the proper computational tool. For realistic problems, the tool of interest is usually a digital computer. The engineer must also be able to evaluate and use correctly available commercial software packages.

Since simulation relies upon a scientific rather than empirical approach to engineering, it has served to stimulate developments in interdisciplinary areas such as bioengineering and environmental engineering. Engineers have found that they have been able to make significant contributions to society through the successful simulations of biological and environmental systems. Future fruitful efforts should lie in the modeling of political and social systems. Chemical process simulations have investigated both the steady–state and dynamic behavior of processes.

The tremendous impact that simulation has had on the chemical process industry is due to the following benefits derived:

1. Economic desirability: For design purposes, it is usually cheaper to use simulation techniques incorporating fundamental laboratory data in the mathematical model than it is to build numerous different-sized pilot plants.

2. It is a convenient way to investigate the effects of system parameters and process disturbances upon operation. It is usually a lot easier to develop alternative operating approaches and evaluate these alternatives via a mathematical model than by experimental methods. In order to verify the simulation results some experiments are usually performed, but only the really critical ones are necessary.

3. Simulations are a reasonable way of extrapolating performance and scaling up processes. By incorporating fundamental mechanisms into process simulations, system performance can be predicted in new and different operating regions.

4. Understanding of the significant process behavior and mechanisms: By undertaking the rigors of mathematical modeling the engineer learns much about the process that is being simulated. In order to obtain a successful simulation, the significant process mechanisms must be quantitatively described. By solving the model, useful relations between the process and equipment variables are revealed and can be easily observed.

The general strategy for the simulation of complex processes follows a fairly well-defined path consisting of the commonsense steps given in the accompanying block diagram. Note that information travels in both directions, indicating the adaptive nature of the development of any successful simulation.

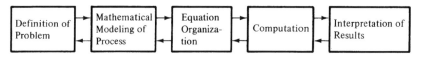

General Strategy of Process Simulation.

DEFINITION OF THE PROBLEM

This is a very important phase of a successful simulation but unfortunately there are very few precise general rules that apply. The real key to problem definition is an imaginative engineer. What is required is creative thought based upon sound engineering training. The engineer must spend sufficient time on this aspect of the problem before proceeding. A good problem definition comes from answering questions such as the following: What do I really want to find out? What are the important consequences of the study? Why should this job be done? What engineering effort should be required? How long should the job take?

MATHEMATICAL MODELING OF THE PROCESS

The engineer is now ready to write the appropriate balance equations and mechanistic relations for the process. Critical laboratory experiments must be designed and performed in order to determine unknown mechanisms and model parameters. Decisions must be made on which effects are important and which ones can be neglected. Order–of–magnitude analysis aids in making these critical simplifying decisions. It is imperative that the engineer be aware of and not overlook nor forget the assumptions made in the development of the mathematical model.

EQUATION ORGANIZATION

Once the mathematical relations have been assembled, they have to be arranged into a solution strategy, that is, decisions have to be made on which variable is to be solved for in each relation. For small problems, we usually perform this function routinely without much thought. However, for large problems care must be taken. Arranging the equations in an *information–flow diagram* is recommended. This block–diagram approach is useful for organizational purposes and illustrates the interrelationships among the equation variables. Also, equations should be arranged so that the solution strategy parallels the logical cause–and–effect relationships of the physical system. This "natural ordering" (see Franks, 1967) of equations usually leads to stable, efficient solution strategies.

COMPUTATION

For obtaining solutions to process simulation problems, the engineer has available several levels of computation—ranging from solution by inspection to analytical and high–speed computer solution. Because of the complexity and nonlinearity of process simulation problems, most solutions require high–speed digital computer solution. Digital computers are particularly useful for solving problems involving numerical manipulations. The FORTRAN language is designed for scientific usage and also has excellent logic capabilities; it is, therefore, used heavily by experienced process engineers. Numerical methods for the solution of sets of algebraic, ordinary differential, and partial differential equations are needed. To ease the programming effort in using numerical methods, generalized scientific subroutines have been written. A particularly useful and well-documented set is that of the NAG library, which is available on both personal computers and workstations. Additional software packages are also now available which have excellent graphical capabilities and ease the programming of specific problems. One popular package is Matlab (Math Works, Inc.; Sherborn, MA). This is a special interactive software package developed for use in the solution of algebraic and dynamic response problems. A number of Toolboxes are also available for use in the solution of specific engineering problems such as process control and process identification.

INTERPRETATION OF RESULTS

The real payoff of the simulation of chemical processes is in the intelligent interpretation of results by the engineer. At this point, the engineer must ascertain whether the model is a valid representation of the actual process or whether it needs revision and updating. The engineer must make sure that the results seem reasonable. Decisions have to be made on whether or not the simulated process achieves the objectives stated in the definition of the problem. Also, reasonable alternatives should be investigated in an effort to improve performance.

LIMITATIONS OF PROCESS SIMULATION

There are some definite limitations of process simulation of which the engineer must be aware. These include the following:

1. Lack of good data and knowledge of process mechanisms: The success of process simulation depends heavily on the basic information available to the engineer.

2. The character of the computational tools: There are certain types of equation sets that still pose a problem for numerical methods. These include some nonlinear algebraic and certain nonlinear partial differential equation sets.

3. The danger of forgetting the assumptions made in modeling the process: This can lead to placing too much significance on the model results.

USEFULNESS OF PROCESS SIMULATION

Computer simulation is playing an increasingly important role in the solution of chemical, biological, energy, and environmental problems. To

Introduction

develop some awareness of this discipline, examine the literature of the past two years, and find a journal article with the application of computer computation and process modeling to the analysis and solution of a chemical engineering problem. Prepare a short, well-written summary of the article, in which you provide the following information:

1. The correctly written journal reference for each article.

2. The name of the author(s), where they are located, and their professional position.

3. The nature of the problem studied.

4. The method of computation and the size of the problem.

5. The value of the result.

Computers in Chemical Engineering, *I&EC Research*, *AIChE Journal*, *Chemical Engineering Communications*, and *Chemical Engineering Science* are good sources.

REFERENCE

Franks, R. G., *Mathematical Modeling in Chemical Engineering*, Wiley, New York (1967).

Chapter 1

DEVELOPMENT OF MACROSCOPIC MASS, ENERGY, AND MOMENTUM BALANCES

In order to mathematically model a physical process, an engineer must write appropriate conservation equations for the process and then incorporate various mechanistic rate and equilibrium relations into those equations. Macroscopic conservation balances are written about a finite control volume and give rise to volume integral expressions for the basic principles of the conservation of mass, the conservation of energy, and the conservation of momentum. Since macroscopic balances are written over a finite control volume, no spatial gradients of the dependent variables appear in the conservation relations. Dependent variables such as temperature and concentration are therefore not differential functions of the spatial independent variables within the control volume, but represent average values over the control volume. The only differential independent variable is time. Therefore, by using the macroscopic conservation principles, mathematical models for unsteady–state processes yield sets of ordinary differential equations, while models for steady–state processes yield sets of algebraic equations. This chapter develops macroscopic mass, energy, and momentum balances and illustrates their use via some classical problems. The information–flow diagram is used to arrange the mathematical relations of these illustrations into solution strategies. Even though, for small problems such as these, we usually perform this function routinely without much thought, the information–flow diagram, or block–diagram approach is introduced here so that the reader may develop competency in using the technique before it is really required in the simulation of more complex problems. Analytical techniques are used to solve the problems presented in this chapter. Appendix A gives a review of analytic methods for the solution of ordinary differential equations.

1.1 CONSERVATION OF TOTAL MASS

The conservation principle for total mass which can neither be created nor destroyed is (Bennett and Meyers, 1982):

$$\begin{array}{c}\text{Rate of}\\\text{Accumulation}\\\text{of Mass}\end{array} = \begin{array}{c}\text{Rate of}\\\text{Mass Flow}\\\text{In}\end{array} - \begin{array}{c}\text{Rate of}\\\text{Mass Flow}\\\text{Out}\end{array}$$

which can be written

$$\begin{array}{c}\text{Rate of}\\\text{Accumulation}\\\text{of Mass}\end{array} = - \begin{array}{c}\text{Net Rate}\\\text{of Mass}\\\text{Flow Out}\end{array}$$

This statement is true for all systems except those involving nuclear reactions.

This conservation principle applied to a general control volume is illustrated in Figure 1.1. The control volume is located in the fluid flow field. The velocity at any point on the surface is given by v and the vector normal to the surface is given by n. The angle between the velocity vector v and the normal vector n is α.

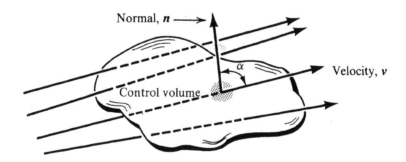

Figure 1.1: Control Volume for Conservation Balance.

The conservation principle for mass becomes

$$\frac{d}{dt}\iiint_V \rho \, dV + \text{net mass flow rate out} = 0 \qquad (1.1.1)$$

To obtain the net mass flow rate out, the point value of mass flow rate must be integrated over the surface of the control volume.

$$\text{Net mass flow rate out} = \iint_A \rho(\boldsymbol{v}\cdot\boldsymbol{n}) \, dA \qquad (1.1.2)$$

where $\boldsymbol{n}$ is the unit normal vector pointing outward from the surface and $\boldsymbol{v}$ is the velocity vector.

The $(\boldsymbol{v}\cdot\boldsymbol{n})$ term is required in order to evaluate the area which is normal to the velocity direction. Therefore, the overall mass balance becomes

$$\frac{d}{dt}\iiint_V \rho\, dV + \iint_A \rho(\boldsymbol{v}\cdot\boldsymbol{n})\, dA = 0 \qquad (1.1.3)$$

The inner product $(\boldsymbol{v}\cdot\boldsymbol{n})$ can also be expressed as

$$(\boldsymbol{v}\cdot\boldsymbol{n}) = v\cos\alpha \qquad (1.1.4)$$

where $v =$ the magnitude of the velocity vector.

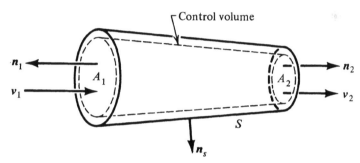

Figure 1.2: Simplified Tapered Tube Geometry.

1.1.1 Tapered Tube Geometry

The most common application of the overall mass balance is to a tapered tube geometry (illustrated in Figure 1.2). Here the velocity is normal to the surfaces A_1 and A_2 and parallel to the side surface, S.

For a system in which the density, ρ, is a constant over the areas A_1 and A_2, equation (1.1.3) becomes equation (1.1.5). The density may vary from plane 1 to 2.

$$\frac{d}{dt}\iiint_V \rho\, dV + \rho_1 \iint_{A_1} v_1 \cos 180°\, dA_1 + \rho_2 \iint_{A_2} v_2 \cos 0°\, dA_2$$

$$+ \iint_S \rho v_3 \cos 90°\, dS = 0 \qquad (1.1.5)$$

Defining the average or bulk velocity as

$$\langle v \rangle = \frac{1}{A}\iint_A v\, dA \qquad (1.1.6)$$

Equation (1.1.5) becomes

$$\frac{d}{dt}\iiint_V \rho\, dV + \rho_2 A_2 \langle v \rangle_2 - \rho_1 A_1 \langle v \rangle_1 = 0 \qquad (1.1.7)$$

Noting that the mass flow rate is defined as

$$w = \rho A \langle v \rangle \qquad (1.1.8)$$

the overall mass balance is

$$\frac{d}{dt}\iiint_V \rho\, dV = w_1 - w_2 \qquad (1.1.9)$$

which is the usual form of the mass balance for this tapered tube system. The lefthand side is the rate of accumulation while the righthand side is rate in minus the rate out.

1.2 CONSERVATION OF COMPONENT i

When considering a mass balance for a component i of a multicomponent mixture, the rate of generation of the component by chemical reaction must be taken into consideration. For the simplified tapered tube geometry of Figure 1.2, a mass balance for component i becomes

$$\begin{array}{c}\text{Rate of}\\ \text{Accumulation}\end{array} = \text{Rate In} - \text{Rate Out} + \begin{array}{c}\text{Rate of}\\ \text{Generation}\end{array}$$

$$\frac{d}{dt}\iiint_V \rho_i\, dV = w_{i_1} - w_{i_2} + \iiint_V r_i\, dV \qquad (1.2.1)$$

where r_i is the rate of generation of species i per unit volume.

1.3 METHOD OF WORKING PROBLEMS

It is suggested in working problems involving applications of the macroscopic mass balances that the balance equations be developed for the individual case under study as the first step. In order to check the development, the general equations can be simplified to make sure that the same describing equation set results.

Macroscopic Mass, Energy, and Momentum Balances

EXAMPLE 1.1 Mixing Tank:

Water is flowing into a well–stirred tank at 150 kg/hr and methanol ($MeOH$) is being added at 30 kg/hr. The resulting solution is leaving the tank at 120 kg/hr. Because of effective stirring, the concentration of the outlet solution is the same as that within the tank. There are 100 kg of fresh water in the tank at the start of the operation, and the rates of input and output remain constant thereafter. Calculate the outlet concentration (mass fraction of methanol) after 1 hr. (See Figure 1.3.)

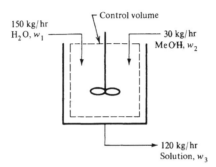

Figure 1.3: Example 1.1.

In order to solve this system with two chemical species, two independent material balances must be written: one overall balance, and one methanol balance. Applying the conservation principle to the system control volume shown as a dotted line in Figure 1.3, the overall mass balance is

$$\begin{array}{c} \text{Rate of} \\ \text{Accumulation} \\ \text{of Total Mass} \end{array} = \begin{array}{c} \text{Rate of} \\ \text{Total} \\ \text{Mass In} \end{array} - \begin{array}{c} \text{Rate of} \\ \text{Total} \\ \text{Mass Out} \end{array}$$

$$\left[\frac{d}{dt} \iiint \rho \, dV \right] = [w_{\text{in}}] - [w_{\text{out}}] \tag{1.3.1}$$

Since the tank is well stirred, the density is constant throughout the volume of the tank and may be removed from inside the volume integral. Therefore

$$\iiint \rho \, dV = \rho \iiint dV = \rho V \tag{1.3.2}$$

or simply $\rho V = M$, the total mass in the tank.

For the particular tank of Figure 1.3, the overall mass balance becomes

$$\frac{dM}{dt} = w_1 + w_2 - w_3 \tag{1.3.3}$$

Using the data given in the problem we have,

$$\frac{dM}{dt} = 150 \text{ kg/hr} + 30 \text{ kg/hr} - 120 \text{ kg/hr} \tag{1.3.4}$$

or
$$\frac{dM}{dt} = 60 \text{ kg/hr} \tag{1.3.5}$$

A methanol balance on the system gives

| Rate of Accumulation of Methanol | = | Rate of Mass of Methanol In | − | Rate of Mass of Methanol Out | + | Rate of Generation of Methanol |

$$\left[\frac{dM_{MeOH}}{dt}\right] = [w_{MeOH \text{ in}}] - [w_{MeOH \text{ out}}] + [0] \tag{1.3.6}$$

Letting ω be the mass fraction of methanol, we have

$$\frac{d(M\omega_3)}{dt} = w_1 \omega_1 + w_2 \omega_2 - w_3 \omega_3 \tag{1.3.7}$$

which for this system is

$$\frac{d(M\omega_3)}{dt} = 30 \text{ kg/hr} - 120\omega_3 \text{ kg/hr} \tag{1.3.8}$$

Equations (1.3.5) and (1.3.8) along with the appropriate initial conditions form the system model. Those equations can be arranged in block–diagram fashion to form the information–flow diagram for the system as shown in Figure 1.4.

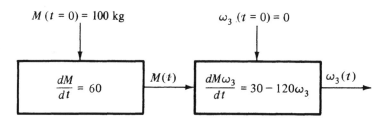

Figure 1.4: Information–Flow Diagram for Example 1.1.

We can easily see from the information–flow diagram exactly what information is required in order to solve each equation and where that information comes from. It also shows us how each equation is used in a simultaneous set of equations. For this example, the overall mass balance is needed in order to compute the total mass transient response. The methanol balance is needed in order to compute the exit methanol concentration response, $\omega_3(t)$. We see from the information–flow diagram that in order to compute this methanol concentration response we need an initial methanol concentration, $\omega_3(t=0)$, and also the current value for the total mass of the tank, $M(t)$. Notice that this solution strategy follows a commonsense cause–and–effect relationship in that the overall mass balance is used to compute the total mass and the methanol balance is used to compute methanol concentration. The transient response is obtained by solving the system equations.

Macroscopic Mass, Energy, and Momentum Balances

Analytical solutions are readily available for equations (1.3.5) and (1.3.8). The overall mass balance is a single equation with a single unknown so it can be solved independent of the component mass balance. It is a separable equation. Separating the variables yields

$$dM = 60 \, dt \tag{1.3.9}$$

Integrating both sides gives,

$$M = 60 \, t + C_1 \tag{1.3.10}$$

where C_1 is a constant of integration which is evaluated from the initial condition of 100 kg of material in the tank at $t = 0$. Using this initial condition gives

$$M = 60 \, t + 100 \tag{1.3.11}$$

Now knowing the solution for the overall mass balance, we seek a solution for the component balance. It can also be rearranged for solution by separation of variables. Expanding the left–hand side of equation (1.3.8) gives

$$\omega_3 \frac{dM}{dt} + M \frac{d\omega_3}{dt} = 30 - 120\omega_3 \tag{1.3.12}$$

Substituting the analytic solution for the mass balance (equation (1.3.8)) and the mass balance (equation (1.3.5)) into equation (1.3.12) and rearranging gives

$$\int_0^{\omega_3} \frac{d\omega_3}{(30 - 180\omega_3)} = \int_0^t \frac{dt}{(100 + 60 \, t)} \tag{1.3.13}$$

Integrating both sides yields

$$-\frac{1}{180} \ln(30 - 180\omega_3) \Big|_0^{\omega_3} = \frac{1}{60} \ln(100 + 60 \, t) \Big|_0^t \tag{1.3.14}$$

or

$$\ln\left[\frac{30}{30 - 180\omega_3}\right] = 3 \ln\left[\frac{100 + 60 \, t}{100}\right] \tag{1.3.15}$$

which can be expressed as

$$\omega_3 = \frac{1}{6}\left[1 - \frac{1}{(1 + 0.6 \, t)^3}\right] \tag{1.3.16}$$

Plots of the solutions for M and ω_3 are given in Figure 1.5. At $t = 1$ hour, $\omega_3 = 0.126$.

Some interpretation of these results can be made. From equation (1.3.16) we see that the steady–state methanol mass fraction is

$$\omega_3 = \frac{1}{6} = 0.1667 \tag{1.3.17}$$

However, there is a problem with the model in that the total mass of the tank is an ever–increasing function of time. Sooner or later the tank will fill.

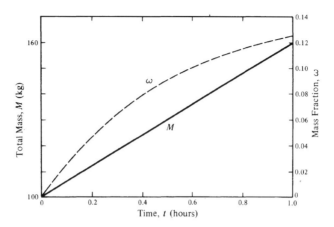

Figure 1.5: **Transient Response for Example 1.1.**

Therefore, there is actually a constraint on the size of the tank or total mass contained in the tank. This constraint limits the range overwhich the model is applicable.

EXAMPLE 1.2 Stirred Tank Reactor:

We want to determine the dynamic response of component A in a continuous stirred tank reactor when the volume of the tank is V (cm^3), the inlet and outlet total volumetric flow rate is F (cm^3/min), the inlet concentration is constant at C_0 (g mol/cm^3), and the initial concentration of component A in the tank is zero. Component A undergoes a first–order reaction in the tank:

$$r = -k\,C \qquad \text{(g mol/min cm}^3\text{)} \qquad (1.3.18)$$

and the rate constant decays according to

$$k = k_0 - at^2 \qquad \text{(1/min)} \qquad (1.3.19)$$

The species conservation balance of component A around the tank is

Rate Of Accumulation of Species A	=	Rate In of Species A	−	Rate Out of Species A	+	Rate of Generation of Species A

$$V\frac{dC}{dt} = C_0 F - CF - kCV \qquad (1.3.20)$$

Dividing by V and using equation (1.3.19) for the rate constant decay gives

$$\frac{dC}{dt} + \left[\frac{F}{V} + k_0 - at^2\right] C = \frac{C_0 F}{V} \qquad (1.3.21)$$

This is a linear first–order differential equation with a nonconstant coefficient and can be solved by seeking an integrating factor as described in

Appendix A, equation (A–34). Details of the solution are left as an exercise for the interested reader. However, the integrating factor is

$$\mu = e^{\left(\frac{F}{V}+k_0\right)t - \frac{at^3}{3}} \tag{1.3.22}$$

and the solution is

$$C = \frac{(F/V)C_0}{\left(\frac{F}{V}+k_0-at^2\right)} + \left[C_0 - \frac{(F/V)C_0}{\left(\frac{F}{V}+k_0\right)}\right] e^{-\left(\frac{F}{V}+k_0\right)t + \frac{at^3}{3}} \tag{1.3.23}$$

There is one important limitation to this model. This is the fact that the rate constant k defined by equation (1.3.19) must be positive. This constraint should be introduced into the model. The solution, equation (1.3.23), is valid as long as this constraint is not violated. Figure 1.6 gives the solution response for different values of the decay parameter a. As the decay parameter increases, the dynamic response of the tank is slowed.

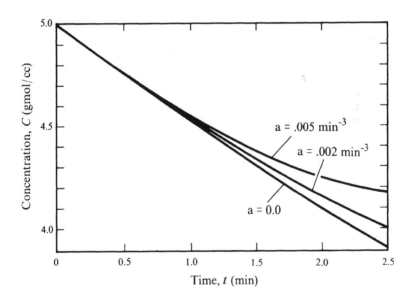

Figure 1.6: Transient Response for Example 1.2.
$F = 10$ cc/min $V = 1000$ cc $k_0 = 0.1$ (min^{-1})

1.4 CONSERVATION OF TOTAL ENERGY

The total energy of a system is composed of internal energy, kinetic energy, and potential energy. The internal energy per unit mass is given by $\hat{U}$.

The kinetic energy per unit mass is computed by integrating the momentum of the particles of the system over their rate of change (velocity) from rest to the velocity of the system,

$$\int_0^v v \, dv = \frac{v^2}{2} \tag{1.4.1}$$

The potential energy per unit mass is obtained by integrating the gravitational acceleration over the system height. If the acceleration of gravity is constant, then the result is gZ as shown in equation (1.4.2)

$$\int_0^Z g \, dZ = gZ \tag{1.4.2}$$

The total energy per unit mass, $\hat{E}$, is the sum of the internal energy per unit mass, the kinetic energy per unit mass and the potential energy per unit mass

$$\hat{E} = \hat{U} + v^2/2 + gZ \tag{1.4.3}$$

It is worth noting at this point that, in working problems, it is imperative that the units in an equation be consistent. Particular emphasis should be placed on keeping force and mass units consistent.

Before the conservation principle for total energy is applied to the general control volume illustrated in Figure 1.1, various terms will be discussed. The net energy change due to the net flow of material out of the system is given by

$$\iint_A \rho \hat{E} (\boldsymbol{v} \cdot \boldsymbol{n}) \, dA \tag{1.4.4}$$

The net rate of heat energy into the system is Q.

The net rate of work done by the system is given by $\dot{W}$. Usually the rate of work is divided into the rate of shaft work or mechanical work, $\dot{W}_s$ the rate of work due to compression; $p\hat{V}$; and the rate of work required to overcome viscous forces on the control volume, $\dot{W}_\eta$. $\hat{V}$ is the specific volume. Applying this over the control volume, we have

$$\dot{W} = \dot{W}_s + \iint_A \rho p \hat{V} (\boldsymbol{v} \cdot \boldsymbol{n}) \, dA - \dot{W}_\eta \tag{1.4.5}$$

The rate of accumulation of energy is given by

$$\frac{d}{dt} \iiint_V \rho \hat{E} \, dV \tag{1.4.6}$$

Therefore, the total energy balance, or the equation for the conservation of energy, becomes

$$\text{Rate of Accumulation of Total Energy} = \text{Rate of Energy Into the System} - \text{Rate of Energy Out of the System} + \text{Rate of Generation of Energy}$$

$$\left[\frac{d}{dt}\iiint_V \rho \hat{E}\, dV\right] = [Q] - \left[\iint_A \rho(\hat{E} + p\hat{V})(\boldsymbol{v}\cdot\boldsymbol{n})\, dA\right] + \left[-\dot{W}_s + \dot{W}_\eta\right] \tag{1.4.7}$$

Introducing the enthalpy per unit mass, $\hat{H}$, defined as $\hat{H} = \hat{U} + p\hat{V}$, equation (1.4.7) becomes

$$\frac{d}{dt}\iiint_V \rho \hat{E}\, dV = -\iint_A \rho\left[\hat{H} + \frac{v^2}{2} + gZ\right](\boldsymbol{v}\cdot\boldsymbol{n})\, dA + Q - \dot{W}_s + \dot{W}_\eta \tag{1.4.8}$$

In working any problem, the internal energy per unit mass $\hat{U}$ on the enthalpy per unit mass can be related to the temperature of the system by the heat capacity (C_p or C_v). The heat capacity at constant volume is given by

$$C_v = \left(\frac{dQ}{dT}\right)_{\hat{V}} = \left(\frac{\partial \hat{U}}{\partial T}\right)_{\hat{V}} \tag{1.4.9}$$

The differential internal energy per unit mass, $d\hat{U}$, can be expressed in trms of $d\hat{V}$ and dT as,

$$\hat{U} = \int_0^{\hat{U}} d\hat{U} = \int_{\hat{V}_r}^{V}\left(\frac{\partial \hat{U}}{\partial \hat{V}}\right)_T d\hat{V} + \int_{T_r}^{T}\left(\frac{\partial \hat{U}}{\partial T}\right)_{\hat{V}} dT \tag{1.4.10}$$

where $\hat{V}_r$ and T_r are the reference specific volume and reference temperature, respectively.

By using the thermodynamic relation for $(\partial \hat{U}/\partial \hat{V})_T$ and equation (1.4.9) for $(\partial \hat{U}/\partial T)_{\hat{V}}$ (Denbigh, 1963), the internal energy can therefore be computed as

$$\hat{U} = \int_{\hat{V}_r}^{\hat{V}}\left[-p + T\left(\frac{\partial p}{\partial T}\right)_{\hat{V}}\right] d\hat{V} + \int_{T_R}^{T} C_v\, dT \tag{1.4.11}$$

The heat capacity at constant pressure is given by

$$C_p = \left(\frac{dQ}{dT}\right)_p = \left(\frac{\partial \hat{H}}{\partial T}\right)_p \tag{1.4.12}$$

The differential enthalpy, $d\hat{H}$, can be expanded in terms of dp and dT; therefore, the enthalpy $\hat{H}$ is given as

$$\hat{H} = \int_0^{\hat{H}} d\hat{H} = \int_{p_r}^{p} \left(\frac{\partial \hat{H}}{\partial p}\right)_T dp + \int_{T_r}^{T} \left(\frac{\partial \hat{H}}{\partial T}\right)_p dT \qquad (1.4.13)$$

Again using the thermodynamic relation for $\left(\frac{\partial \hat{H}}{\partial p}\right)_T$ and equation (1.4.12) for $\left(\frac{\partial \hat{H}}{\partial T}\right)_p$, the enthalpy expression becomes

$$\hat{H} = \int_{p_r}^{p} \left[\hat{V} - T\left(\frac{\partial \hat{V}}{\partial T}\right)_p\right] dp + \int_{T_r}^{T} C_p dT \qquad (1.4.14)$$

with p_r as the reference pressure.

1.4.1 Tapered Tube Geometry

For the simple geometry of flow in a tapered tube (Figure 1.2), the energy balance of equation (1.4.8) can be simplified. First, $W_\eta = 0$ because the velocity on the surface of the control volume is zero. It should be noted that the shear force is not zero at the surface, S. However, the viscous work is zero since it is the product of the system velocity times the viscous force. For the viscous work to be nonzero, the velocity at the control boundary surface, S, must be nonzero. However, the no–slip condition states that the velocity at the surface S is zero. A second simplification occurs because surface area term is nonzero only at the entrance and exit of the tapered tube. The tapered tube surface area term therefore becomes

$$\int\int_A \rho(\boldsymbol{v}\cdot\boldsymbol{n})\left(\frac{v^2}{2}+gZ+\hat{H}\right) dA = \left(\frac{w_2}{2}\frac{\langle v_2^3\rangle}{\langle v_2\rangle}+w_2 g Z_2+w_2\hat{H}_2\right)$$
$$-\left(\frac{w_1}{2}\frac{\langle v_1^3\rangle}{\langle v_1\rangle}+w_1 g Z_1+w_1\hat{H}_1\right) \qquad (1.4.15)$$

where $<\ >$ indicate the average quantity as defined by equation (1.1.6); and ρ, Z, and $\hat{H}$ are assumed constant over the entrance and exit areas. The resulting tapered tube energy relation is therefore

Rate of Accumulation of Total Energy	=	Rate of Energy Into the System	−	Rate of Energy Out of the System	+	Rate of Generation of Energy

$$\left[\frac{d}{dt}\int\int\int \rho\hat{E}\,dV\right] = \left[w_1\left(\frac{1}{2}\frac{\langle v_1^3\rangle}{\langle v_1\rangle}+gZ_1+\hat{H}_1\right)+Q\right]$$
$$-\left[w_2\left(\frac{1}{2}\frac{\langle v_2^3\rangle}{\langle v_2\rangle}+gZ_2+\hat{H}_2\right)\right]+[-\dot{W}_s] \qquad (1.4.16)$$

EXAMPLE 1.3 Constant Density and Ideal Gas Cases for $\hat{U}$ and $\hat{H}$:

We want to evaluate $\hat{U}$ and $\hat{H}$ for the case of an incompressible fluid (ρ = constant) and an ideal gas. First let's assume that the density is constant. Since $\hat{V} = \frac{1}{\rho}$, then the $\int_{\hat{V}_r}^{\hat{V}}$ term in expression for $\hat{U}$ (1.4.11) is zero since $\hat{V} = \hat{V}_r$ when ρ is a constant. This means that

$$\hat{U} = \int_{T_r}^{T} C_v dT \tag{1.4.17}$$

Also when ρ is constant $C_v = C_p$ so equation (1.4.17) can also be expressed as

$$\hat{U} = \int_{T_r}^{T} C_p dT \tag{1.4.18}$$

The enthalpy per unit mass $\hat{H}$ can also be simplified since $\left(\frac{\partial \hat{V}}{\partial T}\right)_p$ is zero when ρ is a constant. Equation (1.4.14) becomes

$$\hat{H} = \int_{p_r}^{p} \hat{V} dp + \int_{T_r}^{T} C_p dT \tag{1.4.19}$$

or

$$\hat{H} = \hat{V}(p - p_r) + \int_{T_r}^{T} C_p dT \tag{1.4.20}$$

For the case of an ideal gas $\left(\frac{\partial p}{\partial T}\right)_{\hat{V}} = \frac{R}{\hat{V}(MW)}$

$$\hat{U} = \int_{\hat{V}_r}^{\hat{V}} \left[-p + \frac{TR}{\hat{V}(MW)}\right] dV + \int_{T_r}^{T} C_v dT \tag{1.4.21}$$

Since $\frac{TR}{\hat{V}(MW)} = p$ from the ideal gas equation of state

$$\hat{U} = \int_{T_4}^{T} C_v dT \tag{1.4.22}$$

For an ideal gas $\left(\frac{\partial \hat{V}}{\partial T}\right)_p = \frac{R}{p(MW)}$ which gives for the enthalpy per unit mass

$$\hat{H} = \int_{T_4}^{T} C_p dT \tag{1.4.23}$$

1.5 METHOD OF WORKING PROBLEMS

It is suggested that, in developing equations involving the macroscopic energy balance, the general equation (1.4.8) or (1.4.16) be simplified for the case under consideration. This approach will help to include all significant energy mechanisms in the final model.

EXAMPLE 1.4 Heating of a Tank:

A dilute solution at 20°C is added to a well-stirred tank at the rate of 180 kg/hr. A heating coil having an area of 0.9 m² is located in the tank and contains steam condensing at 150°C. The heated liquid leaves at 120 kg/hr and at the temperature of the solution in the tank. There is 500 kg of solution at 40°C in the tank at the start of the operation. The overall heat–transfer coefficient is 342 kg/hr m² °C and the heat capacity of water is 1 k cal/kg °C. Calculate the outlet temperature after 1 hr. (See Figure 1.7.)

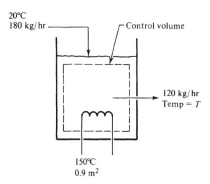

Figure 1.7: Example 1.4.

Applying the overall energy balance equation, (1.4.16), to this system, we will assume an incompressible fluid such that $\hat{E} = \hat{U} = \hat{H}$. This means that $\hat{U}$ is related to the temperature by $\hat{U} = C_p(T - T_r)$, where T_r is the reference temperature, and that the kinetic energy and potential energy terms are negligible. The energy balance is therefore

| Rate of Accumulation of Total Energy | = | Rate of Energy Into the System | − | Rate of Energy Out of the System | + | Rate of Generation of Energy |

$$\left[\frac{d}{dt} \iiint_V \rho \hat{U}\, dV\right] = \left[w_1 \hat{H}_1 + Q\right] - \left[w_2 \hat{H}_2\right] + [0] \quad (1.5.1)$$

Integrating the internal energy term over the volume element gives

$$\iiint_V \rho \hat{U}\, dV = C_p(T - T_r)M \quad (1.5.2)$$

where M is the total mass of the system. The rate of heat transferred from the coil to the tank is given by Newton's law of cooling,

$$Q = UA(T_s - T) \quad (1.5.3)$$

where U is the overall heat–transfer coefficient. Therefore equation (1.5.1) becomes

$$C_p \frac{d}{dt} M(T - T_r) = 180 C_p(T_1 - T_r) + UA(T_s - T) - 120 C_p(T - T_r) \quad (1.5.4)$$

The general energy balance has therefore been reduced to equation (1.5.4). It has four significant terms. The first is the rate of accumulation of internal energy within the control volume, the second is the rate of sensible energy entering the system through the inlet flow to the system, the third is the rate of heat transferred to the system through the heating coil, and the final term is the rate of sensible energy leaving the system with the outlet stream. Assuming that $T_r = 0°C$ gives

$$\frac{d(MT)}{dt} = 180(20) + 342(0.9)(150 - T) - 120T \qquad (1.5.5)$$

The energy balance equation, (1.5.5), must now be solved simultaneously with the overall mass balance given by equation (1.5.6)

$$\begin{matrix} \text{Rate of} \\ \text{Accumulation} \\ \text{of Total Mass} \end{matrix} = \begin{matrix} \text{Rate of} \\ \text{Mass Into} \\ \text{the System} \end{matrix} - \begin{matrix} \text{Rate of} \\ \text{Mass Out of} \\ \text{the System} \end{matrix}$$

$$\left[\frac{dM}{dt}\right] = [180] - [120] \qquad (1.5.6)$$

These two equations can be placed in an information–flow diagram as shown in Figure 1.8.

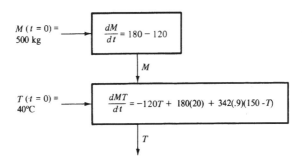

Figure 1.8: Information–Flow Diagram for Example 1.4.

Again the information–flow diagram is useful since it illustrates the relationships between the describing equations. Also the cause-and-effect usage of the equations should be noted. The mass balance is used to solve for the total mass of the system and the energy balance for the temperature of the system.

This set of differential equations can be solved analytically by separation of variables. It is left as an exercise for the interested reader. The transient response of the system obtained is shown in Figure 1.9. The temperature of the tank after 1 hr is 77.3°C. The dynamic response of the system with a 50 percent reduction in the heat transfer area is also shown. The temperature after 1 hr is lowered to 58.7°C.

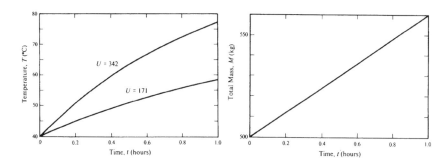

Figure 1.9: Transient Response for Example 1.4.

EXAMPLE 1.5 Temperature Response of Slurry Reactor:
A slurry reactor shown in Figure 1.10 is used to carry out a catalytic reaction. This reactor is a well–stirred tank with catalyst particles dispersed throughout. The rate of such reactions is usually zero order, that is, they do not depend upon the reactant as product concentrations. The reaction rate is, however, a function of the system temperature

$$r = a + bT + cT^2 \qquad \text{(g mol/sec cm}^3\text{)} \qquad (1.5.7)$$

We have a steam–heated stirred tank in which an endothermic slurry reaction is being carried out. Develop an analytical solution to the temperature response of the tank when there is a step change in the steam temperature, T_s.

An overall mass balance on the reactor shows that the inlet mass flow rate, w_1, is equal to the outlet mass flow rate, w_2. A total energy balance around the system gives

Rate of Accumulation of Total Energy	=	Rate of Energy Into the System	−	Rate of Energy Out of the System	+	Rate of Generation of Total Energy

$$M\,C_p\,\frac{dT_2}{dt} = w\,C_p(T_1 - T_r) + U\,A(T_s - T_2) - w\,C_p(T_2 - T_r) - H\,r\,V \qquad (1.5.8)$$

where
- M = total mass of the tank (g)
- C_p = heat capacity (cal/g °C)
- U = heat–transfer coefficient (cal/cm² min °C)
- A = heat–transfer area (cm²)
- H = endothermic heat of reaction (cal/g mol)
- T_r = reference temperature

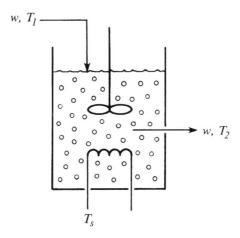

Figure 1.10: Slurry Reactor.

The energy equation can be rearranged as

$$\frac{dT_2}{dt} = C_0 + C_1 T_2 + C_2 T_2^2 \tag{1.5.9}$$

with

$$C_0 = \left[\frac{wT_1}{M} + \frac{UAT_s}{C_p M} - \frac{HVa}{MC_p}\right]$$

$$C_1 = \left[-\frac{w}{M} - \frac{UA}{MC_p} - \frac{HVb}{MC_p}\right]$$

$$C_2 = \left[-\frac{HVc}{MC_p}\right]$$

This is a separable equation with solution

$$t = \frac{2}{\sqrt{4C_2 C_0 - C_1^2}} \tan^{-1}\left[\frac{2C_2 T_2 + C_1}{\sqrt{4C_2 C_0 - C_1^2}}\right] + A \tag{1.5.10}$$

where A is a constant of integration given by the initial tank temperature, which is evaluated by solving the steady–state version of equation (1.5.8) with the old value of the steam temperature before the step change.

1.6 MECHANICAL ENERGY BALANCE

For **isothermal** problems of fluid dynamics, it is better to use a modified version of the energy balance called the mechanical energy energy balance. We will now develop this isothermal version of the energy balance.

The total energy balance was given in equation (1.4.8) as

$$\frac{d}{dt}\iiint_V \rho \hat{E}\, dV + \iint_A \rho(\boldsymbol{v}\cdot\boldsymbol{n})\left(\hat{H}+\frac{v^2}{2}+gZ\right)dA = Q - \dot{W}_s + \dot{W}_\eta \quad (1.6.1)$$

The left–hand side of equation (1.6.1) describes a change of state. The right–hand side of this equation describes the irreversible processes which contribute to the change of state. We will now define a hypothetical reversible process which can cause the same change of state as

$$Q - \dot{W}_s + \dot{W}_\eta = Q_{rev} - \dot{W}_{rev} \quad (1.6.2)$$

The term $\dot{W}_{rev}$ is composed of the actual rate of shaft work, $\dot{W}_s$, and the irreversible losses, E_v

$$\dot{W}_{rev} = \dot{W}_s + E_v \quad (1.6.3)$$

where E_v = the rate of mechanical energy irreversibly converted to thermal energy within and on the surface of the control volume.

Equation (1.6.2) therefore becomes

$$Q - \dot{W}_s + \dot{W}_\eta = Q_{rev} - \left(\dot{W}_s + E_v\right) \quad (1.6.4)$$

From thermodynamics Q_{rev} is given by (Smith and Van Ness, 1959)

$$Q_{rev} = T\frac{dS}{dt} \quad (1.6.5)$$

where S is the total entropy of the system. For an isothermal process, equation (1.6.5) can be written as

$$Q_{rev} = \frac{d(TS)}{dt} \quad (1.6.6)$$

The change in TS with time is the rate of change of TS of the contents of the macroscopic system plus the net rate of TS leaving due to the flow of material across the boundaries of the system; therefore

$$\frac{d(TS)}{dt} = \frac{d}{dt}\iiint_V \rho T\hat{S}\, dV + \iint_A \rho(\boldsymbol{v}\cdot\boldsymbol{n})T\hat{S}\, dA \quad (1.6.7)$$

where $\hat{S}$ is the entropy per unit mass.

If we substitute equations (1.6.7) and (1.6.4) into equation (1.6.1), we get a new form of the energy balance:

$$\frac{d}{dt}\iiint_V \rho\left(\hat{U} - T\hat{S} + \frac{v^2}{2} + gZ\right)dV$$

$$= -\iint_A \rho(\boldsymbol{v}\cdot\boldsymbol{n})\left(\hat{H} - T\hat{S} + \frac{v^2}{2} + gZ\right)dA - E_v - \dot{W}_s \quad (1.6.8)$$

By definition we also have the following

$$\hat{A} = \hat{U} - T\hat{S} \tag{1.6.9}$$

where $\hat{A}$ is the Helmholtz free energy per unit mass, and

$$\hat{G} = \hat{H} - T\hat{S} \tag{1.6.10}$$

where $\hat{G}$ is the Gibbs free energy per unit mass.

Using these two relations, we get the general form of the mechanical energy equation which is given in equation (1.6.11)

$$\begin{matrix} \text{Rate of} \\ \text{Accumulation of} \\ \text{Mechanical Energy} \end{matrix} = \begin{matrix} \text{Rate of} \\ \text{Mechanical} \\ \text{Energy In} \end{matrix} - \begin{matrix} \text{Rate of} \\ \text{Mechanical} \\ \text{Energy Out} \end{matrix} + \begin{matrix} \text{Rate of} \\ \text{Generation of} \\ \text{Mech. Energy} \end{matrix}$$

$$\left[\frac{d}{dt} \iiint_V \rho \left(\hat{A} + \frac{v^2}{2} + gZ \right) dV \right] = [0]$$

$$- \left[\iint_A \rho(\boldsymbol{v} \cdot \boldsymbol{n}) \left(\hat{G} + \frac{v^2}{2} + gZ \right) dA \right] + \left[-\dot{W}_s - E_v \right] \tag{1.6.11}$$

1.6.1 Tapered Tube Geometry

When applying the general mechanical energy equation to tapered tube systems, we have

$$\begin{matrix} \text{Rate of} \\ \text{Accumulation of} \\ \text{Mechanical Energy} \end{matrix} = \begin{matrix} \text{Rate of} \\ \text{Mechanical} \\ \text{Energy In} \end{matrix} - \begin{matrix} \text{Rate of} \\ \text{Mechanical} \\ \text{Energy Out} \end{matrix} + \begin{matrix} \text{Rate of} \\ \text{Generation of} \\ \text{Mech. Energy} \end{matrix}$$

$$\left[\frac{d}{dt} \iiint_V \rho \left(\hat{A} + \frac{v^2}{2} + gZ \right) dV \right] = \left[w_1 \left(\frac{\langle v_1^3 \rangle}{2 \langle v_1 \rangle} + gZ_1 + \hat{G}_1 \right) \right]$$

$$- \left[w_2 \left(\frac{\langle v_2^3 \rangle}{2 \langle v_2 \rangle} + gZ_2 + \hat{G}_2 \right) \right] + \left[-E_v - \dot{W}_s \right] \tag{1.6.12}$$

Here, ρ, Z, and $\hat{G}$ are assumed constant over the inlet and outlet flow areas. To calculate the Gibbs free energy, $\hat{G}$, we can use the following identity (Denbigh, 1963):

$$\hat{G} = - \int_{T_r}^{T} \hat{S} \, dT + \int_{p_r}^{p} \frac{1}{\rho} \, dp \tag{1.6.13}$$

which for an isothermal system is simply

$$\hat{G} = \int_{p_r}^{p} \frac{1}{\rho} \, dp \tag{1.6.14}$$

The Helmholtz free energy $\hat{A}$ is given as (Denbigh, 1963),

$$\hat{A} = -\int_{T_r}^{T} \hat{S}\, dT - \int_{\hat{V}_r}^{\hat{V}} p\, d\hat{V} \qquad (1.6.15)$$

which reduces to

$$\hat{A} = -\int_{\hat{V}_r}^{\hat{V}} p\, d\hat{V} \qquad (1.6.16)$$

The friction loss, E_v, can also be expressed in terms of the head loss due to friction, h_L, as

$$h_L = \frac{E_v}{w} \qquad (1.6.17)$$

At steady state, the mass flow rate is a constant. The mechanical energy equation for tapered tube systems therefore reduces to

| Rate of Accumulation of Mechanical Energy | = | Rate of Mechanical Energy In | − | Rate of Mechanical Energy Out | + | Rate of Generation of Mech. Energy |

$$0 = \left[\left(\frac{\langle v_1^3\rangle}{2\langle v_1\rangle} + gZ_1\right)\right] - \left[\left(\frac{\langle v_2^3\rangle}{2\langle v_2\rangle} + gZ_2\right)\right] - \int_{p_1}^{p_2} \frac{1}{\rho}\, dp + \left[-h_L - \hat{W}_s\right] \qquad (1.6.18)$$

where $\hat{W}_s$ is the work performed per unit mass and is related to the rate of work, $\dot{W}_s$, by

$$\hat{W}_s = \dot{W}_s/w \qquad (1.6.19)$$

Note that each term of equation (1.6.18) has the units of length times acceleration (length^2/time^2). The quantities are usually designated as "heads," namely heads due to velocity, $\langle v^3\rangle/2\langle v\rangle$; due to elevation, gZ; due to pressure,

$$\int_{p_1}^{p_2} \frac{dp}{\rho}$$

and due to friction, h_L. Equation (1.6.18) is commonly known as the Bernoulli equation.

When shaft work is taken out of the system by a turbine, the friction head and shaft work can be expressed as

$$h_L + \hat{W}_s = h_f + \frac{\hat{W}_s}{\eta_t} \qquad (1.6.20)$$

where h_f is the friction head loss not including that of the turbine and η_t is the turbine efficiency.

For shaft work put into the system by a pump,

$$h_L + \hat{W}_s = h_f - \eta_p \hat{W}_s \qquad (1.6.21)$$

where η_p is the pump efficiency.

EXAMPLE 1.6 Analysis of a Piping Network:

We want to compute the horsepower needed to pump water in the piping system shown in Figure 1.11. Water is to be delivered to the upper tank at a rate of 5.7×10^{-3} m^3/s. All of the piping is 10.16 cm (4-in.) ID smooth circular pipe. The pump efficiency is unity.

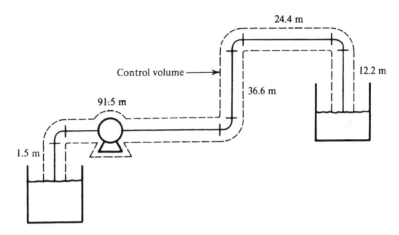

Figure 1.11: Piping System for Example 1.6.

The mechanical energy balance equation (1.6.11) for this system becomes

$$\begin{array}{c}\text{Rate of}\\\text{Accumulation of}\\\text{Mechanical Energy}\end{array} = \begin{array}{c}\text{Rate of}\\\text{Mechanical Energy}\\\text{Into System}\end{array} - \begin{array}{c}\text{Rate of}\\\text{Mechanical Energy}\\\text{Out of System}\end{array}$$

$$+ \begin{array}{c}\text{Rate of}\\\text{Generation of}\\\text{Mechanical Energy}\end{array}$$

$$[0] = [wgZ_1] - [wgZ_2] + [-E_v - \dot{W}_s] \quad (1.6.22)$$

or in terms of the friction head h_f,

$$wg(Z_2 - Z_1) + wh_f + w\eta_p \hat{W}_s = 0 \quad (1.6.23)$$

The friction head for straight pipe is given in terms of the Fanning friction factor, f, as (Bird et al., 1960, p. 214)

$$h_f = 2\langle v \rangle^2 \left(\frac{L}{D}\right) f \quad (1.6.24)$$

The friction–factor–Reynolds–number correlation can be expressed for smooth pipe as (Bird et al., 1960, p. 186)

$$f = 0.0791 \, \text{Re}^{-1/4} \quad (1.6.25)$$

Equivalent lengths of straight pipe can be calculated for the four 90° elbows and the sudden contraction entrance and exit but will be neglected here. The Reynolds number is defined as

$$\text{Re} = \frac{D\rho\langle v\rangle}{\mu} = 7.11 \times 10^4 \qquad (1.6.26)$$

The information–flow diagram for this system is given in Figure 1.12. Performing the computations gives $\dot{W}_s = 14.5\ W$ (2 hp).

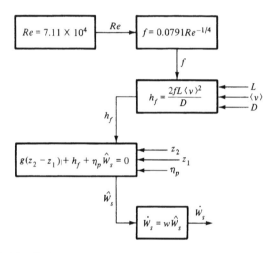

Figure 1.12: Information–Flow Diagram for Example 1.6.

1.7 CONSERVATION OF MOMENTUM

The rate of change of momentum is equivalent to a force as stated by Newton's second law.

Force = rate of change of momentum

where momentum is defined as the product of the mass times the velocity.
Force and velocity are both vector quantities in Euclidian three space.

$$\boldsymbol{f} = \begin{bmatrix} f_x \\ f_y \\ f_z \end{bmatrix} \qquad \boldsymbol{v} = \begin{bmatrix} v_x \\ v_y \\ v_z \end{bmatrix}$$

Therefore when applying the conservation principle to momentum, we will have three component balances corresponding to the three mutually orthogonal directions in Euclidian three space.

The rate of accumulation of the x component of momentum is

$$\frac{d}{dt}\iiint_V \rho v_x\, dV \qquad (1.7.1)$$

The net rate of the x component of momentum out of the general control volume (Figure 1.1) is

$$\iint_A (\rho v_x)(\boldsymbol{v}\cdot\boldsymbol{n})dA \qquad (1.7.2)$$

The rate of generation of momentum can be given by the sum of the x component of the forces acting on the system, $\sum f_x$. Two types of forces make up the force f_x. They are body forces and surface forces. The main body force of consequence is that due to gravity. The gravity force, f_{xg}, is the x directed force of gravity acting on the total mass in the control volume,

$$f_{xg} = \iiint_V \rho g_x\, dV \qquad (1.7.3)$$

There are three specific surface forces that are important in chemical engineering problems. These are pressure forces, shear or viscous forces, and structural or resultant forces. The pressure force f_{xp} is the x directed force caused by the integral of the pressure acting on the surface of the control volume,

$$f_{xp} = \iint_A p n_x\, dA \qquad (1.7.4)$$

where p is the pressure and n_x is the x component of the normal vector. The force f_{xd} is the integrated x directed drag, friction or shear force. This force is normally defined in terms of the friction factor as (Bird et al., 1960)

$$f_{xd} = f \cdot \left(\frac{KE}{\text{vol}}\right) \cdot \text{(wetted surface area)}$$

where f is the Fanning friction factor that is correlated for different geometries as a function of the Reynolds number.

The structural force, or resultant of the forces acting on the system is r_x. This force appears when the control volume cuts through solid objects such as the containing wall of a pipe.

The conservation principle for the rate of change of the x component of momentum therefore becomes

$$\begin{array}{c}\text{Rate of}\\\text{Accumulation of}\\\text{Momentum}\end{array} = \begin{array}{c}\text{Rate of}\\\text{Momentum}\\\text{Into System}\end{array} - \begin{array}{c}\text{Rate of}\\\text{Momentum Out}\\\text{of System}\end{array} + \begin{array}{c}\text{Rate of}\\\text{Generation of}\\\text{Momentum}\end{array}$$

$$\left[\frac{d}{dt}\iiint_V \rho v_x\, dV\right] = [0] - \left[\iint_A (\rho v_x)(\boldsymbol{v}\cdot\boldsymbol{n})\, dA\right]$$
$$+ [r_x + f_{xp} + f_{xd} + f_{xg}] \qquad (1.7.5)$$

Using vector notation to express all components, we have

$$\frac{d}{dt} \iiint_V \rho \boldsymbol{v}\, dV = - \iint_A \rho \boldsymbol{v}(\boldsymbol{v} \cdot \boldsymbol{n}) dA + \boldsymbol{r} + \boldsymbol{f}_p + \boldsymbol{f}_d + \boldsymbol{f}_g \quad (1.7.6)$$

1.7.1 Tapered Tube Geometry

Applying the momentum balance to the simplified tapered tube geometry of Figure 1.2, where we have only flow in and out in the x direction, gives

| Rate of Accumulation of Momentum | = | Rate of Momentum Into System | − | Rate of Momentum Out of System | + | Rate of Generation of Momentum |

$$\left[\frac{d}{dt} \iiint_V \rho v_x\, dV \right] = \left[\frac{w_1 \langle v_x^2 \rangle_1}{\langle v_x \rangle_1} \right] - \left[\frac{w_2 \langle v_x^2 \rangle_2}{\langle v_x \rangle_2} \right] \quad (1.7.7)$$
$$+ [r_x + f_{xp} + f_{xd} + f_{xg}]$$

1.7.2 Comparison Between Mechanical Energy and Momentum Balances

Both the mechanical energy equation, (1.6.12), and momentum balance equation, (1.7.7), are used for the analysis of isothermal flow problems. Most applications fall into one of three classes:

1. Calculate viscous frictional losses for a given flow.

2. Calculate the force for a given flow.

3. Calculate the pressure change for a given flow.

The first class of problems, calculating the viscous frictional losses, involves the use of the mechanical energy equation since these terms appear explicitly in that equation. The second class of problems, calculating a force, involves the use of the momentum balance since forces enter the momentum balance explicitly. Class three problems, calculating the pressure, can be approached by using either the momentum or mechanical energy equations since pressure terms appear in both equations. The choice of which equation to use depends upon what type of information is specified. If viscous frictional losses are given, then the mechanical energy balance is used. If force information is specified, then the momentum balance is used.

EXAMPLE 1.7 Horizontal Converging Nozzle:

Water is flowing at 9.45×10^{-3} m^3/s through a horizontal converging nozzle. The upstream ID is 7.62 cm (3 in.), and the downstream ID is 1 in. Calculate the resultant force on the nozzle when it discharges to the atmosphere. Consider that the nozzle is attached at its upstream end and that frictional forces are negligible.

The control volume for this system as shown in Figure 1.13 includes the walls of the pipe since we are interested in computing the structural or resultant force.

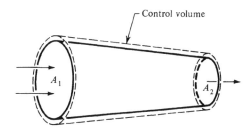

Figure 1.13: Horizontal Converging Nozzle.

The momentum balance equation (1.7.7) for this system is

| Rate of Accumulation of Momentum | = | Rate of Momentum Into System | − | Rate of Momentum Out of System | + | Rate of Generation of Momentum |

$$[0] = [w\langle v_x \rangle_1] - [w\langle v_x \rangle_2] + [r_x + f_{xp}] \qquad (1.7.8)$$

The pressure force is given as difference between the pressure p_1 acting on area A_1 and the sum of the atmospheric pressure, p_a, acting on the area A_2 and the difference of the areas $(A_1 - A_2)$. Essentially, the atmospheric pressure acts on the entire area A_1 when the control volume includes the walls of the pipe.

$$f_{xp} = A_1 p_1 - \left(A_2 p_a + p_a(A_1 - A_2) \right) \qquad (1.7.9)$$

or

$$f_{xp} = A_1(p_1 - p_a) \qquad (1.7.10)$$

where p_a is the atmospheric pressure. We will use equation (1.7.8) to compute the resultant force r_x.

To compute the pressure drop, we use the mechanical energy balance equation (1.6.18), which becomes at steady state and with negligible viscous frictional losses

| Rate of Accumulation of Mechanical Energy | = | Rate of Mechanical Energy In | − | Rate of Mechanical Energy Out | + | Rate of Generation of Mech. Energy |

$$[0] = \left[\frac{\langle v_x \rangle_1^2}{2} \right] - \left[\frac{\langle v_x \rangle_2^2}{2} \right] + \left[-\frac{p_2 - p_1}{\rho} \right] \qquad (1.7.11)$$

The information flow diagram for this system is shown in Figure 1.14. The numerical answer for the problem is $r_x = -625\ N$. Since the resultant force on the nozzle is negative, it is acting in the direction opposite to the flow. The pipe wall is therefore in tension.

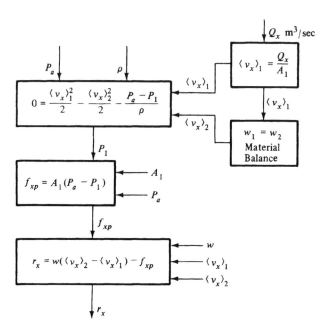

Figure 1.14: Information–Flow Diagram for Example 1.7.

EXAMPLE 1.8 Jet Ejector:

A jet ejector or jet pump is a device with no moving parts that is widely used for moving liquids between tanks and lifting corrosive liquids. The basic configuration is shown in Figure 1.15. The high velocity fluid or jet is introduced into a slow flowing or suction line. The mixing is rather chaotic, but a few diameters downstream, the flow is again uniform. The downstream pressure is increased from the upstream suction pressure. We wish to determine the downstream pressure. Either the mechanical energy or momentum balance can be used in principle. However, in this application, there is no easy way to compute the viscous frictional losses. This means that the mechanical energy equation should not be used. Instead, we can assume that because the distance from planes 1 and 2 is short, the drag force term in the momentum balance will not be important. We will therefore use the momentum balance for this problem. At steady state with the gravity, drag, and resultant forces set to zero, equation (1.7.7) becomes

$$0 = \frac{w_1 \langle v_1^2 \rangle}{\langle v_1 \rangle} - \frac{w_2 \langle v_2^2 \rangle}{\langle v_2 \rangle} + f_{xp} \qquad (1.7.12)$$

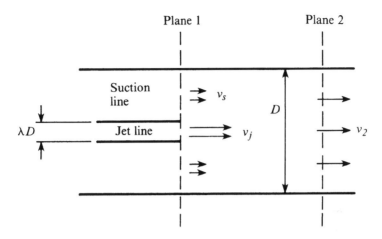

Figure 1.15: Jet Ejector.

The pressure force f_{xp} is computed at the control volume surfaces to be

$$f_{xp} = p_1 A - p_2 A \qquad (1.7.13)$$

A mass balance around the control volume gives

$$w_1 = w_2 \qquad (1.7.14)$$

which for a constant area A yields

$$\langle v_1 \rangle = \langle v_2 \rangle \qquad (1.7.15)$$

The average inlet velocity is computed from the definition of the average velocity as

$$\langle v_1 \rangle = \frac{\int\int v\, dA}{\int\int dA} = \frac{\int\int v_s\, dA_s + \int\int v_j\, dA_j}{A} \qquad (1.7.16)$$

or

$$\langle v_1 \rangle = \frac{\frac{v_s \pi (D^2 - \lambda^2 D^2)}{4} + \frac{v_j \pi \lambda^2 D^2}{4}}{\pi D^2 / 4} \qquad (1.7.17)$$

which simplifies to

$$\langle v_1 \rangle = v_s(1 - \lambda^2) + v_j \lambda^2 \qquad (1.7.18)$$

The average of the inlet velocity squared can be computed in a like manner as

$$\langle v_1^2 \rangle = v_s^2(1 - \lambda^2) + v_j^2 \lambda^2 \qquad (1.7.19)$$

The pressure rise from the momentum balance is therefore

$$p_2 - p_1 = \rho(\langle v_1^2 \rangle - \langle v_2^2 \rangle) \qquad (1.7.20)$$

Using the expressions for the appropriate averaged velocities and some algebraic manipulations, we get

$$p_2 - p_1 = \rho \lambda^2 (1 - \lambda^2)(v_j - v_s)^2 \qquad (1.7.21)$$

PROBLEMS

1.1. It is customary to replace $\langle v^2 \rangle / \langle v \rangle$ by $\langle v \rangle$ in the momentum balance equation (1.7.7). Using the 1/7 power law for the velocity profile in turbulent flow in a pipe, i.e.,

$$\frac{v_z}{v_{z,\,\text{max}}} = \left(1 - \frac{r}{R}\right)^{1/7}$$

calculate the errors in making thise assumption for turbulent flow. What are the errors if the flow is laminar?

1.2. Calculate the pressure drop for air at 38°C and 1 atm flowing at 250 kg/hr through a bed of 1.27 cm (1/2-in.) spheres. The bed is 10 cm in diameter and 20 cm high and has a porosity of 0.38. Consult *Perry's Chemical Engineering Handbook* for an appropriate correlation to calculate h_L.

1.3. A research team is designing a flow system for a nuclear reactor to study corrosion problems. The equipment as constructed consists of 70 m (total equivalent length) of 2.54 cm (1–in.) ID stainless-steel pipe. Molten bismuth is pumped from a melt tank maintained at 350°C and 5 microns absolute pressure, through a test section included in the 70 m and back to the melt tank. If the bismuth velocity is 0.305 m/s, how much theoretical power ($\eta_p = 1.0$) must be supplied to a pump placed in the line? Assume a constant temperature of 350°C is maintained by suitable insulation.

Liquid bismuth properties: Viscosity = 1.28 centipoise
                             Density = 9.8×10^3 kg/m^3

1.4. A conical tank of height 30 cm and radius 10 cm is initially filled with fluid. After 6 hr the height of the fluid is 27 cm. If the fluid evaporates at a rate proportional to the surface area exposed to the air, find a formula for the volume of fluid in the tank as a function of time.

1.5. Analysis of the flow through an ejection molder (see Figure 1.16). In your analysis assume: (i) isothermal conditions; (ii) steady state; (iii) $\rho = \text{constant} = 1.2$ g/cm^3; and (iv) Newtonial fluid with $\mu = 3.16 \times 10^{-3}$ g/cm sec.

  a. Calculate the average velocity in the nozzle. The nozzle has a diameter of 0.198 cm and is 7.6 cm long. Neglect entrance effects. Assume turbulent flow. The pressure at the nozzle entrance is 7000 kPa.

  b. What is the volumetric flow rate through the nozzle?

  c. What is the ram speed?

  d. What is the rate of work required to move the ram? Do not neglect frictional terms through the nozzle.

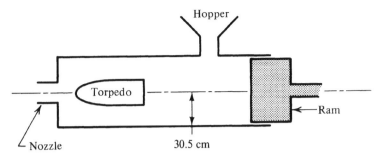

Figure 1.16: Problem 1.5.

1.6. The waste acid from a nitrating process contains 23 percent HNO_3, 57 percent H_2SO_4, and 20 percent H_2O by weight. This acid is to be concentrated to contain 27 percent HNO_3 and 60 percent H_2SO_4 by addition of concentrated sulfuric acid containing 93 percent H_2SO_4 and concentrated nitric acid of 90 percent HNO_3. Calculate the weights of waste and concentrated acids needed to obtain 1000 kg/hr of the desired mixture.

1.7. A centrifugal pump is being used to supply 6.3×10^{-3} m³/s of water to the condensers on a gasoline stabilizer unit. The pump suction is 14 kPa and the discharge pressure 1400 kPa. The pump operates adiabatically. The temperature of the water in and out of the pump is carefully measured and is found to rise from 16 to 16.19°C. Calculate the fraction of the energy supplied to the pump that is dissipated through friction. The suction and discharge lines to the pump are the same size.

1.8. Evaporation and Bypass in Orange Juice Concentration

In a process for concentrating 1000 kg/min of freshly extracted orange juice containing 12.5 wt percent solids, the juice is strained, yielding 800 kg/min of strained juice and 200 kg/min of pulpy juice. The strained juice is concentrated in a vacuum evaporator to give an evaporated juice of 58 percent solids. The 200 kg/min of pulpy juice is bypassed around the evaporator and mixed with the evaporated juice in a mixer to improve the flavor. This final concentrated juice contains 42 wt percent solids. Calculate the concentration of solids in the strained juice, the flow rate of the final concentrated juice, and the concentration of the solids in the pulpy juice bypassed.

1.9. An artificial kidney is a device that removes water and waste metabolites from blood. In one such device, the *hollow fiber hemodialyzer*, blood flows from an artery through the insides of a bundle of hollow cellulose acetate fibers, and dialyzing fluid, which consists of water and various dissolved salts, flows on the outside of the fibers. Water and waste metabolites—principally urea, creatinine, uric acid, and phosphate ions—pass through the fiber walls into the dialyzing fluid, and the purified blood is returned to a vein. (See Figure 1.17.)

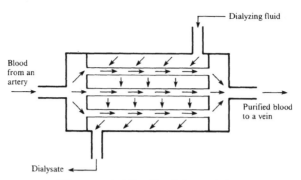

Figure 1.17: Problem 1.9.

Suppose that at some time during a dialyzation, the arterial and venous blood conditions are as follows:

	Arterial (Entering) Blood	Venous (Exiting) Blood
Flow rate	150.0 ml/min	149.2 ml/min
Urea (H_2NHONH_2) concentration	1.90 mg/ml	1.75 mg/ml

a. Calculate the rates at which urea and water are being removed from the blood.

b. If the dialyzing fluid enters at a rate of 1500 ml/min, calculate the concentration of urea in the exiting dialysate. (Neglect the urea volume.)

c. Suppose we want to reduce the patient's urea level from an initial value of 2.7 mg/ml to a final value of 1.1 mg/ml. If the total blood volume is 5 ℓ and the average rate of urea removal is that calculated in (a), how long must the patient be dialyzed? (Neglect the loss in total blood volume due to the removal of water in the dialyzer.)

1.10. Two tanks of 10000–ℓ capacity each are arranged so that when water is fed into the first, an equal quantity of solution overflows from the first to the second tank and likewise from the second to some point out of the system. Agitators keep the contents of each tank uniform in concentration. To start, let each of the tanks be full of a solution of concentration C_0 kg/ℓ. Run water into the first tank at 50 ℓ/min, and let the overflows function as described. Calculate the time required to reduce the concentration in the first tank to $C_0/10$. Calculate the concentration of the second tank at that time.

1.11. Find analytic solutions to:

a. $t\dfrac{dx}{dt} + (1+t)x = e^{-t}$ when $x = 0$ at $t = 1$

b. $(26 + t^2)\, dt = t\, dx$ when $x = 0$ at $t = 2$

c. $tx\, dt = t^2 dx - x^2 dt$ when $x = 1$ at $t = 1$

d. $\dfrac{dx}{dt} = \sec\dfrac{x}{t} + \dfrac{x}{t}$ when $x = \pi$ at $t = 2$

1.12. The following data have been generated in your laboratory for the decomposition of the very pungent sulfuryl chloride.

$$SO_2Cl_2 \to SO_2 + Cl_2$$

Time (min)	3.4	15.7	28.1	41.1	54.5	68.3	96.3
Total Pres.(mm Hg)	325	335	345	355	365	375	395

The reaction was carried out in a closed, constant volume reactor at 279.2°C with an initial pressure of 322 mm Hg.

You are requested to develop a rate expression for this reaction. Try both first– and second–order rate laws to determine which one best fits your data. The sulfuryl chloride can be considered as an ideal gas. Since two moles of gas are generated per mole consumed, the total pressure of the reaction vessel can be related to concentration.

1.13. A cylindrical tank 1.52 m in diameter and 7.62 m high contains cottonseed oil having a density of 917 kg/m³. The tank is opened to the atmosphere. A discharge hole of 15.8 mm diameter is located in the center of the bottom of the tank. The surface of the liquid is located 6.1 m above the drainage hole. The drainage hole is opened, draining the liquid level from $H = 6.1$ m to $H = 4.57$ m. Calculate the time in seconds to do this. Be very careful in defining your control volume and making any simplifying assumptions. One you can make is the following: because of the large tank diameter, accumulation of kinetic energy may be neglected.

1.14. Develop analytic solutions to:

$$x'' + 2x' + 10x = 3t^2$$

$$x'' + 4x' + 5x = 2e^t + te^t$$

1.15. Startup of an Equilibrium Still.

Consider the case of starting the equilibrium still shown in Figure 1.18. The still is purifying benzene and toluene from a small amount of essentially nonvolatile impurity. It is initially charged with 20 kg mol of feed stock of composition $x_F = 0.32$ mol fraction benzene. Feed is supplied at the rate of 10 kg mol/hr, and the heat is adjusted so that the total moles of liquid in the still remain constant at 20 kg mol.

Compute the time required for the composition of overhead product y_D to fall to .4 mol fraction benzene. No liquid stock is removed from the still during this period.

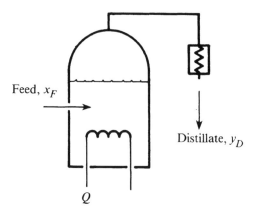

Figure 1.18: Problem 1.15.

You may assume that the relative volatility of benzene and toluene is constant at $\alpha = 2.48$. The definition of this relative volatility is

$$\frac{y_{benzene}}{y_{toluene}} = \alpha \frac{x_{benzene}}{x_{toluene}}$$

or

$$\frac{y_b}{1 - y_b} = \alpha \frac{x_b}{1 - x_b}$$

Hint: Using one component material balance and an overall material balance will ease the solution to this problem.

1.16. Consecutive Reversible Reactions in a Constant Volume Batch Reactor.

Solve for the dynamic response of species A for the following set of reversible reactions taking place in a constant volume batch reactor.

$$A \underset{k_2}{\overset{k_1}{\rightleftarrows}} B$$

$$B \underset{k_4}{\overset{k_3}{\rightleftarrows}} C$$

Assume 1 kg mol of A is present at the start. The forward rate per unit volume associated with the first reaction is

$$r_1 = k_1 C_A$$

The other rate terms are also given by the stoichiometry of the problem as

$$r_2 = k_2 C_B$$
$$r_3 = k_3 C_B$$
$$r_4 = k_4 C_C$$

Derive an analytical solution for the dynamic response of the number of kg moles of A (N_A) as a function of time. Hint: Write two component material balances and an overall balance. Through substitution and differentiation of the A component material balance, get one differential equation which is only a function of N_A and its derivatives. Solve the differential equation. Be sure to evaluate the integration constants using known boundary condition information.

1.17. Analysis of a Turbojet Engine

Figure 1.19 shows a turbojet engine moving through still air at a constant velocity, v_0. As the air enters the diffuser, it slows down and the pressure rises. The compressor further increases the pressure as the air is forced into the combustion chamber, where fuel is added and combustion takes place. The combustion mixture passes through the turbine driving the compressor and expands in the exhaust nozzle. It leaves the nozzle close to atmospheric pressure and at a higher velocity than the entering air.

a. Compute the velocity of exit gas.

b. Compute the thrust (force) that the engine develops.

Data:
1. Temperature rise due to reaction is $200°C$
2. Ambient air temperature = $0°C$
3. ρ_{air} at $0°C = 1.394 \times 10^{-3}$ g/cm^3
4. Cross section of inlet and outlet = 0.1 sq m
5. Rate of fuel consumption = 20 g/sec
6. Inlet gas velocity = v_0 = 100 cm/sec

1.18. Air flowing at a constant mass rate of 0.1 kg/s is heated from $20°C$ to $90°C$. The heating is accomplished in a vertical heat exchanger of constant cross-sectional area. The exchanger is 4 m long. The pressure of the airstream entering at the bottom of the exchanger is 120 kN/m^2, and a 10 kN/m^2 pressure drop occurs through the apparatus. The average linear velocity of the inlet air is 5 m/s. Assuming air to behave as an ideal gas and to have a heat capacity of $c_p = 1000$ J/kg°C, calculate the net heat input to the air in J/s.

REFERENCES

Bennett, C. O. and Meyers, J. E., *Momentum, Heat, and Mass Transfer*, Third Ed., McGraw-Hill, New York (1982).

Bird, R. B., Stewart, W. E., and Lightfoot, E. N., *Transport Phenomena*, Wiley, New York (1960).

Denbigh, K., *The Principle of Chemical Equilibrium*, Cambridge Univ. Press (1963).

Smith, J. M. and Van Ness, H. C., *Introduction to Chemical Engineering Thermodynamics*, Second Ed., McGraw-Hill, New York (1959).

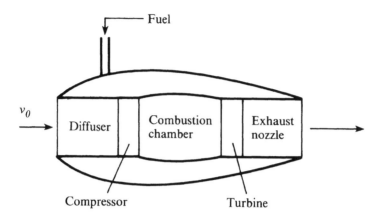

Figure 1.19: Problem 1.17.

Chapter 2

STEADY-STATE LUMPED SYSTEMS

The study of stoichiometry and unit operations concerns itself mainly with the application of steady–state macroscopic balances to chemical process problems. When considering small subsystems of chemical plants, the number of describing relations are small and the development of a computational strategy is not difficult. Usually the relations can be solved directly by partitioning the equations, that is, solving each equation of the equation set for a single unknown variable in a sequential manner. As equation sets become coupled, namely, as each relation involves more of the unknown variables, the probability that an equation set can be partitioned decreases. When an equation set cannot be partitioned, the equations must be solved simultaneously or an iterative scheme devised.

In this chapter, numerical methods for simultaneously solving sets of linear and nonlinear algebraic equations are first discussed. Matrix techniques are used for linear equations. The mathematical notion of a matrix as a linear transformation is presented as well as the computer techniques such as Gaussian elimination for solving matrix algebra. A number of methods are presented for the solution of a single nonlinear equation and sets of nonlinear equations. The use of scientific software library routines such as the NAG routines and MATLAB is discussed. Algorithms for structuring sets of algebraic equations into simple computational strategies are presented.

2.1 METHODS

By a solution to a set of m simultaneous equations in n unknowns we mean those values of the unknowns, $x_1, x_2 \cdots, x_n$ that satisfy

$$\begin{aligned} f_1(x_1, x_2, \cdots, x_n) &= 0 \\ f_2(x_1, x_2, \cdots, x_n) &= 0 \\ &\vdots \\ f_m(x_1, x_2, \cdots, x_n) &= 0 \end{aligned} \quad (2.1.1)$$

or, in vector notation,

$$\boldsymbol{f}(\boldsymbol{x}) = 0 \quad \text{when} \quad \boldsymbol{x} \ \epsilon \ R^n \ \text{ and } \ \boldsymbol{f} \ \epsilon \ R^m \ , \quad (2.1.2)$$

that is, $\boldsymbol{x}$ is contained in an n dimensional real vector space and $\boldsymbol{f}$ in an m dimensional real vector space.

In general, no solution, a number of solutions, or an infinite number of solutions may exist. We will be concerned with the case where $m = n$ and we seek a single, real, physically meaningful solution.

Systems of linear equations arise with great regularity in material balance problems. An example of a two-dimensional linear system is

$$2x_1 + 3x_2 = 8 \qquad x_1 - x_2 = -1 \quad (2.1.3)$$

Systems of nonlinear equations arise in many problems, including calculations of dewpoints and material balances for recycle processes. An example of a two-dimensional system of nonlinear equations is

$$x_1^2 + x_2^2 = 5 \qquad x_1^3 + x_2^3 x_1 = 9 \quad (2.1.4)$$

2.1.1 Partitioning Equations

Although we can approach the problem of solving n equations in n unknowns directly, it may be easier to break the equations into a solution strategy where each equation can be used to solve for a single variable sequentially. For example, consider the following three equations in three unknowns:

$$f_1(x_1, x_3) = 0 \qquad f_2(x_1) = 0 \qquad f_3(x_2, x_3) = 0 \quad (2.1.5)$$

Rather than solve all three at one time, we can solve first for variable x_1 from relation f_2; then, using this value of x_1, solve for variable x_3 from relation f_1; and, finally, solve for variable x_2 from relation f_3. The equations have been partitioned into a sequential solution strategy.

2.1.2 Tearing Equations

Suppose we wish to solve the following set of four equations in four unknowns

$$f_1(x_2, x_3) = 0 \qquad f_3(x_1, x_2, x_3, x_4) = 0$$
$$f_2(x_2, x_3, x_4) = 0 \qquad f_4(x_1, x_2, x_4) = 0 \qquad (2.1.6)$$

These equations cannot be partitioned into a sequential solution strategy. We can solve the four relations simultaneously. However, an alternative approach to the problem is available. Suppose that we estimate variable x_3. Then, variable x_2 could be solved from relation f_1. With x_3 estimated and x_2 known, variable x_4 could be solved from relation f_2. Knowing x_2, x_3, and x_4, variable x_1 could then be solved from f_3. Finally, we can check our estimate of x_3 by relation f_4. If f_4 is zero, we would be finished. If not, a new value of x_3 would have to be assumed.

This procedure is called tearing the equations. The tearing (the picking of the iterative or recycle variable) is not unique, and various criteria for what we mean by the best tear are possible. Generally the best tear is dictated not only by the structure of the equations but by the way in which the variables enter into the equations. It is not always clear whether it is more efficient to solve all equations at once or to use tearing techniques.

2.1.3 Simultaneous Solution

For very large sets of algebraic equations, typical in plant design problems, it is worthwhile to seek sequential computational schemes. However, for small sets of equations (less than 50 equations to a set), simultaneous solution to the describing equations becomes practical and advantageous. This is especially true when the equations are linear and matrix techniques can be applied.

2.2 SIMULTANEOUS SOLUTION OF LINEAR EQUATIONS

Linear equations have the form (m equations in n unknowns)

$$\begin{aligned} a_{11}x_1 + a_{12}x_2 + &\cdots + a_{1n}x_n = b_1 \\ a_{21}x_1 + a_{22}x_2 + &\cdots + a_{2n}x_n = b_2 \\ &\vdots \\ a_{m1}x_1 + a_{m2}x_2 + &\cdots + a_{mn}x_n = b_m \end{aligned} \qquad (2.2.1)$$

In matrix notation this may be expressed

$$\boldsymbol{A}\boldsymbol{x} = \boldsymbol{b} \qquad (2.2.2)$$

The matrix $\boldsymbol{A}$ transforms the vector space X, defined by the vector $\boldsymbol{x}$, into a subspace of the vector space B, defined by the vector $\boldsymbol{b}$. In order to solve the equation $\boldsymbol{Ax} = \boldsymbol{b}$, we seek the solution vector $\boldsymbol{x}$. Using the definition of the inverse of $\boldsymbol{A}$, $\boldsymbol{A}^{-1}\boldsymbol{A} = \boldsymbol{I}$, we have

$$\boldsymbol{x} = \boldsymbol{A}^{-1}\boldsymbol{b} \tag{2.2.3}$$

If the rank $\boldsymbol{A}$ is equal to the number of unknowns, the solution is unique. If the rank is less than the number of unknowns, there are an infinite number of solutions (Carnahan et al., 1969).

Although a large number of techniques exist for solving a well-defined set of linear equations (one having a unique solution), the most efficient methods are those based on the method of Gaussian elimination.

In order to understand Gaussian elimination, consider the following set of three linear equations:

$$a_{11}x_1 + a_{12}x_2 + a_{13}x_3 = b_1 \tag{2.2.4}$$

$$a_{21}x_1 + a_{22}x_2 + a_{23}x_3 = b_2 \tag{2.2.5}$$

$$a_{31}x_1 + a_{32}x_2 + a_{33}x_3 = b_3 \tag{2.2.6}$$

As a first step, let us replace the second equation, (2.2.5), by the result of adding to it the first equation, (2.2.4), multiplied by $(-a_{21}/a_{11})$. This gives in place of equation (2.2.5)

$$a_{22}^*x_2 + a_{23}^*x_3 = b_2^* \tag{2.2.7}$$

Note that we have eliminated the x_1 term. Similarly, we replace the third equation, (2.2.6), by the result of adding it to the first equation, (2.2.4), multiplied by $(-a_{31}/a_{11})$. The resulting equation set is

$$a_{11}x_1 + a_{12}x_2 + a_{13}x_3 = b_1 \tag{2.2.4}$$

$$a_{22}^*x_2 + a_{23}^*x_3 = b_2^* \tag{2.2.7}$$

$$a_{32}^*x_2 + a_{33}^*x_3 = b_3^* \tag{2.2.8}$$

We now replace the new third equation, (2.2.8), by the result of adding it to the new second equation, (2.2.7), multiplied by (a_{32}^*/a_{22}^*). The resulting equation set is

$$a_{11}x_1 + a_{12}x_2 + a_{13}x_3 = b_1 \tag{2.2.4}$$

$$a_{22}^*x_2 + a_{23}^*x_3 = b_2^* \tag{2.2.7}$$

$$a_{33}''x_3 = b_3'' \tag{2.2.9}$$

We may now use the process of back substitution to solve for x_1, x_2, and x_3. Namely, we solve for x_3 from equation (2.2.9); then, knowing x_3, solve for x_2 from equation (2.2.7); and finally, knowing x_2 and x_3, solve for x_1 from equation (2.2.4).

The Gaussian elimination procedure can therefore be considered as two operations, a forward pass and a backward pass. The objective of the forward pass is to transform the original matrix into an upper-triangular matrix. The backward pass calculates the unknown variables using the back substitution procedure.

An alternative method of Gauss–Jordan elimination is possible. It is a one pass method that eliminates the need for the back substitution phase of the Gaussian elimination procedure. Essentially, the Gauss–Jordan method transforms the original A matrix into an identity matrix through row operations. This allows for the direct solution of the unknown x vector. While the Gauss–Jordan method will involve more computational effort for solving a single problem, it is more efficient when we have several right–hand side vectors b that need solution with the same A matrix. The Gauss–Jordan procedure is illustrated below with R_i standing for the original ith row and R_i^n being the new ith row.

Step 1

Original A Matrix		Original b vector	Operation
2	3	-2	$R_1^n = \dfrac{R_1}{2}$
2	5	7	$R_2^n = R_2 - 2R_1^n$

Resultant A Matrix		Resultant b vector
1	$\dfrac{3}{2}$	-1
0	2	9

Step 2

Starting A Matrix		Original b vector	Operation
1	$\dfrac{3}{2}$	-1	$R_1^n = R_1 - \dfrac{3}{2} R_2^n$
0	2	9	$R_2^n = \dfrac{R_2}{2}$

Resultant A Matrix		Resultant b vector
1	0	$-\dfrac{31}{4}$
0	1	$\dfrac{9}{2}$

The solution to the problem is therefore

$$x = \begin{pmatrix} -\dfrac{31}{4} \\ 9 \\ 2 \end{pmatrix}$$

Most computer programs that perform Gaussian elimination do so by operating with elementary lower triangular matrices in order to generate the upper triangular matrix needed for back substitution. The steps are called factorization of the A matrix or LU (Lower triangular, Upper triangular) decomposition.

The forward pass of Gaussian elimination results in an upper triangular matrix, U, which has the general form

$$U = \begin{bmatrix} 1 & u_{12} & u_{13} & \cdots & u_{1n} \\ 0 & 1 & u_{23} & \cdots & u_{2n} \\ 0 & 0 & 1 & \cdots & u_{3n} \\ \vdots & \vdots & \vdots & \ddots & \vdots \\ 0 & 0 & 0 & \cdots & 1 \end{bmatrix} \quad (2.2.10)$$

This upper triangular matrix, U, can be related to the original matrix A using a lower triangular matrix L given as

$$L = \begin{bmatrix} \ell_{11} & 0 & 0 & \cdots & 0 \\ \ell_{21} & \ell_{22} & 0 & \cdots & 0 \\ \ell_{31} & \ell_{32} & \ell_{33} & \cdots & 0 \\ \vdots & \vdots & \vdots & \ddots & \vdots \\ \ell_{n1} & \ell_{n2} & \ell_{n3} & \cdots & \ell_{nn} \end{bmatrix} \quad (2.2.11)$$

In order that $LU = A$, we have

$$\begin{bmatrix} \ell_{11} & 0 & 0 & \cdots & 0 \\ \ell_{21} & \ell_{22} & 0 & \cdots & 0 \\ \ell_{31} & \ell_{32} & \ell_{33} & \cdots & 0 \\ \vdots & \vdots & \vdots & \ddots & \vdots \\ \ell_{n1} & \ell_{n2} & \ell_{n3} & \cdots & \ell_{nn} \end{bmatrix} \begin{bmatrix} 1 & u_{12} & u_{13} & \cdots & u_{1n} \\ 0 & 1 & u_{23} & \cdots & u_{2n} \\ 0 & 0 & 1 & \cdots & u_{3n} \\ \vdots & \vdots & \vdots & \ddots & \vdots \\ 0 & 0 & 0 & \cdots & u_{nn} \end{bmatrix}$$

$$= \begin{bmatrix} a_{11} & a_{12} & a_{13} & \cdots & a_{1n} \\ a_{21} & a_{22} & a_{23} & \cdots & a_{2n} \\ a_{31} & a_{32} & a_{33} & \cdots & a_{3n} \\ \vdots & \vdots & \vdots & \ddots & \vdots \\ a_{n1} & a_{n2} & a_{n3} & \cdots & a_{nn} \end{bmatrix} \quad (2.2.12)$$

Equation (2.2.12) can be expressed in terms of the following algebraic relations that allow for the calculation of the elements of the upper triangular matrix u_{ij} and the elements of this lower triangular matrix ℓ_{ij}.

$$\ell_{i1} = a_{i1} \qquad i = 1, \cdots, n \qquad (2.2.13)$$

$$u_1 = \frac{a_1}{\ell_{11}} \qquad j = 2, \cdots, n \qquad (2.2.14)$$

$$\ell_{ij} = a_{ij} - \sum_{k=1}^{j-1} \ell_{ik} u_{kj} \qquad \begin{array}{l} j = 2, \cdots, n-1 \\ i = j, \cdots, n \end{array} \qquad (2.2.15)$$

$$u_{ij} = \frac{a_{ij} - \sum_{i=1}^{j-1} \ell_{ik} u_{kj}}{\ell_{ii}} \qquad \begin{array}{l} i = 2, \cdots, n-1 \\ j = i+1, \cdots, n \end{array} \qquad (2.2.16)$$

Equations (2.2.15) and (2.2.16) are done sequentially getting a new column of L and then a new row of U on each pass.

$$\ell_{nn} = a_{nn} - \sum_{k=1}^{n-1} \ell_{nk} u_{kn} \qquad (2.2.17)$$

Once the A matrix is decomposed into an upper triangular matrix, U, and a lower triangular matrix, L, they can be used to solve the set of equations

$$Ax = b \qquad (2.2.18)$$

using a Gaussian elimination procedure. The solution is obtained by using a forward pass and then a backward pass. The forward pass is given by the equations

$$Le = b \qquad (2.2.19)$$

or

$$e_1 = \frac{b_1}{\ell_{11}}$$

$$e_i = \frac{b_i - \sum_{j=1}^{i-1} \ell_{ij} e_j}{\ell_{ii}} \qquad i = 2, \cdots, n \qquad (2.2.20)$$

The backward pass gives the solution

$$Ux = e \qquad (2.2.21)$$

or

$$x_n = e_n$$

$$x_i = e_i - \sum_{j=i+1}^{n} u_{ij} x_j \qquad i = n-1, \cdots, 1 \qquad (2.2.22)$$

If any of the pivot elements, a_{ii}, of $\mathbf{A}$ are zero, the Gaussian elimination procedure and the $\mathbf{LU}$ factorization Gaussian elmination procedure will require a division by zero. To avoid this, partial pivoting strategies are employed. Partial pivoting means at the kth step we interchange the rows of the matrix so that the largest remaining element in the kth column is used as the pivot element (Carnahan et al., 1969; Ayyub and McCuen, 1996).

Some computer programs for the solution of sets of algebraic equations also allow for iterative improvement. First, a residual, r_i, is computed using the current solution vector $\mathbf{x}_{i-1}$ as

$$\mathbf{r}_i = \mathbf{A}\mathbf{x}_{i-1} - \mathbf{b} \qquad (2.2.23)$$

Then a refinement vector, $\mathbf{p}_i$, is computed

$$\mathbf{p}_i = \mathbf{A}^{-1}\mathbf{r}_i \qquad (2.2.24)$$

Finally, a new improved value of the solution vector is given by

$$\mathbf{x}_i = \mathbf{x}_{i-1} + \mathbf{p}_i \qquad (2.2.25)$$

If $||\mathbf{p}_i|| < \epsilon ||\mathbf{x}_i||$, then the procedure is stopped. Here ϵ is a small number and $||\cdot||$ is the euclidian norm of the vector defined as $(\mathbf{p} \cdot \mathbf{p})^{1/2}$ or $(\mathbf{x} \cdot \mathbf{x})^{1/2}$.

The inverse of $\mathbf{A}$, $\mathbf{A}^{-1}$, can easily be computed from the $\mathbf{LU}$ factorization. Since

$$\mathbf{A}\mathbf{A}^{-1} = \mathbf{I} \qquad (2.2.26)$$

then the nth column vector of the $\mathbf{A}^{-1}$ matrix can be computed from the following n equations

$$\mathbf{A}\mathbf{a}_1^{-1} = \begin{pmatrix} 1 \\ 0 \\ 0 \\ \vdots \\ 0 \end{pmatrix} \qquad (2.2.27)$$

$$\vdots \quad \vdots \quad \vdots$$

$$\mathbf{A}\mathbf{a}_n^{-1} = \begin{pmatrix} 0 \\ 0 \\ 0 \\ \vdots \\ 1 \end{pmatrix} \qquad (2.2.28)$$

The right hand side of these equations is just the n column vectors of the identity matrix. Since $\mathbf{A}$ has been decomposed into $\mathbf{U}$ and $\mathbf{L}$, then these n solutions for $\mathbf{a}_1^{-1}$ to $\mathbf{a}_n^{-1}$ can be solved using the forward pass and backward pass equations given by (2.2.19) to (2.2.22). Finally $\mathbf{A}^{-1}$ is synthesized by

$$\mathbf{A}^{-1} = \begin{bmatrix} \mathbf{a}_1^{-1} \cdots \mathbf{a}_n^{-1} \end{bmatrix} \qquad (2.2.29)$$

2.2.1 MATLAB Software

MATLAB is an integrated technical computing environment that combines numeric computation, advanced graphics and visualization, and a high-level programming language. The MATLAB software was originally developed to be a matrix laboratory (MATrix LABoratory). Its capabilities have expanded greatly in recent years and is today a leading tool for engineering computation. Because MATLAB commands are similar to the way we express engineering concepts in mathematics, writing computer programs in MATLAB is much quicker than writing computer code in languages such as FORTRAN or C. Also, MATLAB provides excellent graphics capability that is easy to use. The MATLAB software consists of a basic package of mathematical routines as well as optional **toolboxes** that cover specific engineering application areas such as process control, neural networks, optimization, signal processing and symbolic mathematics. It also has a toolbox that makes available many of the NAG (Numerical Algorithms Group) routines. These routines are some of the best implementations of numerical methods that are currently available. MATLAB now offers a convenient computing environment for the solution of process simulation problems and will be used extensively in this book.

Several excellent books have been written that assist a new user of MATLAB. These include those by Etter (1993), Biran and Breiner (1995), and the Math Works, Inc. (1995). To begin MATLAB, select MATLAB from your program manager menu. You start programming when you see the MATLAB prompt ($\gg$). MATLAB uses two windows: a command window to enter commands and data and to print results, and a graphics window to generate plots. An important command is the abort command ($\wedge$C or control C). Often a mistake has been made and extraneous output is being displayed or you find yourself in a seemingly endless loop. At this point you want to abort the run. MATLAB commands are usually entered on separate lines. Multiple statements can be entered on the same line if separated by semicolons. A line can be continued to the next line by using ... at the end of the line. Comments can be entered on a line following following a percent sign (%). The `help` command is an important resource. When you enter the command `help`, a list of help topics is displayed (See Appendix B, Table B.1). You can then select the topic on which you require information. Specific topics are then displayed. You can then enter `help` on the specific topic to learn details about it.

In addition to executing commands entered through the keyboard, you can also use commands stored as files. These files all end in the extension .m and are called M-files because of the file name extension. Long programs are usually created in a text editor and stored as an M-file that can be executed in a MATLAB session. Short programs are usually created and run directly on-line. Since MATLAB is an interpreter language, you must have a correct executable statement before the next statement can be entered.

Appendix B gives reference tables on aspects of MATLAB that will be used in this book.

2.2.2 Linear Algebra Routines Available in MATLAB

Essentially two operations are available in MATLAB for solving linear algebraic equations

$$\boldsymbol{Ax} = \boldsymbol{b} \tag{2.2.30}$$

The first is the use of the inverse operation inv (Table B.18)

$$\gg \quad \text{x = inv (A) * b} \tag{2.2.31}$$

A more efficient method is the use of the \ or / operation which solves the set of equations using $\boldsymbol{LU}$ factorization and Gaussian elimination (Table B.18).

$$\gg \quad \text{x = A\textbackslash b} \tag{2.2.32}$$

The \ operation is the left division and / is right division. In this case, left division is used since the matrix $\boldsymbol{A}$ is to the left of vector $\boldsymbol{b}$. You can learn about these operations in MATLAB by entering help slash. The right division $\boldsymbol{B}/\boldsymbol{A}$ computes $(\boldsymbol{A}^T/\boldsymbol{B}^T)^T$ again using $\boldsymbol{LU}$ factorization and Guassian eliminination.

The $\boldsymbol{LU}$ factorization of the $\boldsymbol{A}$ matrix is available using the lu function in MATLAB (Table B.18).

$$\gg \quad [\text{L}, \text{U}, \text{P}] = \text{lu (A)} \tag{2.2.33}$$

where $\boldsymbol{L}$ is the lower triangular matrix with unity down the diagonal, $\boldsymbol{U}$ is an upper triangular matrix, and $\boldsymbol{P}$ a permutation matrix that keeps track of any row shifting. The result of this algorithm is

$$\boldsymbol{PA} = \boldsymbol{LU} \tag{2.2.34}$$

EXAMPLE 2.1 Six-Plate Absorption Column:

It is desired to develop the steady–state tray compositions for a six–plate absorption column. It can be assumed that a linear equilibrium relation holds between liquid (x_m) and vapor (y_m) on each plate:

$$y_m = ax_m + b \tag{2.2.35}$$

The inlet composition to the column x_0 and y_7 are specified along with the liquid (L) and gas (G) phase flow rates (moles/time). The system is shown schematically in Figure 2.1.

To solve the problem, a material balance is written on a representative tray, n, shown in Figure 2.2.

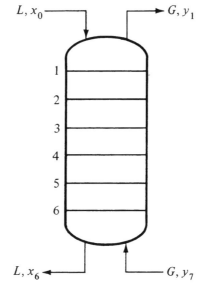

Figure 2.1: Absorption Column.

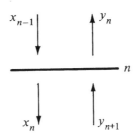

Figure 2.2: Typical Tray.

Applying the macroscopic mass balance yields

$$\text{Rate of mass in} = \text{Rate of mass out}$$
$$Lx_{n-1} + Gy_{n+1} = Lx_n + Gy_n \qquad (2.2.36)$$

Using the linear equilibrium relation in the mass balance gives

$$Lx_{n-1} + G(ax_{n+1} + b) = Lx_n + G(ax_n + b) \qquad (2.2.37)$$

or

$$Lx_{n-1} - (L + Ga)x_n + Gax_{n+1} = 0 \qquad (2.2.38)$$

The entire set of equations for the six-plate column is

$$\begin{aligned}
-(L+Ga)x_1 + Gax_2 &= -Lx_0 \quad \text{(where } x_0 \text{ is specified)} \\
Lx_1 - (L+Ga)x_2 + Gax_3 &= 0 \\
Lx_2 - (L+Ga)x_3 + Gax_4 &= 0 \\
Lx_3 - (L+Ga)x_4 + Gax_5 &= 0 \\
Lx_4 - (L+Ga)x_5 + Gax_6 &= 0 \\
Lx_5 - (L+Ga)x_6 &= -Ga\tfrac{y_7-b}{a} \quad \text{(where } y_7 \text{ is specified)}
\end{aligned} \qquad (2.2.39)$$

The simultaneous solution can be expressed in matrix notation as

$$\begin{bmatrix}
-(L+Ga) & Ga & & & & \\
L & -(L+Ga) & Ga & & & \\
 & L & -(L+Ga) & Ga & & \\
 & & L & -(L+Ga) & Ga & \\
 & & & L & -(L+Ga) & Ga \\
 & & & & L & -(L+Ga)
\end{bmatrix}$$

$$\begin{bmatrix} x_1 \\ x_2 \\ x_3 \\ x_4 \\ x_5 \\ x_6 \end{bmatrix} = \begin{bmatrix} -Lx_0 \\ 0 \\ 0 \\ 0 \\ 0 \\ -Ga\tfrac{y_7-b}{a} \end{bmatrix} \qquad (2.2.40)$$

The matrix equation is

$$\boldsymbol{Ax} = \boldsymbol{b} \qquad (2.2.41)$$

We want to compute the solution vector $\boldsymbol{x}$.

A typical set of parameters for this problem is

$$\begin{aligned}
a &= 0.72 & G &= 66.7 \text{ kg mol/min} \\
b &= 0 & L &= 40.8 \text{ kg mol/min}
\end{aligned}$$

We would like to solve for two cases. The first is when the liquid feed is pure ($x_0 = 0$) and the gas feed is 0.2 kg mol solute/kg mol inert ($y_7 = 0.2$). The second case is $x_0 = 0$ and $y_7 = 0.3$. Figure 2.3 shows a MATLAB file lin2_1.m that solves this problem. This file is available from the world wide web under: http://optimal.colorado.edu/~ramirez/chen4580.html. This M-file first defines the constants for the problem. It then generates the $\boldsymbol{A}$ matrix and two $\boldsymbol{b}$ vectors each associated with the two cases of interest for $y_7 = .2$ and $y_7 = .3$. Solutions to the problem are obtained both using in inv function and the Gaussian elimination procedure using the left division operation \. When you run this M-file, you will obtain the results shown as comment lines at the end of this file. Note that to the accuracy given in the default output the results are equivalent using either approach.

Steady–State Lumped Systems

Figure 2.3: MATLAB M-file `lin2_1.m` **to solve EXAMPLE 2.1**

```
% This is the M-file for solving the set of linear equations
% described in EXAMPLE 2.1.

%
% read in process parameters:
%
a = 0.72;
b = 0;
G = 66.7;
L = 40.8;

%
% read in initial conditions
%
x0 = 0; y7(1) = 0.2; y7(2) = 0.3;

%
% read in matrices A and B:
%
A = zeros(6);
A(1,1) = -(L+G*a);
A(2,1) = L;
A(1,2) = G*a;
A(2,2) = -(L+G*a);
A(3,2) = L;
A(2,3) = G*a;
A(3,3) = -(L+G*a);
A(4,3) = L;
A(3,4) = G*a;
A(4,4) = -(L+G*a);
A(5,4) = L;
A(4,5) = G*a;
A(5,5) = -(L+G*a);
A(6,5) = L;
A(5,6) = G*a;
A(6,6) = -(L+G*a);

B1 = zeros(6,1);
B1(1,1) = -L*x0;
B1(6,1) = -G*a*(y7(1)-b)/a;

B2 = zeros(6,1);
B2(1,1) = -L*x0;
B2(6,1) = -G*a*(y7(2)-b)/a;

%
% solve for the tray compositions:
%

%
% if you want to display the results with long format, use:
```

```
% (see "help format" for more info.)
%
%format long

%
% use matrix inversion:
%
x1 = inv(A)*B1
x2 = inv(A)*B2

%
% for better numerical accuracy, use Gaussian elimination:
%
x1 = A\B1
x2 = A\B2


%
%=========
% results
%=========
%>> linear2_1
%

% results using matrix inversion

%x1 =
%
%      0.0614
%      0.1136
%      0.1579
%      0.1955
%      0.2275
%      0.2547


%x2 =

%      0.0921
%      0.1703
%      0.2368
%      0.2933
%      0.3413
%      0.3820
%
% results using matrix division

%x1 =

%      0.0614
%      0.1136
%      0.1579
%      0.1955
```

Steady–State Lumped Systems

```
%       0.2275
%       0.2547

%x2 =

%       0.0921
%       0.1703
%       0.2368
%       0.2933
%       0.3413
%       0.3820
```

EXAMPLE 2.2 Material Balances for Alcohol Distillation:

A company plans to make commercial alcohol by the distillation process shown in Figure 2.4. The process contains two distillation columns, both having reflux ratios of three to one. A feed of 10000 kg/hr is 80 percent (wt) water, 10 percent alcohol, and 10 percent organic material. The distillate from the first column is 60 percent alcohol while that from the second column is 95 percent. The bottom stream from the first column contains 80 percent of the organic material feed while the rest leaves the bottom of the second column. No alcohol is present in either of the bottom streams. Solve for the amount of each component in each stream.

Component material balances are written for the system. The component material balances written around a control volume which includes both still 1 and condenser 1 are

$$W_2 = W_7 + W_6 \quad (7)$$
$$A_2 = A_7 \quad (10)$$
$$R_2 = R_7 + R_6 \quad (11)$$

The reflux specifications are

$$W_5 - 3W_7 = 0 \quad (4)$$
$$A_5 - 3A_7 = 0 \quad (5)$$
$$R_5 - 3R_7 = 0 \quad (6)$$

and the material balances describing the split leaving condenser 1 can be expressed as

$$W_3 - \tfrac{4}{3}W_5 = 0 \quad (1)$$
$$A_3 - \tfrac{4}{3}A_5 = 0 \quad (2)$$
$$R_3 - \tfrac{4}{3}R_5 = 0 \quad (3)$$

Material balances written around a control volume which includes still 2, condenser 2, and heat exchanger 1 are

$$W_7 = W_{13} + W_{12} \quad (16)$$
$$A_7 = A_{13} \quad (19)$$
$$R_7 = R_{12} \quad (17)$$

The reflux specifications are

$$W_{11} - 3W_{13} = 0 \quad (14)$$
$$A_{11} - 3A_{13} = 0 \quad (15)$$

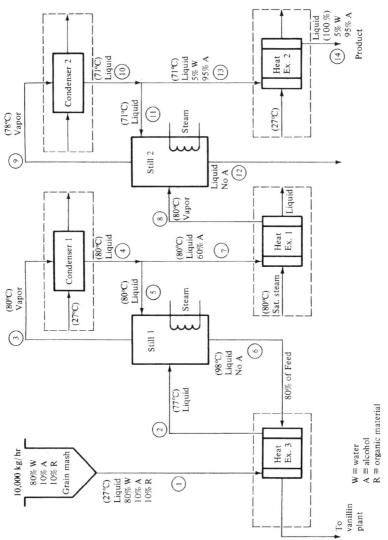

Figure 2.4: Alcohol Distillation Process for Example 2.2

Steady–State Lumped Systems

and the material balances describing the split leaving condenser 2 can be expressed as

$$W_9 - \tfrac{4}{3}W_{11} = 0 \quad (12)$$
$$A_9 - \tfrac{4}{3}A_{11} = 0 \quad (13)$$

From the problem specifications we also have

$$W_2 = 8000 \text{ kg/hr water feed}$$
$$A_2 = 1000 \text{ kg/hr alcohol feed}$$
$$R_2 = 1000 \text{ kg/hr organic material feed}$$

From the percent weight specifications, three independent constraining relations are obtained:

$$W_7 + R_7 - \tfrac{2}{3}A_7 = 0 \quad (9)$$
$$W_{13} - \tfrac{5}{95}A_{13} = 0 \quad (18)$$
$$R_6 - 0.8 \; R_2 = 0 \quad (8)$$

These equations can be expressed in matrix notation as

$$\boldsymbol{Ax = b}$$

Care should be taken in arranging the equations so that the diagonal elements all have nonzero entries. This procedure helps to organize the problem and allows the engineer to catch any redundant dependent equations that may have been written (can't put a value on the diagonal). It also aids in assuring that problems will not arise in using partial pivoting algorithms. It is convenient to start with the variables of the first stream and continue in the order of streams. For this example we started with stream 3 and ended with stream 13. The equations are then numbered so as to fill the diagonal elements of the matrix. (See Figure 2.5.)

Figure 2.6 gives the MATLAB file lin2_2.m that solves this material balance problem. Again the file is available from the world wide web. The program uses the left division Gaussian elimination solution technique \ to solve for the component flow rates of each stream. The results that you will obtain by running the file are given as comment lines. The variables have been identified and the units given in these comment lines, although that information will not appear when you run the program.

Figure 2.6: MATLAB M-file lin2_2.m to Solve Example 2.2

```
% This is the M-file for solving the set of linear equations
% described in EXAMPLE 2.2.
%
% read in matrices A and B:
%
A = zeros(19);
A(1,1) = 1; A(1,4) = -1.33;
A(2,2) = 1; A(2,5) = -1.33;
A(3,3) = 1; A(3,6) = -1.33;
```

$$
\begin{bmatrix}
1 & 0 & 0 & -1.33 & & & & & & & & & & & & & & & \\
0 & 1 & 0 & 0 & -1.33 & & & & & & & & & & & & & & \\
0 & 0 & 1 & 0 & 0 & -1.33 & & & & & & & & & & & & & \\
0 & 0 & 0 & 1 & 0 & 0 & -3 & 0 & & & & & & & & & & & \\
0 & 0 & 0 & 0 & 1 & 0 & 0 & 0 & -3 & & & & & & & & & & \\
0 & 0 & 0 & 0 & 0 & 1 & 0 & 1 & 0 & & -3 & & & & & & & & \\
0 & 0 & 0 & 0 & 0 & 0 & 1 & 0 & 0 & -0.67 & 1 & & & & & & & & \\
0 & 0 & 0 & 0 & 0 & 0 & 0 & 1 & 0 & 1 & 0 & & & & & & & & \\
0 & 0 & 0 & 0 & 0 & 0 & 0 & 0 & 1 & 0 & 1 & -1.33 & 0 & & & & & & \\
0 & 0 & 0 & 0 & 0 & 0 & 0 & 0 & 0 & 1 & 0 & 0 & -1.33 & & & & & & \\
0 & 0 & 0 & 0 & 0 & 0 & 0 & 0 & 0 & 0 & 1 & 0 & 0 & & & & & & \\
0 & 0 & 0 & 0 & 0 & 0 & 0 & 0 & 0 & 0 & 0 & 1 & 0 & -3 & 0 & 0 & 0 & -3 & 0 \\
0 & 0 & 0 & 0 & 0 & 0 & 0 & 0 & 0 & 0 & 0 & 0 & 1 & 0 & -3 & 0 & 0 & 0 & -3 \\
0 & 0 & 0 & 0 & 0 & 0 & 0 & 0 & -1 & 0 & 0 & 0 & 0 & 1 & 0 & 1 & 0 & & \\
0 & 0 & 0 & 0 & 0 & 0 & 0 & 0 & 0 & -1 & 0 & 0 & 0 & 0 & 1 & 0 & 1 & & \\
0 & 0 & 0 & 0 & 0 & 0 & 0 & 0 & 0 & 0 & 0 & 0 & 0 & 0 & 0 & 1 & 0 & 1 & -0.05 \\
0 & 0 & 0 & 0 & 0 & 0 & 0 & 0 & 0 & 0 & 0 & 0 & 0 & 0 & 0 & 0 & 1 & 0 & 1 \\
-1 & 0 & 0 & 0 & 0 & 0 & 0 & 0 & 0 & 0 & 0 & 0 & 0 & 0 & 0 & 0 & 0 & 1 & 0 \\
0 & -1 & 0 & 0 & 0 & 0 & 0 & 0 & 0 & 0 & 0 & 0 & 0 & 0 & 0 & 0 & 0 & 0 & 1 \\
\end{bmatrix}
\begin{bmatrix} W_3 \\ A_3 \\ R_3 \\ W_5 \\ A_5 \\ R_5 \\ W_6 \\ R_6 \\ W_7 \\ A_7 \\ R_7 \\ W_9 \\ A_9 \\ W_{11} \\ A_{11} \\ W_{12} \\ R_{12} \\ W_{13} \\ A_{13} \end{bmatrix}
=
\begin{bmatrix} 0.00 \\ 0.00 \\ 0.00 \\ 0.00 \\ 0.00 \\ 0.00 \\ 8000. \\ 800. \\ 0.00 \\ 1000. \\ 1000. \\ 0.00 \\ 0.00 \\ 0.00 \\ 0.00 \\ 0.00 \\ 0.00 \\ 0.00 \\ 0.00 \end{bmatrix}
$$

	Stream	kg/hr

x vector, b vector

Figure 2.5: Matrix Notation for Example 2.2

Steady–State Lumped Systems

```
A(4,4) = 1; A(4,9) = -3;
A(5,5) = 1; A(5,10) = -3;
A(6,6) = 1; A(6,11) = -3;
A(7,7) = 1; A(7,9) = 1;
A(8,8) = 1;
A(9,9) = 1; A(9,10) = -0.67; A(9,11) = 1;
A(10,10) = 1;
A(11,8) = 1; A(11,11) = 1;
A(12,12) = 1; A(12,14) = -1.33;
A(13,13) = 1; A(13,15) = -1.33;
A(14,14) = 1; A(14,18) = -3;
A(15,15) = 1; A(15,19) = -3;
A(16,9) = -1; A(16,16) = 1; A(16,18) = 1;
A(17,11) = -1; A(17,17) = 1;
A(18,18) = 1; A(18,19) = -0.05;
A(19,10) = -1; A(19,19) = 1;

b = zeros(19,1);
b(7,1) = 8000;
b(8,1) = 800;
b(10,1) = 1000;
b(11,1) = 1000;

%
% solve for the component flow rate of each stream using Gaussian
% elimination:
%
x = A\b

%=========
% results
%=========

%>> linear2_2

%x =

%    1.0e+03 *

%     1.8753   = W3 (kg/hr)
%     3.9900   = A3 (kg/hr)
%     0.7980   = R3 (kg/hr)
%     1.4100   = W5 (kg/hr)
%     3.0000   = A5 (kg/hr)
%     0.6000   = R5 (kg/hr)
%     7.5300   = W6 (kg/hr)
%     0.8000   = R6 (kg/hr)
%     0.4700   = W7 (kg/hr)
%     1.0000   = A7 (kg/hr)
%     0.2000   = R7 (kg/hr)
%     0.1995   = W9 (kg/hr)
%     3.9900   = A9 (kg/hr)
%     0.1500   = W11(kg/hr)
```

```
%     3.0000   = A11(kg/hr)
%     0.4200   = W12(kg/hr)
%     0.2000   = R12(kg/hr)
%     0.0500   = W13(kg/hr)
%     1.0000   = A13(kg/hr)
```

It is also of interest to determine the *LU* factorization of this $\boldsymbol{A}$ matrix. The MATLAB file `lua.m` (Figure 2.7) can be run after running the file `lin2_2.m` to determine its triangular factorization. Note that we have first used the help facility to learn specific details on the MATLAB function `lu`.

Figure 2.7: MATLAB M-file `lua.m` to determine a *LU* factorization.

```
help lu

% LU      Factors from Gaussian elimination.
%         [L,U] = LU(X) stores a upper triangular matrix in U and a
%         "psychologically lower triangular matrix", i.e. a product
%         of lower triangular and permutation matrices, in L , so
%         that X = L*U.
%
%         [L,U,P] = LU(X) returns lower triangular matrix L, upper
%         triangular matrix U, and permutation matrix P so that
%         P*X = L*U.
%
%         By itself, LU(X) returns the output from LINPACK'S ZGEFA
%         routine

[l,u,p] = lu(A)
```

2.2.3 Other Matrix Capabilities in MATLAB

2.2.3.1 *Singular Value Decomposition*

The singular value decomposition (SVD) of an $m \times n$ matrix $\boldsymbol{A}$ is given by

$$\boldsymbol{A} = \boldsymbol{U}\boldsymbol{S}\boldsymbol{V}^T \qquad (2.2.42)$$

where $\boldsymbol{U}$ is an orthogonal $m \times m$ matrix, $\boldsymbol{V}$ an orthogonal $n \times n$ matrix and $\boldsymbol{S}$ is a real diagonal $m \times n$ matrix. The elements of the diagonal matrix are called the signular values of $\boldsymbol{A}$. Normally they are arranged in decreasing value so that $s_1 > s_2 > \cdots > s_n$. An orthogonal matrix has the property that $\boldsymbol{U}\boldsymbol{U}^T = \boldsymbol{I}$. In MATLAB we have the function `svd`.

One application of SVD is in determining the rank of a matrix. In practice MATLAB determines rank (using function `rank` from the SVD by counting the number of singular values greater than some tolerance value. In addition, SVD allows us to compute the condition number of a matrix using the function `cond`. The condition number in MATLAB is the ratio of the largest singular value to the smallest. A well–conditioned matrix has a condition number around unity. A very large condition number implies an ill–conditioned matrix. The norm of a matrix uses SVD. The function `norm` provides several norms (see the `help norm` output). The usual of ℓ_2 norm is the largest singular value.

2.2.3.2 *The Pseudo–Inverse*

If $\boldsymbol{A}$ is an $m \times n$ rectangular matrix such that $m > n$, then the system

$$\boldsymbol{A}\boldsymbol{x} = \boldsymbol{b} \tag{2.2.43}$$

is an over–determined system of equations. We cannot invert $\boldsymbol{A}$ since it is not a square matrix. If we pre–multiply equation (2.2.43) by $\boldsymbol{A}^T$, then we have

$$\boldsymbol{A}^T\boldsymbol{A}\boldsymbol{x} = \boldsymbol{A}^T\boldsymbol{b} \tag{2.2.44}$$

Since $\boldsymbol{A}^T\boldsymbol{A}$ is a square matrix, we can take its inverse so that

$$\boldsymbol{x} = (\boldsymbol{A}^T\boldsymbol{A})^{-1}\boldsymbol{A}^T\boldsymbol{b} \tag{2.2.45}$$

The pseudo–inverse of $\boldsymbol{A}$ is therefore

$$\boldsymbol{A}^+ = (\boldsymbol{A}^T\boldsymbol{A})^{-1}\boldsymbol{A}^T \tag{2.2.46}$$

Usually the pseudo–inverse, $\boldsymbol{A}^+$, is computed using SVD since

$$\boldsymbol{A}^T\boldsymbol{A} = (\boldsymbol{V}\boldsymbol{S}^T\boldsymbol{U}^T)(\boldsymbol{U}\boldsymbol{S}\boldsymbol{V}^T) \tag{2.2.47}$$

or

$$\boldsymbol{A}^T\boldsymbol{A} = \boldsymbol{V}\boldsymbol{S}^T\boldsymbol{S}\boldsymbol{V}^T \tag{2.2.48}$$

since $\boldsymbol{U}^T\boldsymbol{U} = \boldsymbol{I}$. The pseudo–inverse, $\boldsymbol{A}^+$ is therefore

$$\boldsymbol{A}^+ = (\boldsymbol{V}\boldsymbol{S}^T\boldsymbol{S}\boldsymbol{V}^T)^{-1}\boldsymbol{V}\boldsymbol{S}^T\boldsymbol{U}^T \tag{2.2.49}$$

or

$$\boldsymbol{A}^+ = (\boldsymbol{V}^T)^{-1}(\boldsymbol{S}^T\boldsymbol{S})^{-1}\boldsymbol{V}^{-1}\boldsymbol{V}\boldsymbol{S}^T\boldsymbol{U}^T \tag{2.2.50}$$

since $(\boldsymbol{V}^T)^{-1} = (\boldsymbol{V}^T)^T = \boldsymbol{V}$, we have

$$\boldsymbol{A}^+ = \boldsymbol{V}(\boldsymbol{S}^T\boldsymbol{S})^{-1}\boldsymbol{S}^T\boldsymbol{U}^T \tag{2.2.51}$$

Here $\boldsymbol{V}$ is an $m \times m$ matrix, $\boldsymbol{U}$ an $n \times n$ matrix and $\boldsymbol{S}$ an $n \times m$ diagonal matrix. When $\boldsymbol{A}$ is rank deficient, then $\boldsymbol{S}^T\boldsymbol{S}$ cannot be inverted because of very small or zero singular values. In this case, we only take the r non–zero singular values so that $\boldsymbol{S}$ becomes an $r \times r$ matrix where r is the rank of $\boldsymbol{A}$. The MATLAB function `pinv` computes the pseudo–inverse.

2.2.3.3 Sparse Matrices

MATLAB has extensive sparse matrix facilities. Sparse matrices arise in many engineering problems including those of material and energy balance calculations as illustrated earlier. It is more efficient to define matrices as sparse when appropriate. If you are not sure if a matrix is sparse, then you can use the function issparse (A) which returns a value of 1 if A is sparse and 0 if it is not sparse. To define a matrix as sparse, we use the function sparse (rowpos, colpos, val, m,n) where rowpos are the positions of the nonzero row elements, colpos are the positions of the non–zero column elements, val are the values of the non–zero elements, m is the number of rows, and n the number of columns. For example, if

```
>> colpos = [1 2 1 2 5 3 4 3 4 5];
>> rowpos = [1 1 2 2 2 4 4 5 5 5];
>> val = [12 -4 7 3 -8 -13 11 2 7 -4];
>> a = sparse (rowpos,colpos,val,5,5)
```

The result is

```
a =

   (1,1)       12
   (2,1)        7
   (1,2)       -4
   (2,2)        3
   (4,3)      -13
   (5,3)        2
   (4,4)       11
   (5,4)        7
   (2,5)       -8
   (5,5)       -4
```

It is important to note that matrix operations such as *, +, -, \ produce sparse results if **both** operands are sparse. You can convert a full matrix to a sparse one by using the command

$$b=sparse(a);$$

and a sparse matrix to a full one by the command

$$a=full(b);$$

There is significant efficiency in properly using sparse matrix operations when appropriate.

Steady–State Lumped Systems 67

2.3 SOLUTION OF NONLINEAR EQUATIONS

2.3.1 Solving a Single Nonlinear Equation in One Unknown

There are many problems which require the finding of the root of a single nonlinear equation in one unknown. For example, the heat capacity of carbon dioxide is given as a function of temperature as

$$C_p = 1.716 - 4.257 \times 10^{-6} \, T - \frac{15.04}{\sqrt{T}} \qquad (2.3.1)$$

where the units of C_p are (kJ/kg K) and the unit of temperature T is (K). Suppose we want to determine the temperature which yields a value of the heat capacity of 1 (kJ/kg K). We therefore seek a solution to the implicit nonlinear equation

$$f(T) = 0.716 - 4.257 \times 10^{-6} \, T - \frac{15.04}{\sqrt{T}} = 0 \qquad (2.3.2)$$

The MATLAB function cp1.m (Figure 2.8) gives a program that defines this $f(T)$ function.

Figure 2.8: Heat Capacity Example $f(T) = 0$; M-file cp1.m

```
%
% This defines the nonlinear relationship between heat capacity
% and temperature (Eq. 2.3.2). The format is suitable for the
% bisection and tangent methods:
%
function f = heat(T)
f = 0.716-4.257*T*10^(-6)-15.04/sqrt(T);
```

In general we are seeking the value of the unknown variable, x, which makes the following equation valid

$$f(x) = 0 \qquad (2.3.3)$$

Often equation (2.3.3) can be rearranged so that the root x appears uniquely as

$$x = g(x) \qquad (2.3.4)$$

For the heat capacity example, this would be

$$T = \left[0.716 - \frac{15.04}{\sqrt{T}} \right] \bigg/ 4.257 \times 10^{-6} \qquad (2.3.5)$$

The MATLAB File cp2.m (Figure 2.9) gives a program that defines this explicit function.

Figure 2.9: Heat Capacity as $x = g(x)$; M-file cp2.m

```
%
%
% This defines the nonlinear relationship between heat capacity
% and temperature (Eq. 2.3.5). The format is suitable for the
% Wegstein method and is a divergent expression using direct
% substution

function f = cp(T)

f = (0.716-(15.04/sqrt(T)))/(4.257*10^(-6));
```

A number of methods for finding the root (roots) of equations of the implicit form of equation (2.3.3) or the explicit form of equation (2.3.4) have been developed. In general, they fall into derivative–free or derivative–based categories. Derivative–free methods are usually more stable, less sensitive to initial guesses, and converge less rapidly than derivative–based methods.

2.3.1.1 Half–Interval (Bisection)

A straightforward derivative–free method is that of halving the interval of uncertainty. Consider a monotonic function $f(x)$, shown in Figure 2.10, which is continuous from $x = a$ to $x = b$ where the values $f(a)$ and $f(b)$ have opposite signs. For the equation

$$f(x) = 0 \qquad (2.3.6)$$

the root of the equation lies at $x = R$.

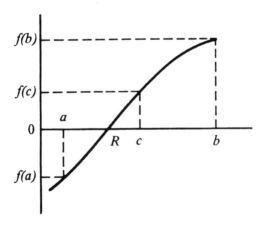

Figure 2.10: Half–Interval or Bisection Method.

For a case such as this, fundamental calculus says that there will be one real root in this interval. Note that we can determine the root graphically by plotting the function $f(x)$ and estimating where it intersects the x-axis.

The bisection or half–interval method is a systematic trial–and–error solution consisting of the following steps:

1. Choose points $x = a$ and $x = b$ so that the interval $a < x < b$ encompasses one and only one real root.

2. Determine the midpoint of the interval, point c. If as in this case, $f(c) > 0$, then the root must lie between the midpoint of the original interval, c, and the lower end of the interval, a. If $f(c) < 0$, then the root must be between c and b.

3. Repeat the previous step and calculate $f(x)$ at the midpoint of the new interval at each step of the iteration and discard the half of the interval that does not contain the root.

The result is that the interval containing the root will decrease by a factor of two each iteration. Thus by repeated application of this algorithm, the root can be determined to any degree of accuracy.

For our example of equation (2.3.1),

$$f(T) = 0.716 - 4.257 \times 10^{-6}\, T - \frac{15.04}{\sqrt{T}} = 0 \qquad (2.3.2)$$

the method gives:

Iteration	$a = T_L$	$b = T_R$	$f(T_L)$	$f(T_R)$	c
1	400.	600.	−0.038	0.099	500.
2	400.	500.	−0.038	0.041	450.
3	400.	450.	−0.038	0.0051	425.
4	425.	450.	−0.015	0.0051	437.5
5	437.5	450.	−0.0049	0.0051	443.75
6	437.5	443.75	−0.0049	0.00014	440.625
7	440.625	443.75	−0.0024	0.00014	

MATLAB has the function `fzero` which performs this bisection algorithm. The M-file `bisec.m` (Figure 2.11) uses `fzero` to calculate the root for this heat capacity example.

Figure 2.11: MATLAB M-file bisec.m to determine the root of Equation (2.3.2) using the bisection method.

```
% This is an example showing how to solve a nonlinear equation by
% calling the fzero function (bisection algorithm) in Matlab.
%
% pick initial guess
%
init = 400;

fzero('cp1', init)

%
% more arguments can be added to the fzero function call: the
% third argument is user specified tollerance; a nonzero fourth
% argument allows user to trace the convergence/divergence. Use
% "help fzero" for more information.
%


%
% =========
%   results
% =========
%
% >> bisec
%
% ans =
%
% 443.5712
```

2.3.1.2 Linear Inverse Interpolation (Regula Falsi)

This method is similar to bisection except that the new estimate of the root, c, shown in Figure 2.12, is found by linear interpolation between the points $(b, f(b))$ and $(a, f(a))$.

By the principle of similar triangles we have,

$$\frac{c - a}{0 - f(a)} = \frac{b - c}{f(b) - 0} \tag{2.3.7}$$

Solving for c gives

$$c = \frac{a\, f(b) - b\, f(a)}{f(b) - f(a)} \tag{2.3.8}$$

Then c will replace a according to Figure 2.12 since $f(c) < 0$.

Applying this method to the same example gives

Steady–State Lumped Systems

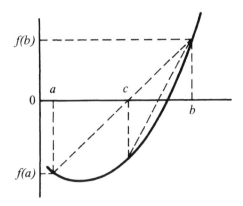

Figure 2.12: Regula Falsi Method.

Iteration	$a = T_1$	$b = T_R$	$f(T_1)$	$f(T_R)$	c	$f(c)$
1	400.	600.	−0.038	0.099	455.5	0.0093
2	400.	455.5	−0.038	0.0093	444.5	0.00076
3	400.	444.5	−0.038	0.00076	443.6	0.000062
4	400.	443.6	−0.038	0.000062	443.58	0.000005

As typical, the convergence using Regula Falsi is more rapid than that by bisection.

MATLAB does not have a routine that implements the Regula Falsi algorithm. We have developed such an algorithm and it is given in the M-file `regfals.m` (Figure 2.13). The file also gives as an example of the use of the routine the solution to equation (2.3.2). Note that just as in the bisection algorithm, the initial two guesses must be such that one gives a positive function evaluation and the other a negative function evaluation. This allows for the new root to always be an interpolation between the previous values.

Figure 2.13: MATLAB file `regfals.m` that implements the Regula Falsi algorithm.

```
function c = regfals(Func, a, b, tol, max)
%
% REGFALS   Find a zero of a nonlinear equation of one variable
%           using the regula falsi algorithm. Func is the name of
%           an M-file that contains a real-valued function of a
%           single real variable, and a and b are the initial
%           guesses.  Note that a bad formulation of the problem
```

```
%              may cause divergence.
%
%              An optional fourth argument sets the tolerance for
%              the convergence. The presence of an optional fifth
%              argument specifies the maximum iteration number.
%
%
% pre-processing, set default tolerance and maximum iteration
% number

if nargin<4, tol=1e-10; max=1000; end
if nargin==4, max=1000; end

%
% start the regula falsi algorithm
%

i = 1;
TOL_FLAG = 1;
while ( (i<max) & TOL_FLAG ),
  fb = feval(Func,b);
  fa = feval(Func,a);
  c = (a*fb-b*fa)/(fb-fa);
  fc = feval(Func,c);
  if (sign(fc)==sign(fa)),
    a = c;
  else
    b = c;
  end
  if (abs(feval(Func,c))<tol),
    TOL_FLAG = 0;
  end
  t(i) = i;
  x(i) = c;
  i = i + 1;
end

%
% plot to see the convergence/divergence of the process
%

plot(t, x, 'x');
title('Convergence/Divergence of the Regula Falsi Algorithm')
xlabel('iteration number')

%
% examples of calling this function:
%
%        >> regfals('cp1', 400, 600)
%        >> regfals('cp1', 400, 600, 0.000001)
%        >> regfals('cp1', 400, 600, 0.000001, 1000)
%
```

```
return

end
```

2.3.1.3 Direct Substitution

A direct substitution strategy can be used for equations of the form

$$x = g(x) \tag{2.3.4}$$

Graphically, the solution exists when the curve $g(x)$ crosses the diagonal as shown in Figure 2.14.

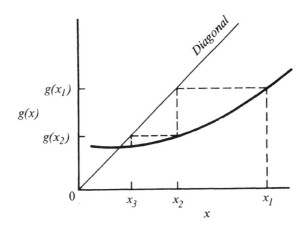

Figure 2.14: Direct Substitution (Convergence).

The direct substitution convergence of x_1, x_2, etc. is illustrated in the figure. Not all functions $g(x)$ will converge using direct substitution as shown in Figure 2.15.

The difference between a convergent and divergent system is the slope of $g(x)$ at the solution condition. When the slope $\frac{dg(x)}{dx} > 1$, then the method diverges. However, when the slope $\frac{dg(x)}{dx} < 1$, then the method converges. For the example of equation (2.3.5)

$$T = g(T) = \frac{0.716 - 15.04/\sqrt{T}}{4.257 \times 10^{-6}} \tag{2.3.5}$$

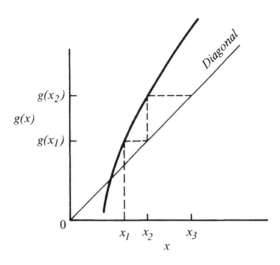

Figure 2.15: Direct Substitution (Divergence).

the slope is

$$\frac{dg(T)}{dt} = \frac{7.52}{4.257 \times 10^{-6}} T^{-1.5} \tag{2.3.9}$$

Assuming a couple of values of T gives

$$\frac{dg}{dT}(600) = 120$$

$$\frac{dg}{dT}(400) = 220$$

Therefore, it appears that the direct substitution algorithm will be unstable. Actual implementation is as follows.

Iteration	T	$g(T)$
1	600	23960
2	23960	145638
	Divergent	
1	400	−8457
	Divergent	

Sometimes it is possible to reformulate a divergent $g(x)$ function to obtain a new convergent expression. For our example, an alternative

Steady–State Lumped Systems

formulation is

$$\sqrt{T} = \frac{15.04}{0.716 - 4.257 \times 10^{-6}\,T} \qquad (2.3.10)$$

or

$$T = \left[\frac{15.04}{0.716 - 4.257 \times 10^{-6}\,T}\right]^2 \qquad (2.3.11)$$

Using this formulation, the slope is

$$\frac{dg(T)}{dT} = \frac{1.93 \times 10^{-3}}{(0.716 - 4.257 \times 10^{-6}\,T)^3} \qquad (2.3.12)$$

and

$$\frac{dg(600)}{dT} = 0.0053$$

$$\frac{dg(400)}{dT} = 0.0053$$

Therefore, with this formulation, the direct substitution algorithm should be stable:

Iteration	T	$g(T)$
1	600.	444.4
2	444.4	443.576
3	443.576	443.571
1	400.	443.34
2	443.34	443.57

2.3.1.4 Wegstein Method

Use of the Wegstein accelerator, another substitution algorithm, allows for convergence even when the direct substitution scheme is unstable. This method is represented in Figure 2.16. The Wegstein algorithm can also accelerate the convergence of a normally stable situation. An initial guess is used (x_1) to generate a second estimate $x_2 = g(x_1)$. Note in the figure that x_2 is diverging from the solution. However, at this point we project a straight line through the points (x_1, $g(x_1)$) and (x_2, $g(x_2)$) to the diagonal line, that is

$$\frac{y - g(x_1)}{x - x_1} = \frac{g(x_1) - g(x_2)}{x_1 - x_2} \qquad (2.3.13)$$

at the point $y = x = x_3$; then

$$x_3 = \frac{x_1 g(x_2) - x_2 g(x_1)}{x_1 - x_2 - g(x_1) + g(x_2)} \qquad (2.3.14)$$

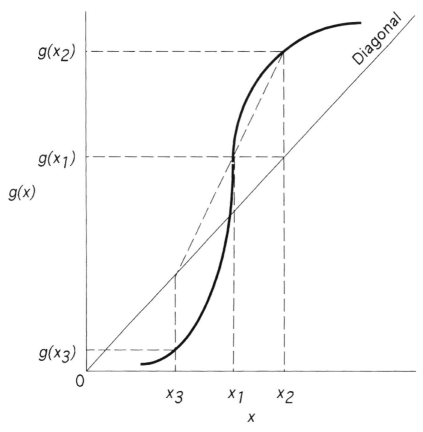

Figure 2.16: **Wegstein Accelerator**

At the next iteration, x_2 replaces x_1 and x_3 replaces x_2 and in equation (2.3.14).

Now, we apply the Wegstein formula to the unstable form of the example:

$$T = g(T) = \frac{0.716 - 15.04/\sqrt{T}}{4.257 \times 10^{-6}} \qquad (2.3.5)$$

Iteration	T_1	$g(T) = T_2$	T_3
1	600.	23959.	−4965.
			cannot proceed
1	400.	−8457.	cannot proceed
1	445.	713.08	442.94
2	713.08	442.94	443.85
3	442.94	443.85	443.571

Steady-State Lumped Systems

Because of the problem of generating negative temperatures and taking the square-root of negative numbers, the Wegstein method gives convergence only for guesses close to the actual solution.

The MATLAB M-file wegstein.m (Figure 2.17) has been created in order to solve root solving problems using the Wegstein acceleration method. The M-file cp2 formulates the explicit unitable form of the problem (equation (2.3.5)). Convergence is only possible for guesses close to the actual root. For example, a initial guess of $T = 468K$ will converge as shown in Figure 2.18, but an initial guess of 469 K will result in the necessity of taking the square root of a negative number which results in complex solutions which are not physically real. It is suggested that the reader study the convergence properties of this function by running the problem with different initial guesses for the temperature root.

Figure 2.17: MATLAB M-file wegstein.m for Implementation of the Wegstein Accelerator

```
function w = wegstein(Func, x, tol, max)
%
%WEGSTEIN   Find a zero of a nonlinear equation of one variable
%           using the wegstein algorithm. Func is the name of an
%           M-file that contains a real-valued function of a
%           single real variable, and x is the initial guess.
%           Note that a bad formulation of the problem may cause
%           divergence.
%           An optional third argument sets the tolerance for the
%           convergence. The presence of an optional fourth
%           argument specifies the maximum iteration number.
%

%
% pre-processing, set default tolerance and maximum iteration
% number.
%
if nargin<3, tol=0.000001; max=1000; end
if nargin==3, max=1000; end
t=[1:max+1]';

%
% initialization
%
x1 = x;
x2 = feval(Func, x1);

%
% start wegstein algorithm
%
for i=1:max,
    x3(i) =(x1*feval(Func,x2)-x2*feval(Func,x1))...
           /(x1-x2-feval(Func,x1)+feval(Func,x2));
    if(abs(x3(i)-x2) < tol)
```

```
            T = x3(i)
            iteration = i
            t = [1:iteration+1]';
            break;
        end
        x1 = x2;
        x2 = x3(i);
end
%
% plot to see convergence/divergence
%
x3 = [x x3];
plot(t, x3, '-')
title('Convergence/Divergence of the Wegstein Algorithm')
xlabel('iteration number+1')

%
% examples of calling this function:
%
%       >> wegstein('cp2', 450)
%       >> wegstein('cp2', 450, 0.000001)
%       >> wegstein('cp2', 450, 0.000001, 1000)
%
% all these function calls in Matlab gives following results:
%
%       T = 443.5712
%       iteration = 7
%
% together with a graph window.
```

Now, we apply the Wegstein method to the reformulated problem:

$$T = \left[\frac{15.04}{0.716 - 4.257 \times 10^{-6}\, T} \right]^2 \quad (2.3.11)$$

Iteration	T_1	$g(T_1) = T_2$	T_3
1	600.	444.40	443.571
1	400.	443.34	443.571

The acceleration over the direct substitution method is apparent since convergence is achieved in one iteration from both initial guesses.

The M-file cp3.m (Figure 2.19) has the convergent path formulation of equation (2.3.11). It can be used with the M-file wegstein.m to obtain solutions to the problem. The algorithm converges rapidly and from all initial guesses for the temperature T.

The rapidity of convergence of substitution methods depends strongly on the slope of the solution line with respect to the diagonal line. If direct substitution diverges, either the problem must be reformulated or

Steady–State Lumped Systems

the Wegstein method applied. Equations can have multiple roots. All of these algorithms may find different roots depending on their initial guesses. Plotting of the function will be helpful in such cases.

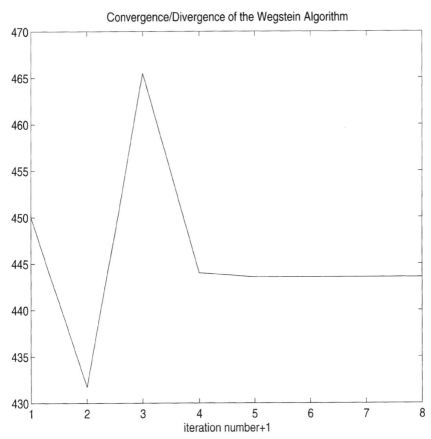

Figure 2.18: Wegstein Convergence

Figure 2.19: MATLAB M-file cp3.m which is Equation (2.3.11)

```
%
% This defines the nonlinear relationship between heat capacity
% and temperature (Eq. 2.3.11). The format is suitable for the
% Wegstein method and is a convergent expression using direct
% substitution
%
function f = cp(T)
f = (15.04/(0.716-4.257*10^(-6)*T))^2;
```

2.3.1.5 Newton Method

This widely used derivative–based method for a single nonlinear algebraic equation is illustrated in Figure 2.20. Here we want to solve equation (2.3.6),

$$f(x) = 0 \tag{2.3.6}$$

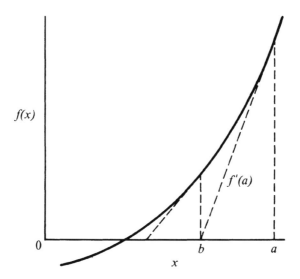

Figure 2.20: Newton Method for a Single Equation.

The slope or derivative of $f(x)$ is computed at an initial guess, a, for the root of $f(x) = 0$. The new value of the root, b, is computed based on a first–order Taylor Series expansion of $f(x)$ about the initial guess, a,

$$f(x) = 0 = f(a) + f'(a)(b - a) + \text{HOT} \tag{2.3.15}$$

or

$$b = a - \frac{f(a)}{f'(a)} \tag{2.3.16}$$

As illustrated in Figure 2.19, the value b is the intersection of the tangent line of slope $f'(a)$ and the $f(x) = 0$ line.

Again, this method is iterative, but it only requires one initial guess. An important advantage of the Newton method is its rapid convergence.

Steady–State Lumped Systems

Two disadvantages are that it requires computation of the derivative and that it can be very sensitive to the initial guess (tends to blow up with poor initial guesses).

Applying Newton's method to the same example gives

$$f(T) = 0.716 - 4.257 \times 10^{-6}\, T - \frac{15.04}{\sqrt{T}} = 0 \qquad (2.3.2)$$

and

$$f'(T) = -4.257 \times 10^{-6} + 7.52 T^{-3/2} \qquad (2.3.17)$$

so

$$T_{\text{new}} = T_{\text{old}} - \frac{\left(0.716 - 4.257 \times 10^{-6}\, T_{\text{old}} - 15.04/\sqrt{T_{\text{old}}}\right)}{\left(4.257 \times 10^{-6} + 7.52 T_{\text{old}}^{-3/2}\right)} \qquad (2.3.18)$$

The results for two different initial guesses are

Iteration	T_{old}	T_{new}
1	600.	404.03
2	404.03	440.87
3	440.87	443.56
4	443.56	443.571
1	400.	440.29
2	440.29	443.55
3	443.55	443.571

Note the rapid convergence using different initial guesses. The MATLAB M-file `tangent.m` (Figure 2.21) has been created in order to implement the Newton method. It uses a finite difference approximation to the first derivative

$$f'(a) = \frac{f(a) - f(a + .001a)}{a - (a + .001a)} \qquad (2.3.19)$$

Figure 2.21: MATLAB M-file `tangent.m` for Solutions Using the Newton Method

```
function s = tangent(Func, x0, tol, max)
%
%TANGENT    Find a zero of a nonlinear equation of one variable
%           using the tangent algorithm. Func is the name of an
%           M-file that contains a real-valued function of a
%           single real variable, and x0 is the initial guess.
%           Note that a poor initial quess may cause divergence.
%
%           An optional third argument sets the tolerance for the
```

```
%              convergence. The presence of an optional fourth
%              argument specifies the maximum iteration number.
%
% pre-processing, set default tolerance and maximum iteration
% number
%
if nargin<4, tol=0.000001; max=1000; end
if nargin==4, max=1000; end
t=[1:max]';

%
% start the tangent algorithm
%
for i=1:max,
    x1 = x0+ .001*x0;
    x2(i) = x0 - (feval(Func,x0)*(x1-x0))/(feval(Func,x1)...
        -feval(Func,x0));
    if(abs(x2(i)-x0) < tol)
        iteration = i
        T = x2(i)
        t = [1:iteration]';
        break;
    end
    x0 = x2(i);
end
%
% plot to see the convergence/divergence of the process
%
plot(t, x2, '-');
title('Convergence/Divergence of the Tangent Algorithm')
xlabel('iteration number')

%
% examples of calling this function:
%
%       >> tangent('cp1', 400)
%       >> tangent('cp1', 400, 0.000001)
%       >> tangent('cp1', 400, 0.000001, 1000)
%
% all these function calls in Matlab gives following results:
%
%       T = 443.5712
%       iteration = 5
%
% together with a graph window.
```

2.3.2 Simultaneous Solution of Nonlinear Algebraic Equations

Whereas the procedures for solving systems of linear equations are straightforward, those for solving sets of nonlinear equations are not nearly so well formulated. The method that will be used for the solution of n simultaneous nonlinear equations in n unknowns is a generalization of the Newton method of the previous section and is called the Newton–Raphson method. This technique tries to improve an initial guess for the solution vector via a linearization procedure.

A set of n equations in n unknowns may be written:

$$\begin{aligned} f_1(x_1, x_2, \cdots x_n) &= 0 \\ f_2(x_1, x_2, \cdots x_n) &= 0 \\ &\vdots \\ f_n(x_1, x_2, \cdots x_n) &= 0 \end{aligned} \qquad (2.3.20)$$

or, in vector notation,

$$\boldsymbol{f}(\boldsymbol{x}) = \boldsymbol{0} \qquad (2.3.21)$$

If there exists a present set of guesses, $\boldsymbol{x}^i$, for the solution of (2.3.21), the function $\boldsymbol{f}$ may be written for any other $\boldsymbol{x}$ as a Taylor series expansion about $\boldsymbol{x}^i$ as follows:

$$\boldsymbol{f}(\boldsymbol{x}) = \boldsymbol{f}(\boldsymbol{x}^i) + \frac{\partial \boldsymbol{f}}{\partial \boldsymbol{x}}(\boldsymbol{x}^i)(\boldsymbol{x} - \boldsymbol{x}^i) + \mathbf{HOT} \qquad (2.3.22)$$

The partial derivative in equation (2.3.22) is called the Jacobian matrix and is given in terms of the following first partials:

$$\frac{\partial \boldsymbol{f}}{\partial \boldsymbol{x}}(\boldsymbol{x}^i) = \begin{bmatrix} \frac{\partial f_1}{\partial x_1} & \frac{\partial f_1}{\partial x_2} & \cdots & \frac{\partial f_1}{\partial x_n} \\ \vdots & \vdots & & \vdots \\ \frac{\partial f_n}{\partial x_1} & \frac{\partial f_n}{\partial x_2} & \cdots & \frac{\partial f_n}{\partial x_n} \end{bmatrix} \qquad (2.3.23)$$

The Jacobian can be evaluated analytically by using analytic first partials for each entry (2.3.23); however, it is usually estimated numerically by

$$\frac{\partial \boldsymbol{f}}{\partial \boldsymbol{x}}(\boldsymbol{x}^i) = \begin{bmatrix} \frac{f_1(x_1^i + \delta x_1) - f_1(x_1^i)}{\delta x_1} & \cdots & \frac{f_1(x_n^i + \delta x_n) - f_1(x_n^i)}{\delta x_n} \\ \vdots & & \vdots \\ \frac{f_n(x_1^i + \delta x_1) - f_n(x_1^i)}{\delta x_1} & \cdots & \frac{f_n(x_n^i + \delta x_n) - f_n(x_n^i)}{\delta x_n} \end{bmatrix} \qquad (2.3.24)$$

Ideally, the next set of guesses, $\boldsymbol{x}^{i+1}$, would be the solution to equation (2.3.21). If this were the case, (2.3.22) becomes, truncating the nonlinear higher-order terms,

$$\boldsymbol{0} = \boldsymbol{f}(\boldsymbol{x}^i) + \frac{\partial \boldsymbol{f}}{\partial \boldsymbol{x}}(\boldsymbol{x}^i)(\boldsymbol{x}^{i+1} - \boldsymbol{x}^i) \qquad (2.3.25)$$

Solving equation (2.3.25) for x^{i+1} gives

$$x^{i+1} = x^i - \left[\frac{\partial f}{\partial x}(x^i)\right]^{-1} f(x^i) \qquad (2.3.26)$$

Efficient numerical routines will not compute the inverse but instead perform a **LU** factorization Gaussian elimination solution on (2.3.25). Because of the linear approximation of equation (2.3.26), x^{i+1} will not be the exact solution to equation (2.3.21); however, equation (2.3.26) can be used iteratively to converge on a solution to equation (2.3.21).

A couple of precautions are in order. Systems of nonlinear equations may have many solutions. Depending on the initial guess, x^o, the Newton–Raphson method may converge to different solutions. In that case, it is wise to make the best initial guesses possible and use physical reasoning in interpreting the solution. Also, the Jacobian matrix may become singular as the solution is approached. If this occurs, solution by the Newton–Raphson technique may be impossible, and other, nonderivative methods should be used.

MATLAB in the Optimization Toolbox has the routine `fsolve`. This routine implements an advanced version of the Newton–Raphson method as described above. However, rather than just a linear extrapolation, it implements a mixed quadratic and cubic extrapolation procedure. To learn more about this function, type

$$\gg \texttt{help} \quad \texttt{fsolve}$$

EXAMPLE 2.3 Exothermic Continuous Stirred Tank Reactor:

It is desired to determine the steady–state temperature and concentration in a continuous stirred tank reactor in which an exothermic reaction is taking place. We want to determine conditions for both maximum and minimum cooling. The reactor is shown in Figure 2.22. The macroscopic material and energy balances are

$$F(C_{A_0} - C_A) - VR = 0 \qquad (2.3.27)$$

and

$$Fc_p\rho(T_0 - T_R) - UA(T_R - T_C) + (-\Delta H)VR = 0 \qquad (2.3.28)$$

The reaction rate expression for a first–order reaction is

$$R = k_0 C_A \exp(-E/RT_R) \qquad (2.3.29)$$

where C_A = concentration of A leaving the reactor
C_{A_0} = feed concentration of A
c_p = heat capacity
T_R = temperature of reactants leaving reactor
T_0 = temperature of feed stream to reactor

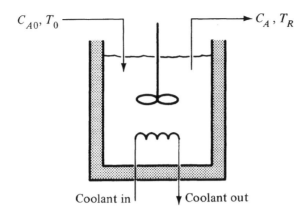

Figure 2.22: Exothermic Continuous Stirred Tank Reactor.

V = volume of reactor
F = volumetric flow rate of input and output streams to reactor
$R(C_A, T)$ = reaction rate (rate/vol) This term is a function of both the reactor temperature and concentration.
For a first–order reaction,
$R = -k_0 C_A \exp(-E/R_g T_R)$
$-\Delta H$ = heat of reaction
U = heat–transfer coefficient
A = area for heat transfer
T_C = coolant temperature

It is useful to formulate the problem in terms of dimensionless variables:

$$\xi = \frac{C_A}{C_{A_0}} \quad \text{dimensionless concentration}$$

$$\eta = \frac{\rho c_p T}{(-\Delta H) C_{A_0}} \quad \text{dimensionless temperature}$$

$$P = \frac{VR}{F C_{A_0}} \quad \text{dimensionless reaction rate}$$

$$U^* = \frac{UA(T_R - T_C)}{F C_{A_0}(-\Delta H)} \quad \text{dimensionless heat transfer term}$$

Using these variables the material and energy balances become

$$0 = 1 - \xi - P \quad (2.3.30)$$

$$0 = (\eta_0 - \eta) - U^* + P \quad (2.3.31)$$

A typical set of numerical parameters is

$$P = \xi \, \exp[50(1/2 - 1/\eta)]$$

$\eta_0 = \eta_c = 1.75$ (feed temperature and coolant temperature)

$$U^* = U_1 (\eta - \eta_c) \quad \text{where} \quad U_1 = \frac{UA}{\rho C_p F}$$

There are two conditions for the dimensionless heat transfer coefficient U_1 that are of interest. These are $U_1 = 0$, which is the condition of no cooling or an adiabatic reactor. The other condition is that of maximum cooling, which corresponds to $U_1 = 2.0$.

An M-file newtrap.m uses fsolve to find the solution to this reactor problem. It uses the M-file cstr.m which defines the equations to be solved (equations (2.3.30) and (2.3.31)). Both files are shown in Figure 2.23. To find the solution when U_1 is zero, the reader should change that variable in cstr.m.

Figure 2.23: M-files newtrap.m and cstr.m to Solve Equations (2.3.30) and (2.3.31)

File newtap.m

```
%
% solve the set of nonlinear equations in Example 2.3 by calling
% fsolve in Matlab.
%

x = fsolve('cstr', [0.2 2.0]');
concentration = x(1)
temperature = x(2)

%=========
% results
%=========
%
%concentration = 0.9676
%temperature = 1.7608
```

File cstr.m

```
function c = cstr(x)
% example for the CSTR in Ex.2.3.
%

eta0 = 1.75; etac = 1.75;
U1 = 2.0;

c(1) = 1 - x(1) - x(1)*exp(50*(0.5-1/x(2)));
c(2) = (eta0-x(2)) - U1*(x(2)-etac) + x(1)*exp(50*(0.5-1/x(2)));
```

2.4 STRUCTURAL ANALYSIS AND SOLUTION OF SYSTEMS OF ALGEBRAIC EQUATIONS

Design equations are predominantly algebraic, often nonlinear, and commonly sparse in that each equation contains only a few of the system variables. The major difference between a simulation study and a design study is in the type of variables that are specified. All input variables and equipment parameters must be specified in a simulation study and the output variables are calculated. The structure for simulation studies is set. This allows simulation studies to be formulated into a modular approach (Motard et al., 1975). Inputs, outputs, or combinations of input and output variables may be known in a design study. Specified variables may change from one solution to another. Equipment parameters and unspecified input and output variables are the result of the calculations.

Analysis of the characteristics of design problems demonstrates a need for a methodology to analyze systems of algebraic equations. This methodology must be oriented at the equation, not modular, level since the calculation path is not fixed for a design problem. In this section we develop the method of structural analysis, which is used to develop simplified solution strategies for large sets of algebraic equations.

Structural analysis is the study of the interrelationships and interactions among the variables that form a set of algebraic equations. The goal is the attainment of the simplest, most efficient calculational path for the set of algebraic equations.

The crux of the ordering problem arises when equations cannot be partitioned and must be solved by either simultaneous or iterative methods. The initial algorithms developed by Sargent and Westerberg (1964), Lee and Rudd (1966), and Christensen and Rudd (1969) sought to minimize the number of iterative variables in a solution strategy. This recognized the fact that convergence problems tend to increase with the size of the set of iterative variables.

Structuring design calculations varies from structuring simulation calculations in two significant ways. The first is the undirected format of design equations. In the analysis of design equations, an order must be assigned to the equations and the equation must be assigned a variable for which to be solved. This assignment is the admissible output set of Steward (1965). The second variation lies in the inability of simulation calculations to circumvent a module. Design calculations often have numerous parallel paths, not all of which must be taken. Each of the parallel paths must be analyzed to determine the path which provides the minimum calculational difficulty.

Parallel paths in design problems arise from the existence of the degrees of freedom in design variables and redundant equations. Steward (1965) and Himmelblau (1967) studied the assignment of admissible output sets for systems of algebraic equations which contained no design variables or redundant equations. The first attempt to develop a mathematical basis for the selection of design variables was the bipartite graph structure of Lee, Christensen and Rudd (1966). The algorithm given in terms of the more convenient occurrence matrix is presented in Rudd and Watson (1968). The algorithm fails for systems of equations with persistent iteration. Christensen (1970) developed an algorithm which handles persistent iteration. The algorithm of Stadtherr et al. (1974) tends to yield nested, implicit iterative loops, both of which are to be avoided for iterative calculations using direct substitution. Ramirez and Vestal (1972) present algorithms which are constructed in two phases. The first phase selects the design variables for the system and determines the minimum number of iterative variables for the system. The second phase selects iterative variables so as to obtain explicit iterative loops if possible. Algorithms by Westerberg and Edie (1971) and Edie and Westerberg (1971) use the maximum eigenvalue principle to select design variables. One of the major limitations of all of these algorithms is that they produce a single combination of design and iterative variables. Book and Ramirez (1976) developed methods of expressing all solution sequences which are acyclic for systems of equations without persistent iteration and all solution sequences with a minimum number of iterative variables in systems of equations with persistent iteration. Friedman and Ramirez (1973) show that convergence of explicit iterative loops by the method of direct substitution is dominated by the order of solution of the equations in the loop. Convergence by direct substitution will occur on the reverse path if the forward path diverges and the reverse path exists.

2.4.1 The Functionality Matrix

Book and Ramirez (1984) developed the use of the functionality matrix for expressing the structure of a system of algebraic equations. The prime requisite of a method for expressing the structure of a system of algebraic equations is that the structure be explicitly expressed. An occurrence matrix is not an explicit representation in that it only expresses the occurrence structure of a system of equations. To remove this deficiency, a new type of occurrence matrix was developed called the functionality matrix. In order to express the functional form of an equation, the different functional forms in which a variable can appear in equations must be defined. Table 2.1 gives the various functional forms which were considered by Book and Ramirez. In Table 2.1, C_i are constants and $f(x)$ is the function of all the other system variables except the explicit variables. The functional forms of Table 2.1 are those which predominate in algebraic design equations. Almost any original equation can be algebraically manipulated to fit these basic forms. The functionality matrix is, therefore, a special occurrence matrix which expresses both equation structure and functionality.

Table 2.1: Functional Forms of Equations

Form Designation	Functional Form	Equation Form
A	Linear	$C_1 \, x_1 + f(\boldsymbol{x}) = 0$
B	Product	$C_1 \, x_1 \, x_2 + f(\boldsymbol{x}) = 0$
D	Triple Product	$C_1 \, x_1 \, x_2 \, x_3 + f(\boldsymbol{x}) = 0$
H	Exponential	$C_1 \, C_2^{x_1} + f(\boldsymbol{x}) = 0$
I	Hyperbolic Sine	$C_1 \, \sinh(x_1) + f(\boldsymbol{x}) = 0$
J	Hyperbolic Cosine	$C_1 \, \cosh(x_1) + f(\boldsymbol{x}) = 0$
K	Hyperbolic Tangent	$C_1 \, \tanh(x_1) + f(\boldsymbol{x}) = 0$
L	Natural Logarithm	$C_1 \, \ln(x_1) + f(\boldsymbol{x}) = 0$
M	Common Logarithm	$C_1 \, \log(x_1) + f(\boldsymbol{x}) = 0$
N	Cubic Form	$C_1 \, x_1^3 + f(\boldsymbol{x}) = 0$
P	Power Form	$C_1 \, x_1^{C_2} + f(\boldsymbol{x}) = 0$
Q	Quadratic Form	$C_1 \, x_1^2 + C_2 \, x_1 + f(\boldsymbol{x}) = 0$
R	Fourth–Order Form	$C_1 \, x_1^4 + f(\boldsymbol{x}) = 0$
S	Second–Order Form	$C_1 \, x_1^2 + f(\boldsymbol{x}) = 0$
T	Sine	$C_1 \, \sin(x_1) + f(\boldsymbol{x}) = 0$
U	Cosine	$C_1 \, \cos(x_1) + f(\boldsymbol{x}) = 0$
V	Tangent	$C_1 \, \tan(x_1) + f(\boldsymbol{x}) = 0$

The example functional matrix of Figure 2.24a demonstrates the effects of the functional forms of equations. The functionality matrix (Figure 2.24a) is a representation of the following equations:

$$C_1 x_2 + C_2 x_3 + C_3 = 0 \tag{2.4.1}$$

$$C_4 x_2 + C_5 x_3 + C_6 x_4^2 + C_7 = 0 \tag{2.4.2}$$

$$C_8 x_1 + C_9 x_3^2 + C_{10} = 0 \tag{2.4.3}$$

If the constants are $C_1 = 1$, $C_2 = 1$, $C_3 = -5$, $C_4 = -2$, $C_5 = 1$, $C_6 = 3$, $C_7 = -2$, $C_8 = 2$, $C_9 = 1$, $C_{10} = -17$, then the desired solution is $x_1 = 4$, $x_2 = 2$, $x_3 = 3$, and $x_4 = 1$.

The matrix of Figure 2.24b is the rearranged matrix (Book and Ramirez, 1976) of Figure 2.24a, and the matrix of Figure 2.24c is the solution mapping matrix (Book and Ramirez, 1976) of the rearranged matrix. The solution mapping matrix contains only those combinations of design variables which result in partitionable set of equations. This problem does not contain persistent iteration. The solution mapping matrix indicates that the selection of x_1, or x_2, or x_3 will reduce the rearranged matrix such that it has an acyclic solution sequence.

The matrix of Figure 2.24d is the acyclic solution sequence for the case when x_1 is chosen as the design variable. This solution sequence encounters some difficulty in the solution of equation (2.4.3) for variable x_3 since multiple roots of ± 3 are available as solutions to the quadratic equation. A physical distinction must be made between these two roots, if the design problem has a single, real solution. It is sometimes quite difficult to distinguish between the two roots unless further calculations are made with both roots. Continuing with the solution sequence and solving equation (2.4.1) for x_2 gives $x_2 = 2$ or 8. This has increased the difficulty in that there are two solutions, each associated with one of the two solutions for x_3. The final step in the solution sequence is to solve equation (2.4.2) for x_4. This gives values of $x_2 = \pm 1$ or $\pm 4i$. It has now been determined that $x_3 = -3$ and $x_2 = 8$ corresponds to an imaginary root. Also, physical distinction must be made between the $\pm$ values for x_4. Clearly, this computational path leads to difficulties.

The rearranged matrix associated with Figure 2.24e corresponds to the selection of x_2 as the design variable. The square term in equation (2.4.3) presents no difficulty since it does not enter as an output variable. A physical distinction must, however, be employed to select the correct root for variable x_4. The rearranged functionality matrix of Figure 2.24e would be considered an easier solution sequence than the sequence represented by the functionality matrix of Figure 2.24d because of the elimination of carrying two solutions through the entire calculations.

The solution sequence represented by the functionality matrix of Figure 2.24f (x_3 is the design variable) is equal in difficulty structurally to

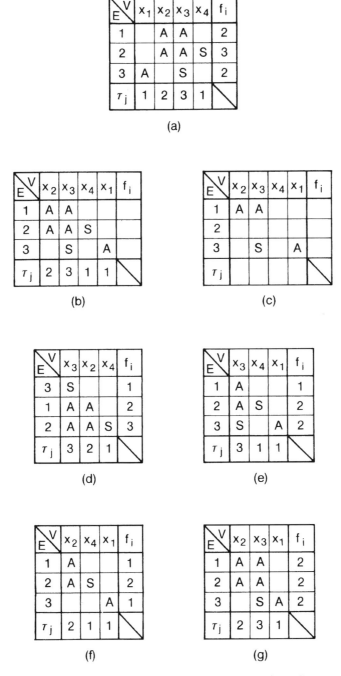

Figure 2.24: Functional Effects on Solution Sequences.

that of Figure 2.24e. This is because only one multiple root appears as an output variable.

The selection of x_4 as a design variable does not yield an acyclic solution sequence. However, proper arrangement of the rows and columns of the matrix results in the functionality matrix of Figure 2.24g. This matrix describes the solution of two linear equations (equations (2.4.1) and (2.4.2)) in two unknowns (x_2 and x_3) followed by the solution of equation (2.4.3), which is linear in variable x_1. Because the solution of simultaneous linear equations is easily obtained, the functionality matrix of Figure 2.24g is considered the simplest to obtain for machine computation. The acyclic calculations involving multiple root functions actually increase the complexity in obtaining the solution. By including information on the functional form of the equation to be solved, the "optimality" of a solution sequence can be analyzed in great depth, even to the point, as the example has shown, of sacrificing acyclicity.

2.4.2 An Optimal Solution Strategy

The explicit representation of a system of equations contained in the functionality matrix allows differentiation among numerous solution strategies. This allows algorithms to be developed for defining an optimal solution strategy. The formulation of a suitable objective function will yield an optimal solution strategy considering available solution methods. For example, if equations are to be solved by hand, direct substitution with a single iterative variable (explicit or implicit) is often an excellent method. Likewise, multiple roots can be easily carried along in hand calculations. However, for implementation on digital computers, general methods for the solution of implicit iterative loops are not well established. With machine computation, methods which employ matrix manipulations are desirable.

For machine computation of algebraic equations describing design or simulation problems, the following heuristic objective function is selected:

1. Assign arbitrarily an acyclic output element to a single equation if the solution for the output element is single–valued. If not possible, go to step 2.

2. For a linear set of equations, assign arbitrarily output variables to that set. If not possible, go to step 3.

3. Assign arbitrarily an acyclic output element to a single equation with multiple roots. If not possible, go to step 4.

4. Assign arbitrarily the minimum number of iterative variables to equations which contain those variables.

Book and Ramirez (1984) present algorithms that use this heuristic objective function. Figures 2.25 and 2.26 present the algorithms. These algorithms result in a rearranged functionality matrix for the system of equations.

There are several useful properties of a rearranged matrix. These are:

1. *A Minimum Difficulty Solution Strategy.* The rearranged functionality matrix provides information on the simplest strategy available. Hierarchy information from the equation ordering algorithm indicates the presence and size of persistent iteration loops or loops that must be solved simultaneously.

2. *Platform Variables.* Platform variables of an equation are new variables of an equation in which there are more than one new variable. Platform variables contain desirable design variable selections. If multiple root variables appear as platform variables, it is useful to select them as a design variable if possible.

3. *Step Equations.* Equations which contain no new variables are step equations. The existence of step equations indicates a system of equations with persistent iteration.

The Equation Ordering Algorithm (Figure 2.25) searches the variable degrees of freedom for values of one. The existence of a one indicates that an acyclic assignment can be made. The entry is located, and the row is eliminated if the functional designation indicates a single–valued functional. If the function designation indicates a multiple–valued function, the column location is stored in MULCOL. If all the variable degrees of freedom are searched and no single–valued acyclic assignment is made, a set of linear simultaneous equations are desired (Level 2 of the objective function). If a linear simultaneous assignment cannot be made, MULCOL is checked to see if an acyclic multiple–root assignment can be made (Level 3 of the objective function). As the variable degrees of freedom are searched, the column with the minimum value greater than one is stored in MINCOL. This variable indicates the equations to be eliminated in assigning the minimum number of iterative variables. Hierarchy pointers store the location of step equations in systems with persistent iteration.

The Variable Group Algorithm (Figure 2.26) searches the equation degrees of freedom. If a value of one is encountered, the equations are "shuffled" to move the equation up into the subgroup being formed. The column (variable) containing the entry is then eliminated. If an equation is outside a cyclic equation set (below the hierarchy pointer), no "shuffling" is done. This avoids moving additional equations into an iterative loop. When no value of one is encountered for the equation degrees of freedom, a new set of platform variables (a new cyclic equation set) is formed by eliminating the columns containing entries in the topmost row. Subgroup pointers store the location of the platform variables.

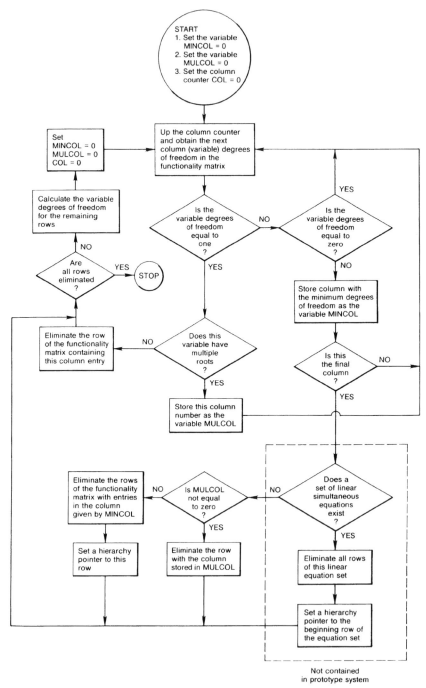

Figure 2.25: Equation Ordering Algorithm.

Steady–State Lumped Systems

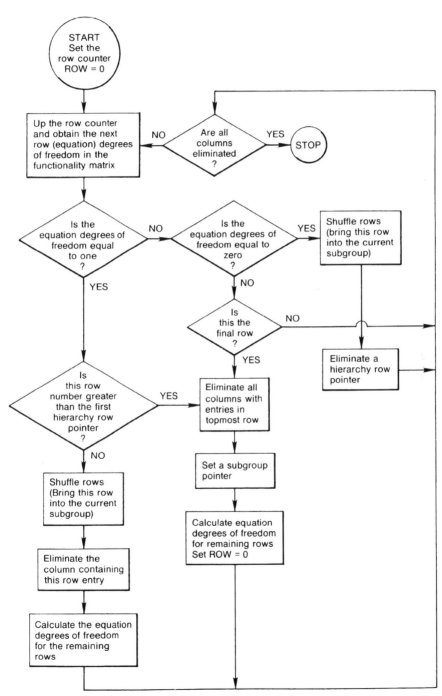

Figure 2.26: Variable Group Algorithm.

2.4.3 Simple Example of Structural Analysis

Figure 2.27a gives a simple functionality matrix for which we want to perform structural analysis using the Equation Ordering Algorithm of Figure 2.25 and the Variable Group Algorithm of Figure 2.26.

Starting with the Equation Ordering Algorithm, we first need to determine the variable degrees of freedom, which are shown in Figure 2.27a. The algorithm seeks the first column for which the variable degrees of freedom are unity. For this example, this is column 3. The row containing this column is eliminated from the Functionality Matrix since the equation is linear and multiple roots do not exist. Then the variable degrees of freedom are recomputed. This step is shown in Figure 2.27b. Following the algorithm, a column frequency of one is found in column 5 for a linear equation, and therefore row 6 is eliminated. Now a one is found in column 9, which eliminates row 1. At this point, a column frequency of one does not exist. According to the algorithm, eliminating the search for linear equations which has not been implemented in the computer software given in the next section, we seek the column with the minimum degrees of freedom, which is column 2 with two degrees of freedom. We then eliminate the rows of the functionality matrix with entries in this column. This results in the elimination of rows 2 and 5. Also row 2 is designated with the hierarchy 1 pointer. This row will be a step equation in the final functionality matrix and will be the end of an iterative loop or set of simultaneous equations. Now we can proceed with the elimination of row 3 because of the column 1 frequency of unity. Finally row 7 is eliminated because of the unity frequency in column 4.

Now we move to the Variable Group Algorithm. First the Functionality Matrix is reconstructed with the rows being in the reverse order as eliminated from application of the Equation Ordering Algorithm. This is shown in Figure 2.27c. Using the Variable Group Algorithm, the row frequencies are computed. The algorithm seeks a row frequency of unity. For this problem none exists at the initial level. Since a unity frequency is not found, we eliminate all columns with entries in the topmost remaining row. Therefore, columns 4, 6, and 8 are eliminated and row 7 starts subgroup 1. Now the equation (row) frequencies are recomputed. The first unity frequency is found in row 3, which results in the elimination of variable 1. Now a row frequency of unity exists (row 1), but it is below the first hierarchy row pointer so we must eliminate all columns with entries in the topmost remaining row. We, therefore, eliminate columns 2 and 7 from row 5. Also row 5 is the first row of subgroup 2. Now an extra row frequency becomes zero. This is row 2. Its hierarchy row pointer is now eliminated. Next, row 1 is eliminated with column 9, then row 6 with column 5, and finally row 4 with column 3.

Finally, the rearranged Functionality Matrix is reconstructed by placing the rows in the order eliminated and the columns in the order eliminated. The rearranged functionality matrix is shown in Figure 2.27d with

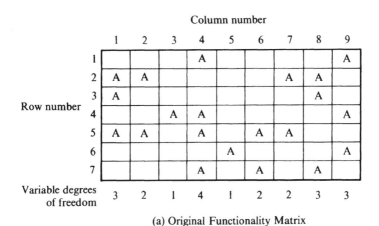

(a) Original Functionality Matrix

Figure 2.27a: Structural Analysis of Algebraic System Structure.

	Column number									
		1	2	3	4	5	6	7	8	9
Row number	1				A					A
	2	A	A					A	A	
	3	A							A	
	4				A	A				A
	5	A	A			A		A	A	
	6						A			A
	7				A		A		A	

	1	2	3	4	5	6	7	8	9	
Original variable degrees of freedom	3	2	1	4	1	2	2	3	3	Row 4 eliminated
Next calculation of variable degrees of freedom	3	2	0	3	1	2	2	3	2	Row 6 eliminated
Next calculation	3	2	0	3	0	2	2	3	1	Row 1 eliminated
Next calculation	3	2	0	2	0	2	2	3	0	Rows 2 and 5 eliminated Row 2 is hierarchy 1
Next calculation	1	0	0	1	0	1	0	2	0	Row 3 eliminated
Next calculation	0	0	0	1	0	1	0	1	0	Row 7 eliminated
Next calculation	0	0	0	0	0	0	0	0	0	

(b) Application of Equation Ordering Algorithm

Figure 2.27b: Structural Analysis of Algebraic System Structure.

98 Computational Methods for Process Simulation

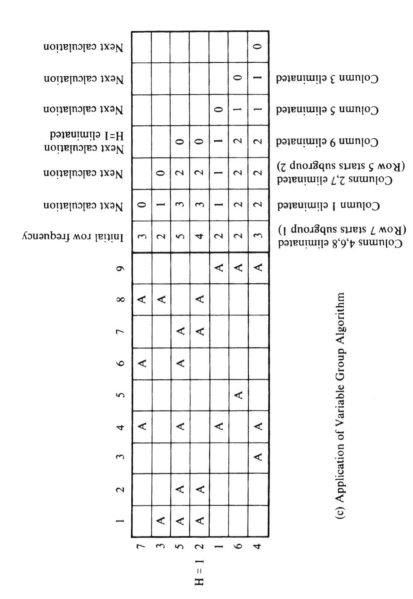

(c) Application of Variable Group Algorithm

Figure 2.27c: Structural Analysis of Algebraic System Structure.

Steady–State Lumped Systems

		DV↓4	DV↓6	8	1	2	7	9	5	3
SG 1	7	A	A	A						
	3			A	A					
	5	A	A		A	A	A			
	2		A	A	A	A				
SG 2	1	A						A		
	6							A	A	
	4	A						A		A

(d) Rearranged Functionality Matrix

Figure 2.27d: Structural Analysis of Algebraic System Structure.

subgroups 1 and 2 indicated. In order to be able to solve the equations in subgroup 1, both design variables must come from this subgroup. The two platform variables for this are variables 4 and 6 and are therefore natural candidates for the design variables. With this choice of design variables subgroup 1 is partitioned so that equation 7 gives variable 8 and equation 3 gives variable 1. In subgroup 2 we find equations 5 and 2 both involving variables 2 and 7. The best solution strategy is to solve these two linear equations in two unknowns simultaneously using matrix methods discussed earlier in this chapter. Finally, the remaining equations are then partitioned so that they each only involve a single unknown.

2.4.4 Computer Implementation of Structural Analysis

We have developed a FORTRAN 90 computer program that implements most of the Equation Ordering Algorithm and the Variable Group Algorithm. This program is fun.f and can be obtained on the world wide web at http://optimal.colorado.edu/~ramirez/chen4580.html. This FORTRAN program uses a fixed format that must be followed. The file fig2_28.matrix (see Figure 2.28) gives the input file that runs the example of section 2.4.3. In order to run this file using program fun, file fig2_26.matrix must be renamed functionality.matrix. Also the FORTRAN program fun.f must be compiled on your machine to get an executable code. The output of running this program is given as file fig2_28 (see Figure 2.26). This current version does not implement the

multicolumn (multiple root) part of the equation ordering algorithm.

Figure 2.28: Computer Implementation of Structural Analysis Input file for Example of Section 2.4.3 (fig2_28.matrix)

```
7           (Number of Equations,I2 format)
9           (Number of Variables, I2 format)
21          (Number of Elements in Functionality Matrix, I2 format)
1, 4,a      (Row, Column, Equation type: I2,1X,I2,1X,A format)
1, 9,a
2, 1,a
2, 2,a
2, 7,a
2, 8,a
3, 1,a
3, 8,a
4, 3,a
4, 4,a
4, 9,a
5, 1,a
5, 2,a
5, 4,a
5, 6,a
5, 7,a
6, 5,a
6, 9,a
7, 4,a
7, 6,a
7, 8,a
```

Output from Program (fig2_28)

	1	2	3	4	5	6	7	8	9
1	-	-	-	a	-	-	-	-	a
2	a	a	-	-	-	-	a	a	-
3	a	-	-	-	-	-	-	a	-
4	-	-	a	a	-	-	-	-	a
5	a	a	-	a	-	a	a	-	-
6	-	-	-	-	a	-	-	-	a
7	-	-	-	a	-	a	-	a	-

	1	2	3	4	5	6	7	8	9
7	-	-	-	a	-	a	-	a	-
3	a	-	-	-	-	-	-	a	-

```
5    a  a  -  a  -  a  a  -  -
2    a  a  -  -  -  -  a  a  -
1    -  -  -  a  -  -  -  -  a
6    -  -  -  -  a  -  -  -  a
4    -  -  a  a  -  -  -  -  a
HIERARCHY  1 = ROW  2

     4  6  8  1  2  7  9  5  3
7    a  a  a  -  -  -  -  -  -
3    -  -  a  a  -  -  -  -  -
5    a  a  -  a  a  a  -  -  -
2    -  -  a  a  a  a  -  -  -
1    a  -  -  -  -  -  a  -  -
6    -  -  -  -  -  -  a  a  -
4    a  -  -  -  -  -  a  -  a
SUB GROUP  1 = ROW  7
SUB GROUP  2 = ROW  5
```

2.4.5 Mixer–Exchanger–Mixer Design

The production of methylamines from methanol and ammonia is economically affected by the ratio of the demands of the three products (monomethylamine, dimethylamine, and trimethylamine). The recycle of trimethylamine will reduce the production of dimethylamine and trimethylamine relative to the production of monomethylamine. Similarly, the dilution of the reaction mixture with water will result in a relative increase in the production of monomethylamine. A mixer-heat exchanger-mixer portion of a methylamine plant is a proposed capital investment which would allow the relative production of the methylamines to be varied to meet changes in demand.

A detailed schematic for the mixer-exchanger-mixer system is shown in Figure 2.29. Trimethylamine recycle enters in stream 4, is cooled in the heat exchanger, and is mixed with water from stream 1 in mixer 1. The trimethylamine–water mixture is used as the cold side fluid in the heat exchanger and is then mixed with the ammonia–methanol stream from the gas absorber in mixer 3. The mixture leaving mixer 3 is the reaction mixture, which feeds into the preheater of the existing plant. A preliminary estimate of the cost of installing the mixer–heat exchanger–mixer system is desired. A generic mixer–exchanger-mixer problem has been studied by Ramirez and Vestal (1972).

The installation is modeled by the 49 equations presented in Table 2.2 all of which fit an acceptable form from Table 2.1. There are four redundant material balance equations in this equation set (equations 12, 13, 14, and 16) and three trivial equations (equations 1, 2, and 4). Each of the 55 variables is assigned a number for convenience (Table 2.3). Removing the redundant and trivial equations (with their variables) reduces the problem to 42 equations and 52 variables. The installation of

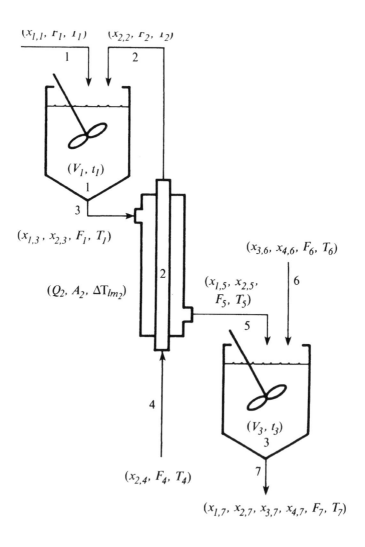

Figure 2.29: Mixer–Exchanger–Mixer System.

Table 2.2: Equations for Mixer–Exchanger–Mixer System

Material Balance Equations		Energy Balance Equations	
1.	$x_{1,1} - 1 = 0$	25.[18]	$F_1 T_1 - z_1 = 0$
2.	$x_{2,2} - 1 = 0$	26.[19]	$F_2 T_2 - z_2 = 0$
3. [1]	$x_{1,3} + x_{2,3} - 1 = 0$	27.[20]	$F_3 T_3 - z_3 = 0$
4.	$x_{2,4} - 1 = 0$	28.[21]	$F_4 T_4 - z_4 = 0$
5. [2]	$x_{1,5} + x_{2,5} - 1 = 0$	29.[22]	$F_5 T_5 - z_5 = 0$
6. [3]	$x_{3,6} + x_{4,6} = 1$	30.[23]	$F_6 T_6 - z_6 = 0$
7. [4]	$x_{1,7} + x_{2,7} + x_{3,7} + x_{4,7} - 1 = 0$	31.[24]	$F_7 T_7 - z_7 = 0$
		32.[25]	$C_1 z_1 + C_2 z_2 - C_3 z_3 = 0$
Flow Balance Equations		33.[26]	$C_4 z_4 - Q_2 - C_2 z_2 = 0$
8. [5]	$F_1 + F_2 - F_3 = 0$	34.[27]	$C_3 z_3 + Q_2 - C_5 z_5 = 0$
9. [6]	$F_2 - F_4 = 0$	35.[28]	$C_5 z_5 + C_6 z_6 - C_7 z_7 = 0$
10. [7]	$F_3 - F_5 = 0$	Equipment Specification Equations	
11. [8]	$F_5 + F_6 - F_7 = 0$	36.[29]	$V_1 - (F_3 t_1) / \rho_3 = 0$
Component Balance Eqns.		37.[30]	$Q_2 - U_2 A_2 \Delta T_{lm_2} = 0$
12.	$x_{1,1} F_1 - y_1 = 0$	38.[31]	$w_1 + T_3 - T_2 = 0$
13.	$x_{1,3} F_3 - y_1 = 0$	39.[32]	$w_2 + T_5 - T_4 = 0$
14.	$x_{1,5} F_5 - y_1 = 0$	40.[33]	$w_2 w_3 - w_1 = 0$
15. [9]	$x_{1,7} F_7 - y_1 = 0$	41.[34]	$w_4 - \ln(w_3) = 0$
16.	$x_{2,2} F_2 - y_2 = 0$	42.[35]	$w_4 \Delta T_{lm_2} + w_2 - w_1 = 0$
17.[10]	$x_{2,3} F_3 - y_2 = 0$	43.[36]	$V_3 - (F_7 t_3) / \rho_7 = 0$
18.[11]	$x_{2,4} F_4 - y_2 = 0$	Capital Cost Estimation Equations	
19.[12]	$x_{2,5} F_5 - y_2 = 0$	44.[37]	$\log(S_1) - 0.515 \log(v_1) - 3.354 = 0$
20.[13]	$x_{2,7} F_7 - y_2 = 0$	45.[38]	$v_1 - V_1 = 0$
21.[14]	$x_{3,6} F_6 - y_3 = 0$	46.[39]	$\log(S_2) - 0.699 \log(A_2) - 2.414 = 0$
22.[15]	$x_{4,7} F_7 - y_3 = 0$	47.[40]	$\log(S_3) - 0.515 \log(v_3) - 3.354 = 0$
23.[16]	$x_{4,6} F_6 - y_4 = 0$	48.[41]	$v_3 - V_3 = 0$
24.[17]	$x_{4,7} F_7 - y_4 = 0$	49.[42]	$S - S_1 - S_2 - S_3 = 0$

Table 2.3: Variable Assignments

Variable	Designated Number	Variable	Designated Number	Variable	Designated Number
$x_{1,1}$	1	y_1	21[14]	A_2	41[31]
$x_{2,2}$	2	y_2	22[15]	V_1	42[32]
$x_{1,3}$	3 [1]	y_3	23[16]	V_3	43[33]
$x_{2,3}$	4 [2]	y_4	24[17]	t_1	44[43]
$x_{2,4}$	5	T_1	25	t_3	45[44]
$x_{1,5}$	6 [3]	T_2	26[18]	w_1	46[45]
$x_{2,5}$	7 [4]	T_3	27[19]	w_2	47[34]
$x_{3,6}$	8 [5]	T_4	28	w_3	48[35]
$x_{4,6}$	9 [6]	T_5	29[20]	w_4	49[36]
$x_{1,7}$	10	T_6	30	v_1	50[37]
$x_{2,7}$	11	T_7	31[21]	v_3	51[38]
$x_{3,7}$	12	z_1	32[22]	S_1	52[39]
$x_{4,7}$	13 [7]	z_2	33[23]	S_2	53[40]
F_1	14 [8]	z_3	34[24]	S_3	54[41]
F_2	15 [9]	z_4	35[25]	S	55[42]
F_3	16[10]	z_5	36[26]		
F_4	17[11]	z_6	37[27]		
F_5	18[12]	z_7	38[28]		
F_6	19[13]	Q_2	39[29]		
F_7	20	ΔT_{lm_2}	40[30]		

the mixer–exchanger–mixer system into the existing plant tends to specify certain of the process variables. The temperatures of the entering streams (T_1, T_4 and T_6) are known. Likewise it is desired to produce a flow rate and composition of the exit stream which meets the proper specifications. Hence, values are set for variables F_7, $x_1 7$, $x_{2,7}$, $x_{3,7}$, and $x_{4,7}$. Only three of the four exit mole fractions are independent and may be specified as known variables. Therefore, of the ten degrees of freedom, seven are essentially specified as known variables via the problem. With seven known variables, three additional variables are required as design variables. The residence times for the two mixers (t_1 and t_3) and the cold end temperature difference of the heat exchanger (w_1) were preliminarily selected as possible design variables. These variables are not necessarily the best design variables for solution simplicity but are ones for which the designer can best assign reasonable values. Engineering experience, flow sheet analysis, and knowledge of the process mathematical model are used to construct the preliminary set of design variables, and are chosen whenever possible in the structural analysis. The preliminary set of design variables could contain any number of variables and does not have to exactly constrain the system.

We now used program fun.f to perform a structural analysis on this problem., First the completely specified 42 equations with 42 variables case is analyzed. The assignment of the equations for this problem is given in Table 2.2 and is indicated by the numbers in [ ]. The variable assignment is given in Table 2.3 with the numbers in [ ]. File mixer.matrix contains the input file necessary for running this problem. Again all files are available on the world wide web under http://optimal.colorado.edu/~ramirez/chen4580.html1. The original 42 by 42 functionality matrix is given in Figure 2.30.

The results of running the equation ordering algorithm is the functionality matrix of Figure 2.31. There are two hierarchies which indicate that there are two iterative variables needed to solve this problem or that some of the equations must be solved simultaneously. The step equations will be located at rows 5 and 3 of the rearranged functionality matrix.

The final results of the equation ordering algorithm is given in Figure 2.32. There is one subgroup which starts at row 25 with two platform variables. This means that one section of the problem must be solved simultaneously. The simultaneous equation set starts at row 25 and goes through row 3. This is a set of 14 equations in 14 unknowns. Due to the presence of b elements (products), this is a set of nonlinear algebraic equations. All other equations can be solved explicitly with one equation in one unknown. We have reduced the computational complexity from one of 42 simultaneous nonlinear equations to one of 14 simultaneous nonlinear equations.

A new functionality matrix can be considered with the three design variables t_1, t_3, and w_1 not specified. This problem now involves 45 variables and 42 equations. There are three degrees of freedom so that these design variables can be chosen using structural analysis to minimize computational complexity. File `mix1.matrix` gives the input file for this problem. The final rearranged functionality matrix for this problem is given in Figure 2.33. There is one small set of three nonlinear equations that must be solved simultaneously (equations 16 to 3). The loop starts with the platform variables associated with the first subgroup and ends with the step equation of equation 3 which is the location of the only hierarchy pointer. To make the rest of the problem acyclic, the design variables should be the platform variables 23 or 24 (z_2 or z_3), 33 or 44 (V_3 or t_3), and 32 or 43 (V_1 or t_1). Therefore, we can keep t_1 and t_3 as design variables and by substituting z_2 or z_3 for w_1, we reduce the computational difficulty of the problem significantly (from 14 simultaneous nonlinear equations to 3 simultaneous nonlinear equations).

2.4.5.1 Nomenclature for Mixer–Exchanger–Mixer Design

$x_{i,j}$ Mole fraction of component i in stream j
F_j Molar flow rate of stream j (g mol/hr)
y_i Molar flow rate of component i (g mol/hr)
T_j Temperature of stream j (°C)
z_j Auxiliary state variable substitution for $F_j\ T_j$
Q_k Heat transferred in unit k (cal/hr)
A_k Heat transfer area for unit k (m^2)
V_k Volume of unit k (m^3)
t_k Residence time for unit k (hr^{-1})
w_1 Auxiliary state variable substitution
v_k Volume of unit k (m^3)
S_k Cost of unit k (dollars)
S Total cost of installation (dollars)
C_j Molar heat capacity of stream j (cal/g mol °C)
U_k Overall heat transfer coefficient of unit k (cal/hr m^2 °C)
f_i Row frequency of row i
ρ_k Molar density of fluid in unit k (g mol/cm^3)
τ_j Column frequency of column j
ΔT_{lm_k} Log mean temperature difference for unit k (°C)

Figure 2.30: Functionality Matrix for Mixer-Exchanger-Mixer Problem With All Design Variables Chosen

	1	2	3	4	5	6	7	8	9	10	11	12	13	14	15	16	17	18	19	20	21	22	23	24	25	26	27	28	29	30	31	32	33	34	35	36	37	38	39	40	41	42
1	a	a	-	-	-	-	-	-	-	-	-	-	-	-	-	-	-	-	-	-	-	-	-	-	-	-	-	-	-	-	-	-	-	-	-	-	-	-	-	-	-	-
2	-	-	a	a	-	-	-	-	-	-	-	-	-	-	-	-	-	-	-	-	-	-	-	-	-	-	-	-	-	-	-	-	-	-	-	-	-	-	-	-	-	-
3	-	-	-	-	a	a	-	-	-	-	-	-	-	-	-	-	-	-	-	-	-	-	-	-	-	-	-	-	-	-	-	-	-	-	-	-	-	-	-	-	-	-
4	-	-	-	-	-	-	a	-	-	-	-	-	-	-	-	-	-	-	-	-	-	-	-	-	-	-	-	-	-	-	-	-	-	-	-	-	-	-	-	-	-	-
5	-	-	-	-	-	-	-	a	a	a	-	-	-	-	-	-	-	-	-	-	-	-	-	-	-	-	-	-	-	-	-	-	-	-	-	-	-	-	-	-	-	-
6	-	-	-	-	-	-	-	-	a	-	a	-	-	-	-	-	-	-	-	-	-	-	-	-	-	-	-	-	-	-	-	-	-	-	-	-	-	-	-	-	-	-
7	-	-	-	-	-	-	-	-	-	a	-	a	-	-	-	-	-	-	-	-	-	-	-	-	-	-	-	-	-	-	-	-	-	-	-	-	-	-	-	-	-	-
8	-	-	-	-	-	-	-	-	-	-	-	a	a	-	-	-	-	-	-	-	-	-	-	-	-	-	-	-	-	-	-	-	-	-	-	-	-	-	-	-	-	-
9	-	-	-	-	-	-	-	-	-	-	-	-	-	a	-	-	-	-	-	-	-	-	-	-	-	-	-	-	-	-	-	-	-	-	-	-	-	-	-	-	-	-
10	-	b	-	-	-	-	-	-	-	b	-	-	-	-	a	-	-	-	-	-	-	-	-	-	-	-	-	-	-	-	-	-	-	-	-	-	-	-	-	-	-	-
11	-	-	-	-	-	-	-	-	-	-	a	-	-	-	a	-	-	-	-	-	-	-	-	-	-	-	-	-	-	-	-	-	-	-	-	-	-	-	-	-	-	-
12	-	-	-	b	-	-	-	-	-	-	-	b	-	-	a	-	-	-	-	-	-	-	-	-	-	-	-	-	-	-	-	-	-	-	-	-	-	-	-	-	-	-
13	-	-	-	-	-	-	-	-	-	-	-	-	-	-	a	-	-	-	-	-	-	-	-	-	-	-	-	-	-	-	-	-	-	-	-	-	-	-	-	-	-	-
14	-	-	-	-	b	-	-	-	-	-	-	-	b	-	-	a	-	-	-	-	-	-	-	-	-	-	-	-	-	-	-	-	-	-	-	-	-	-	-	-	-	-
15	-	-	-	-	-	-	-	-	-	-	-	-	-	-	-	a	-	-	-	-	-	-	-	-	-	-	-	-	-	-	-	-	-	-	-	-	-	-	-	-	-	-
16	-	-	-	-	-	b	-	-	-	-	-	-	b	-	-	-	a	-	-	-	-	-	-	-	-	-	-	-	-	-	-	-	-	-	-	-	-	-	-	-	-	-
17	-	-	-	-	-	-	a	-	-	-	-	-	-	-	-	-	a	-	-	-	-	-	-	-	-	-	-	-	-	-	-	-	-	-	-	-	-	-	-	-	-	-
18	-	-	-	-	-	-	-	a	-	-	-	-	-	-	-	-	-	-	-	-	-	a	-	-	-	-	-	-	-	-	-	-	-	-	-	-	-	-	-	-	-	-
19	-	-	-	-	-	-	-	-	b	-	-	-	-	-	-	-	-	b	-	-	-	-	a	-	-	-	-	-	-	-	-	-	-	-	-	-	-	-	-	-	-	-
20	-	-	-	-	-	-	-	-	-	b	-	-	-	-	-	-	-	-	b	-	-	-	-	a	-	-	-	-	-	-	-	-	-	-	-	-	-	-	-	-	-	-
21	-	-	-	-	-	-	-	-	-	-	a	-	-	-	-	-	-	-	-	-	-	-	-	-	a	-	-	-	-	-	-	-	-	-	-	-	-	-	-	-	-	-
22	-	-	-	-	-	-	-	-	-	-	-	b	-	-	-	-	-	-	-	a	-	-	-	-	-	-	-	-	-	-	-	-	-	-	-	-	-	-	-	-	-	-
23	-	-	-	-	-	-	-	-	-	-	-	-	a	-	-	-	-	-	-	-	-	-	-	-	-	-	a	-	-	-	-	-	-	-	-	-	-	-	-	-	-	-
24	-	-	-	-	-	-	-	-	-	-	-	-	-	-	-	-	-	-	-	-	a	-	-	-	-	-	-	a	-	-	-	-	-	-	-	-	-	-	-	-	-	-
25	-	-	-	-	-	-	-	-	-	-	-	-	-	-	-	-	-	-	-	-	-	a	a	a	-	-	-	-	-	-	-	-	-	-	-	-	-	-	-	-	-	-
26	-	-	-	-	-	-	-	-	-	-	-	-	-	-	-	-	-	-	-	-	-	-	a	-	a	-	-	-	a	-	-	-	-	-	-	-	-	-	-	-	-	-
27	-	-	-	-	-	-	-	-	-	-	-	-	-	-	-	-	-	-	-	-	-	-	-	a	-	a	-	-	a	-	-	-	-	-	-	-	-	-	-	-	-	-
28	-	-	-	-	-	-	-	-	-	-	-	-	-	-	-	-	-	-	-	-	-	-	-	-	-	a	a	a	-	-	-	-	-	-	-	-	-	-	-	-	-	-
29	-	-	-	-	-	-	-	-	-	a	-	-	-	-	-	-	-	-	-	-	-	-	-	-	-	-	-	-	-	-	-	a	-	-	-	-	-	-	-	-	-	-
30	-	-	-	-	-	-	-	-	-	-	-	-	-	-	-	-	-	-	-	-	-	-	-	-	-	-	-	-	a	b	b	-	-	-	-	-	-	-	-	-	-	-
31	-	-	-	-	-	-	-	-	-	-	-	-	-	-	-	-	-	a	a	-	-	-	-	-	-	-	-	-	-	-	-	-	-	-	-	-	-	-	-	-	-	-
32	-	-	-	-	-	-	-	-	-	-	-	-	-	-	-	-	-	-	-	a	-	-	-	-	-	-	-	-	-	-	-	-	-	a	-	-	-	-	-	-	-	-
33	-	-	-	-	-	-	-	-	-	-	-	-	-	-	-	-	-	-	-	-	-	-	-	-	-	-	-	-	-	-	-	-	-	b	b	-	-	-	-	-	-	-
34	-	-	-	-	-	-	-	-	-	-	-	-	-	-	-	-	-	-	-	-	-	-	-	-	-	-	-	-	-	-	-	-	-	-	l	a	-	-	-	-	-	-
35	-	-	-	-	-	-	-	-	-	-	-	-	-	-	-	-	-	-	-	-	-	-	-	-	-	-	-	-	-	b	-	-	-	a	-	b	-	-	-	-	-	-
36	-	-	-	-	-	-	-	-	-	-	-	-	-	-	-	-	-	-	-	-	-	-	-	-	-	-	-	-	-	-	-	-	a	-	-	-	-	-	-	-	-	-
37	-	-	-	-	-	-	-	-	-	-	-	-	-	-	-	-	-	-	-	-	-	-	-	-	-	-	-	-	-	-	-	-	-	-	-	-	m	-	m	-	-	-
38	-	-	-	-	-	-	-	-	-	-	-	-	-	-	-	-	-	-	-	-	-	-	-	-	-	-	-	-	-	-	-	a	-	-	-	-	a	-	-	-	-	-
39	-	-	-	-	-	-	-	-	-	-	-	-	-	-	-	-	-	-	-	-	-	-	-	-	-	-	-	-	-	-	m	-	-	-	-	-	-	-	-	m	-	-
40	-	-	-	-	-	-	-	-	-	-	-	-	-	-	-	-	-	-	-	-	-	-	-	-	-	-	-	-	-	-	-	-	-	-	-	-	-	m	-	-	m	-
41	-	-	-	-	-	-	-	-	-	-	-	-	-	-	-	-	-	-	-	-	-	-	-	-	-	-	-	-	-	-	-	-	a	-	-	-	-	a	-	-	-	-
42	-	-	-	-	-	-	-	-	-	-	-	-	-	-	-	-	-	-	-	-	-	-	-	-	-	-	-	-	-	-	-	-	-	-	-	-	-	-	a	a	a	a

Figure 2.31: Equation Ordering Results for Mixer-Exchanger-Mixer Example

	1	2	3	4	5	6	7	8	9	10	11	12	13	14	15	16	17	18	19	20	21	22	23	24	25	26	27	28	29	30	31	32	33	34	35	36	37	38	39	40	41	42
25	-	-	-	-	-	-	-	-	-	-	-	-	-	-	-	-	-	-	-	-	-	a	a	a	-	-	-	-	-	-	-	-	-	-	-	-	-	-	-	-	-	-
13	-	-	-	-	-	-	-	-	-	-	-	-	-	-	a	-	-	-	-	-	-	-	-	-	-	-	-	-	-	-	-	-	-	-	-	-	-	-	-	-	-	-
11	-	-	-	-	-	-	-	-	-	-	a	-	-	-	a	-	-	-	-	-	-	-	-	-	-	-	-	-	-	-	-	-	-	-	-	-	-	-	-	-	-	-
6	-	-	-	-	-	-	-	-	a	-	a	-	-	-	-	-	-	-	-	-	-	-	-	-	-	-	-	-	-	-	-	-	-	-	-	-	-	-	-	-	-	-
19	-	-	-	-	-	-	-	-	b	-	-	-	-	-	-	-	-	b	-	-	-	-	a	-	-	-	-	-	-	-	-	-	-	-	-	-	-	-	-	-	-	-
31	-	-	-	-	-	-	-	-	-	-	-	-	-	-	-	-	-	a	a	-	-	-	-	-	-	-	-	-	-	-	-	-	-	-	-	-	-	-	-	-	-	-
20	-	-	-	-	-	-	-	-	-	b	-	-	-	-	-	-	-	-	b	-	-	-	-	a	-	-	-	-	-	-	-	-	-	-	-	-	-	-	-	-	-	-
18	-	-	-	-	-	-	-	a	-	-	-	-	-	-	-	-	-	-	-	-	-	a	-	-	-	-	-	-	-	-	-	-	-	-	-	-	-	-	-	-	-	-
5	-	-	-	-	-	-	-	a	a	a	-	-	-	-	-	-	-	-	-	-	-	-	-	-	-	-	-	-	-	-	-	-	-	-	-	-	-	-	-	-	-	-
4	-	-	-	-	-	-	a	-	-	-	-	-	-	-	-	-	-	-	-	-	-	-	-	-	-	-	-	-	-	-	-	-	-	-	-	-	-	-	-	-	-	-
17	-	-	-	-	-	-	a	-	-	-	-	-	-	-	-	-	a	-	-	-	-	-	-	-	-	-	-	-	-	-	-	-	-	-	-	-	-	-	-	-	-	-
15	-	-	-	-	-	-	-	-	-	-	-	-	-	-	-	a	-	-	-	-	-	-	-	-	-	-	-	-	-	-	-	-	-	-	-	-	-	-	-	-	-	-
7	-	-	-	-	-	-	-	-	-	a	-	a	-	-	-	-	-	-	-	-	-	-	-	-	-	-	-	-	-	-	-	-	-	-	-	-	-	-	-	-	-	-
8	-	-	-	-	-	-	-	-	-	-	-	a	a	-	-	-	-	-	-	-	-	-	-	-	-	-	-	-	-	-	-	-	-	-	-	-	-	-	-	-	-	-
16	-	-	-	-	-	b	-	-	-	-	-	-	b	-	-	-	a	-	-	-	-	-	-	-	-	-	-	-	-	-	-	-	-	-	-	-	-	-	-	-	-	-
14	-	-	-	-	b	-	-	-	-	-	-	-	b	-	-	a	-	-	-	-	-	-	-	-	-	-	-	-	-	-	-	-	-	-	-	-	-	-	-	-	-	-
3	-	-	-	-	a	a	-	-	-	-	-	-	-	-	-	-	-	-	-	-	-	-	-	-	-	-	-	-	-	-	-	-	-	-	-	-	-	-	-	-	-	-
36	-	-	-	-	-	-	-	-	-	-	-	-	-	-	-	-	-	-	-	-	-	-	-	-	-	-	-	-	-	-	-	-	a	-	-	-	-	-	-	-	-	-
41	-	-	-	-	-	-	-	-	-	-	-	-	-	-	-	-	-	-	-	-	-	-	-	-	-	-	-	-	-	-	-	-	a	-	-	-	-	a	-	-	-	-
40	-	-	-	-	-	-	-	-	-	-	-	-	-	-	-	-	-	-	-	-	-	-	-	-	-	-	-	-	-	-	-	-	-	-	-	-	-	m	-	-	m	-
22	-	-	-	-	-	-	-	-	-	-	-	b	-	-	-	-	-	-	-	a	-	-	-	-	-	-	-	-	-	-	-	-	-	-	-	-	-	-	-	-	-	-
32	-	-	-	-	-	-	-	-	-	-	-	-	-	-	-	-	-	-	-	a	-	-	-	-	-	-	-	-	-	-	-	-	-	a	-	-	-	-	-	-	-	-
33	-	-	-	-	-	-	-	-	-	-	-	-	-	-	-	-	-	-	-	-	-	-	-	-	-	-	-	-	-	-	-	-	-	b	b	-	-	-	-	-	-	-
34	-	-	-	-	-	-	-	-	-	-	-	-	-	-	-	-	-	-	-	-	-	-	-	-	-	-	-	-	-	-	-	-	-	-	l	a	-	-	-	-	-	-
35	-	-	-	-	-	-	-	-	-	-	-	-	-	-	-	-	-	-	-	-	-	-	-	-	-	-	-	-	-	b	-	-	-	a	-	b	-	-	-	-	-	-
21	-	-	-	-	-	-	-	-	-	-	a	-	-	-	-	-	-	-	-	-	-	-	-	-	a	-	-	-	-	-	-	-	-	-	-	-	-	-	-	-	-	-
26	-	-	-	-	-	-	-	-	-	-	-	-	-	-	-	-	-	-	-	-	-	-	a	-	a	-	-	-	a	-	-	-	-	-	-	-	-	-	-	-	-	-
30	-	-	-	-	-	-	-	-	-	-	-	-	-	-	-	-	-	-	-	-	-	-	-	-	-	-	-	-	a	b	b	-	-	-	-	-	-	-	-	-	-	-
39	-	-	-	-	-	-	-	-	-	-	-	-	-	-	-	-	-	-	-	-	-	-	-	-	-	-	-	-	-	-	m	-	-	-	-	-	-	-	-	m	-	-
29	-	-	-	-	-	-	-	-	-	a	-	-	-	-	-	-	-	-	-	-	-	-	-	-	-	-	-	-	-	-	-	a	-	-	-	-	-	-	-	-	-	-
38	-	-	-	-	-	-	-	-	-	-	-	-	-	-	-	-	-	-	-	-	-	-	-	-	-	-	-	-	-	-	-	a	-	-	-	-	a	-	-	-	-	-
37	-	-	-	-	-	-	-	-	-	-	-	-	-	-	-	-	-	-	-	-	-	-	-	-	-	-	-	-	-	-	-	-	-	-	-	-	m	-	m	-	-	-
42	-	-	-	-	-	-	-	-	-	-	-	-	-	-	-	-	-	-	-	-	-	-	-	-	-	-	-	-	-	-	-	-	-	-	-	-	-	-	a	a	a	a
23	-	-	-	-	-	-	-	-	-	-	-	-	a	-	-	-	-	-	-	-	-	-	-	-	-	-	a	-	-	-	-	-	-	-	-	-	-	-	-	-	-	-
27	-	-	-	-	-	-	-	-	-	-	-	-	-	-	-	-	-	-	-	-	-	-	-	a	-	a	-	-	a	-	-	-	-	-	-	-	-	-	-	-	-	-
28	-	-	-	-	-	-	-	-	-	-	-	-	-	-	-	-	-	-	-	-	-	-	-	-	-	a	a	a	-	-	-	-	-	-	-	-	-	-	-	-	-	-
24	-	-	-	-	-	-	-	-	-	-	-	-	-	-	-	-	-	-	-	-	a	-	-	-	-	-	-	a	-	-	-	-	-	-	-	-	-	-	-	-	-	-
9	-	-	-	-	-	-	-	-	-	-	-	-	-	a	-	-	-	-	-	-	-	-	-	-	-	-	-	-	-	-	-	-	-	-	-	-	-	-	-	-	-	-
12	-	-	-	b	-	-	-	-	-	-	-	b	-	-	a	-	-	-	-	-	-	-	-	-	-	-	-	-	-	-	-	-	-	-	-	-	-	-	-	-	-	-
2	-	-	a	a	-	-	-	-	-	-	-	-	-	-	-	-	-	-	-	-	-	-	-	-	-	-	-	-	-	-	-	-	-	-	-	-	-	-	-	-	-	-
10	-	b	-	-	-	-	-	-	-	b	-	-	-	-	a	-	-	-	-	-	-	-	-	-	-	-	-	-	-	-	-	-	-	-	-	-	-	-	-	-	-	-
1	a	a	-	-	-	-	-	-	-	-	-	-	-	-	-	-	-	-	-	-	-	-	-	-	-	-	-	-	-	-	-	-	-	-	-	-	-	-	-	-	-	-

HIERARCHY 2 = ROW 5
HIERARCHY 1 = ROW 3

Figure 2.32: Final Rearranged Functionality Matrix for Mixer-Exchanger-Mixer Example

	15	11	9	22	23	24	18	19	10	8	7	17	16	12	13	6	5	33	38	41	20	34	35	36	30	25	29	31	40	32	37	39	42	27	26	28	21	14	4	3	2	1
13	a	-	-	-	-	-	-	-	-	-	-	-	-	-	-	-	-	-	-	-	-	-	-	-	-	-	-	-	-	-	-	-	-	-	-	-	-	-	-	-	-	-
11	a	a	-	-	-	-	-	-	-	-	-	-	-	-	-	-	-	-	-	-	-	-	-	-	-	-	-	-	-	-	-	-	-	-	-	-	-	-	-	-	-	-
6	-	a	a	-	-	-	-	-	-	-	-	-	-	-	-	-	-	-	-	-	-	-	-	-	-	-	-	-	-	-	-	-	-	-	-	-	-	-	-	-	-	-
25	-	-	-	a	a	a	-	-	-	-	-	-	-	-	-	-	-	-	-	-	-	-	-	-	-	-	-	-	-	-	-	-	-	-	-	-	-	-	-	-	-	-
19	-	-	b	-	a	-	b	-	-	-	-	-	-	-	-	-	-	-	-	-	-	-	-	-	-	-	-	-	-	-	-	-	-	-	-	-	-	-	-	-	-	-
31	-	-	-	-	-	-	a	a	-	-	-	-	-	-	-	-	-	-	-	-	-	-	-	-	-	-	-	-	-	-	-	-	-	-	-	-	-	-	-	-	-	-
20	-	-	-	-	-	a	-	b	b	-	-	-	-	-	-	-	-	-	-	-	-	-	-	-	-	-	-	-	-	-	-	-	-	-	-	-	-	-	-	-	-	-
18	-	-	-	a	-	-	-	-	-	a	-	-	-	-	-	-	-	-	-	-	-	-	-	-	-	-	-	-	-	-	-	-	-	-	-	-	-	-	-	-	-	-
5	-	-	a	-	-	-	-	-	a	a	-	-	-	-	-	-	-	-	-	-	-	-	-	-	-	-	-	-	-	-	-	-	-	-	-	-	-	-	-	-	-	-
4	-	-	-	-	-	-	-	-	-	-	a	-	-	-	-	-	-	-	-	-	-	-	-	-	-	-	-	-	-	-	-	-	-	-	-	-	-	-	-	-	-	-
17	-	-	-	-	-	-	-	-	-	-	a	a	-	-	-	-	-	-	-	-	-	-	-	-	-	-	-	-	-	-	-	-	-	-	-	-	-	-	-	-	-	-
15	-	-	-	-	-	-	-	-	-	-	-	-	a	-	-	-	-	-	-	-	-	-	-	-	-	-	-	-	-	-	-	-	-	-	-	-	-	-	-	-	-	-
7	-	-	-	-	-	-	-	-	a	-	-	-	-	a	-	-	-	-	-	-	-	-	-	-	-	-	-	-	-	-	-	-	-	-	-	-	-	-	-	-	-	-
8	-	-	-	-	-	-	-	-	-	-	-	-	-	a	a	-	-	-	-	-	-	-	-	-	-	-	-	-	-	-	-	-	-	-	-	-	-	-	-	-	-	-
16	-	-	-	-	-	-	-	-	-	-	-	a	-	-	b	b	-	-	-	-	-	-	-	-	-	-	-	-	-	-	-	-	-	-	-	-	-	-	-	-	-	-
14	-	-	-	-	-	-	-	-	-	-	-	-	a	-	b	-	b	-	-	-	-	-	-	-	-	-	-	-	-	-	-	-	-	-	-	-	-	-	-	-	-	-
3	-	-	-	-	-	-	-	-	-	-	-	-	-	-	-	a	a	-	-	-	-	-	-	-	-	-	-	-	-	-	-	-	-	-	-	-	-	-	-	-	-	-
36	-	-	-	-	-	-	-	-	-	-	-	-	-	-	-	-	-	a	-	-	-	-	-	-	-	-	-	-	-	-	-	-	-	-	-	-	-	-	-	-	-	-
41	-	-	-	-	-	-	-	-	-	-	-	-	-	-	-	-	-	a	a	-	-	-	-	-	-	-	-	-	-	-	-	-	-	-	-	-	-	-	-	-	-	-
40	-	-	-	-	-	-	-	-	-	-	-	-	-	-	-	-	-	-	m	m	-	-	-	-	-	-	-	-	-	-	-	-	-	-	-	-	-	-	-	-	-	-
22	-	-	-	-	-	-	-	-	-	-	-	-	-	b	-	-	-	-	-	-	a	-	-	-	-	-	-	-	-	-	-	-	-	-	-	-	-	-	-	-	-	-
32	-	-	-	-	-	-	-	-	-	-	-	-	-	-	-	-	-	-	-	-	a	a	-	-	-	-	-	-	-	-	-	-	-	-	-	-	-	-	-	-	-	-
33	-	-	-	-	-	-	-	-	-	-	-	-	-	-	-	-	-	-	-	-	-	b	b	-	-	-	-	-	-	-	-	-	-	-	-	-	-	-	-	-	-	-
34	-	-	-	-	-	-	-	-	-	-	-	-	-	-	-	-	-	-	-	-	-	-	l	a	-	-	-	-	-	-	-	-	-	-	-	-	-	-	-	-	-	-
35	-	-	-	-	-	-	-	-	-	-	-	-	-	-	-	-	-	-	-	-	-	a	-	b	b	-	-	-	-	-	-	-	-	-	-	-	-	-	-	-	-	-
21	-	a	-	-	-	-	-	-	-	-	-	-	-	-	-	-	-	-	-	-	-	-	-	-	-	a	-	-	-	-	-	-	-	-	-	-	-	-	-	-	-	-
26	-	-	-	-	a	-	-	-	-	-	-	-	-	-	-	-	-	-	-	-	-	-	-	-	-	a	a	-	-	-	-	-	-	-	-	-	-	-	-	-	-	-
30	-	-	-	-	-	-	-	-	-	-	-	-	-	-	-	-	-	-	-	-	-	-	-	-	b	-	a	b	-	-	-	-	-	-	-	-	-	-	-	-	-	-
39	-	-	-	-	-	-	-	-	-	-	-	-	-	-	-	-	-	-	-	-	-	-	-	-	-	-	-	m	m	-	-	-	-	-	-	-	-	-	-	-	-	-
29	-	-	-	-	-	-	-	-	a	-	-	-	-	-	-	-	-	-	-	-	-	-	-	-	-	-	-	-	-	a	-	-	-	-	-	-	-	-	-	-	-	-
38	-	-	-	-	-	-	-	-	-	-	-	-	-	-	-	-	-	-	-	-	-	-	-	-	-	-	-	-	-	a	a	-	-	-	-	-	-	-	-	-	-	-
37	-	-	-	-	-	-	-	-	-	-	-	-	-	-	-	-	-	-	-	-	-	-	-	-	-	-	-	-	-	-	m	m	-	-	-	-	-	-	-	-	-	-
42	-	-	-	-	-	-	-	-	-	-	-	-	-	-	-	-	-	-	-	a	-	-	-	-	-	-	-	-	a	-	-	a	a	-	-	-	-	-	-	-	-	-
23	-	-	-	-	-	-	-	-	-	-	-	-	-	-	a	-	-	-	-	-	-	-	-	-	-	-	-	-	-	-	-	-	-	a	-	-	-	-	-	-	-	-
27	-	-	-	-	-	a	-	-	-	-	-	-	-	-	-	-	-	-	-	-	-	-	-	-	-	-	a	-	-	-	-	-	-	-	a	-	-	-	-	-	-	-
28	-	-	-	-	-	-	-	-	-	-	-	-	-	-	-	-	-	-	-	-	-	-	-	-	-	-	-	-	-	-	-	-	-	a	a	a	-	-	-	-	-	-
24	-	-	-	-	-	-	-	-	-	-	-	-	-	-	-	-	-	-	-	-	-	-	-	-	-	-	-	-	-	-	-	-	-	-	-	a	a	-	-	-	-	-
9	-	-	-	-	-	-	-	-	-	-	-	-	-	-	-	-	-	-	-	-	-	-	-	-	-	-	-	-	-	-	-	-	-	-	-	-	-	a	-	-	-	-
12	a	-	-	-	-	-	-	-	-	-	-	-	-	b	-	-	-	-	-	-	-	-	-	-	-	-	-	-	-	-	-	-	-	-	-	-	-	-	b	-	-	-
2	-	-	-	-	-	-	-	-	-	-	-	-	-	-	-	-	-	-	-	-	-	-	-	-	-	-	-	-	-	-	-	-	-	-	-	-	-	-	a	a	-	-
10	a	-	-	-	-	-	-	-	b	-	-	-	-	-	-	-	-	-	-	-	-	-	-	-	-	-	-	-	-	-	-	-	-	-	-	-	-	-	-	-	b	-
1	-	-	-	-	-	-	-	-	-	-	-	-	-	-	-	-	-	-	-	-	-	-	-	-	-	-	-	-	-	-	-	-	-	-	-	-	-	-	-	-	a	a

SUB GROUP 1 = ROW 25

Figure 2.33: Rearranged Functionality Matrix with Three Degrees of Freedom

	7	17	16	6	13	5	15	12	11	10	9	8	22	20	34	25	27	14	4	3	2	1	23	24	19	18	45	35	36	30	29	31	40	26	28	21	33	44	38	41	32	43	37	39	42
4	a	-	-	-	-	-	-	-	-	-	-	-	-	-	-	-	-	-	-	-	-	-	-	-	-	-	-	-	-	-	-	-	-	-	-	-	-	-	-	-	-	-	-	-	-
17	a	a	-	-	-	-	-	-	-	-	-	-	-	-	-	-	-	-	-	-	-	-	-	-	-	-	-	-	-	-	-	-	-	-	-	-	-	-	-	-	-	-	-	-	-
15	-	-	a	-	-	-	-	-	-	-	-	-	-	-	-	-	-	-	-	-	-	-	-	-	-	-	-	-	-	-	-	-	-	-	-	-	-	-	-	-	-	-	-	-	-
16	-	a	-	b	b	-	-	-	-	-	-	-	-	-	-	-	-	-	-	-	-	-	-	-	-	-	-	-	-	-	-	-	-	-	-	-	-	-	-	-	-	-	-	-	-
14	-	-	a	-	b	b	-	-	-	-	-	-	-	-	-	-	-	-	-	-	-	-	-	-	-	-	-	-	-	-	-	-	-	-	-	-	-	-	-	-	-	-	-	-	-
3	-	-	-	a	-	a	-	-	-	-	-	-	-	-	-	-	-	-	-	-	-	-	-	-	-	-	-	-	-	-	-	-	-	-	-	-	-	-	-	-	-	-	-	-	-
13	-	-	-	-	-	-	a	-	-	-	-	-	-	-	-	-	-	-	-	-	-	-	-	-	-	-	-	-	-	-	-	-	-	-	-	-	-	-	-	-	-	-	-	-	-
8	-	-	-	-	a	-	-	a	-	-	-	-	-	-	-	-	-	-	-	-	-	-	-	-	-	-	-	-	-	-	-	-	-	-	-	-	-	-	-	-	-	-	-	-	-
11	-	-	-	-	-	-	a	-	a	-	-	-	-	-	-	-	-	-	-	-	-	-	-	-	-	-	-	-	-	-	-	-	-	-	-	-	-	-	-	-	-	-	-	-	-
7	-	-	-	-	-	-	-	a	-	a	-	-	-	-	-	-	-	-	-	-	-	-	-	-	-	-	-	-	-	-	-	-	-	-	-	-	-	-	-	-	-	-	-	-	-
6	-	-	-	-	-	-	-	-	a	-	a	-	-	-	-	-	-	-	-	-	-	-	-	-	-	-	-	-	-	-	-	-	-	-	-	-	-	-	-	-	-	-	-	-	-
5	-	-	-	-	-	-	-	-	-	a	a	a	-	-	-	-	-	-	-	-	-	-	-	-	-	-	-	-	-	-	-	-	-	-	-	-	-	-	-	-	-	-	-	-	-
18	-	-	-	-	-	-	-	-	-	-	-	a	a	-	-	-	-	-	-	-	-	-	-	-	-	-	-	-	-	-	-	-	-	-	-	-	-	-	-	-	-	-	-	-	-
22	-	-	-	-	-	-	-	b	-	-	-	-	-	a	-	-	-	-	-	-	-	-	-	-	-	-	-	-	-	-	-	-	-	-	-	-	-	-	-	-	-	-	-	-	-
32	-	-	-	-	-	-	-	-	-	-	-	-	-	a	a	-	-	-	-	-	-	-	-	-	-	-	-	-	-	-	-	-	-	-	-	-	-	-	-	-	-	-	-	-	-
21	-	-	-	-	-	-	-	-	a	-	-	-	-	-	-	a	-	-	-	-	-	-	-	-	-	-	-	-	-	-	-	-	-	-	-	-	-	-	-	-	-	-	-	-	-
23	-	-	-	-	a	-	-	-	-	-	-	-	-	-	-	-	a	-	-	-	-	-	-	-	-	-	-	-	-	-	-	-	-	-	-	-	-	-	-	-	-	-	-	-	-
9	-	-	-	-	-	-	-	-	-	-	-	-	-	-	-	-	-	a	-	-	-	-	-	-	-	-	-	-	-	-	-	-	-	-	-	-	-	-	-	-	-	-	-	-	-
12	-	-	-	-	-	-	a	b	-	-	-	-	-	-	-	-	-	-	b	-	-	-	-	-	-	-	-	-	-	-	-	-	-	-	-	-	-	-	-	-	-	-	-	-	-
2	-	-	-	-	-	-	-	-	-	-	-	-	-	-	-	-	-	-	a	a	-	-	-	-	-	-	-	-	-	-	-	-	-	-	-	-	-	-	-	-	-	-	-	-	-
10	-	-	-	-	-	-	a	-	-	b	-	-	-	-	-	-	-	-	-	-	b	-	-	-	-	-	-	-	-	-	-	-	-	-	-	-	-	-	-	-	-	-	-	-	-
1	-	-	-	-	-	-	-	-	-	-	-	-	-	-	-	-	-	-	-	-	a	a	-	-	-	-	-	-	-	-	-	-	-	-	-	-	-	-	-	-	-	-	-	-	-
25	-	-	-	-	-	-	-	-	-	-	-	-	a	-	-	-	-	-	-	-	-	-	a	a	-	-	-	-	-	-	-	-	-	-	-	-	-	-	-	-	-	-	-	-	-
20	-	-	-	-	-	-	-	-	-	b	-	-	-	-	-	-	-	-	-	-	-	-	-	a	b	-	-	-	-	-	-	-	-	-	-	-	-	-	-	-	-	-	-	-	-
19	-	-	-	-	-	-	-	-	-	-	b	-	-	-	-	-	-	-	-	-	-	-	a	-	-	b	-	-	-	-	-	-	-	-	-	-	-	-	-	-	-	-	-	-	-
31	-	-	-	-	-	-	-	-	-	-	a	-	-	-	-	-	-	-	-	-	-	-	-	-	a	a	a	-	-	-	-	-	-	-	-	-	-	-	-	-	-	-	-	-	-
33	-	-	-	-	-	-	-	-	-	-	-	-	-	-	b	-	-	-	-	-	-	-	-	-	-	-	a	b	-	-	-	-	-	-	-	-	-	-	-	-	-	-	-	-	-
34	-	-	-	-	-	-	-	-	-	-	-	-	-	-	-	-	-	-	-	-	-	-	-	-	-	-	-	1	a	-	-	-	-	-	-	-	-	-	-	-	-	-	-	-	-
35	-	-	-	-	-	-	-	-	-	-	-	-	-	-	a	-	-	-	-	-	-	-	-	-	-	-	a	-	b	b	-	-	-	-	-	-	-	-	-	-	-	-	-	-	-
26	-	-	-	-	-	-	-	-	-	-	-	-	-	-	-	a	-	-	-	-	-	-	a	-	-	-	-	-	-	-	a	-	-	-	-	-	-	-	-	-	-	-	-	-	-
30	-	-	-	-	-	-	-	-	-	-	-	-	-	-	-	-	-	-	-	-	-	-	-	-	-	-	-	-	-	b	a	b	-	-	-	-	-	-	-	-	-	-	-	-	-
39	-	-	-	-	-	-	-	-	-	-	-	-	-	-	-	-	-	-	-	-	-	-	-	-	-	-	-	-	-	-	-	m	m	-	-	-	-	-	-	-	-	-	-	-	-
27	-	-	-	-	-	-	-	-	-	-	-	-	-	-	-	-	-	-	-	-	-	-	-	a	-	-	-	-	-	-	a	-	-	a	-	-	-	-	-	-	-	-	-	-	-
28	-	-	-	-	-	-	-	-	-	-	-	-	-	-	-	-	a	-	-	-	-	-	-	-	-	-	-	-	-	-	-	-	-	a	a	-	-	-	-	-	-	-	-	-	-
24	-	-	-	-	-	-	-	-	-	-	-	-	-	-	-	-	-	-	-	-	-	-	-	-	-	-	-	-	-	-	-	-	-	-	a	a	-	-	-	-	-	-	-	-	-
36	-	-	-	-	-	-	-	-	-	-	-	-	-	-	-	-	-	-	-	-	-	-	-	-	-	-	-	-	-	-	-	-	-	-	-	-	a	a	-	-	-	-	-	-	-
41	-	-	-	-	-	-	-	-	-	-	-	-	-	-	-	-	-	-	-	-	-	-	-	-	-	-	-	-	-	-	-	-	-	-	-	-	a	-	a	-	-	-	-	-	-
40	-	-	-	-	-	-	-	-	-	-	-	-	-	-	-	-	-	-	-	-	-	-	-	-	-	-	-	-	-	-	-	-	-	-	-	-	-	-	m	m	-	-	-	-	-
29	-	-	-	-	-	-	-	-	-	b	-	-	-	-	-	-	-	-	-	-	-	-	-	-	-	-	-	-	-	-	-	-	-	-	-	-	-	-	-	-	a	b	-	-	-
38	-	-	-	-	-	-	-	-	-	-	-	-	-	-	-	-	-	-	-	-	-	-	-	-	-	-	-	-	-	-	-	-	-	-	-	-	-	-	-	-	a	-	a	-	-
37	-	-	-	-	-	-	-	-	-	-	-	-	-	-	-	-	-	-	-	-	-	-	-	-	-	-	-	-	-	-	-	-	-	-	-	-	-	-	-	-	-	-	m	m	-
42	-	-	-	-	-	-	-	-	-	-	-	-	-	-	-	-	-	-	-	-	-	-	-	-	-	-	-	-	-	-	-	-	a	-	-	-	-	-	-	a	-	-	-	a	a

SUB GROUP 1 = ROW 16
SUB GROUP 2 = ROW 25
SUB GROUP 3 = ROW 36
SUB GROUP 4 = ROW 29

PROBLEMS

2.1. Lapidus and coworkers (1961) have studied the dynamics and control of a six–plate absorber controlled by the inlet feed streams. In their work, they assume a linear equilibrium relationship between liquid (x_m) and vapor (y_m) at each plate:

$$y_m = a x_m + b$$

The two control variables are the inlet liquid composition $u_1 = x_0$ and the inlet vapor composition $\bar{u}_2 = y_7$; $\bar{u}_2$ has an equivalent composition:

$$u_2 = \frac{\bar{u}_2 - b}{a}$$

A convenient set of parameters for this system is

$$a = 0.72 \quad H = 1.0 \text{ kg} \quad G = 66.7 \text{ kg/min}$$
$$b = 0 \quad h = 75 \text{ kg} \quad L = 40.8 \text{ kg/min}$$

where h = inert liquid hold–up on each plate
H = inert vapor hold–up on each plate
L = flow rate of inert liquid absorbent
G = flow rate of inert gas stream

Solve for the steady–state plate composition (x_m) when (i) the liquid feed is pure ($u_1 = 0$) and the gas feed is 0.2 kg solute/kg inert ($u_2 = 0.2$) and (ii) the liquid feed is pure and the gas feed is 0.3 kg solute/kg inert. (See Example 2.1.)

2.2. Koepcke and Lapidus (1961) have studied the dynamics and control of a liquid–liquid countercurrent extraction train with a secondary feed to one of the middle states. Assume that there are a total of six stages with the secondary feed introduced at stage 3. The equilibrium between the extract and raffinate is given by

$$y_m = a x_m$$

For the parameters

$w = 60$ = extract solvent flow rate (kg/min)
$a = 2.2$
$y_7 = 0$ (pure solvent)
$x_0 = 0.04$
$x_F = 0.03$ (secondary feed)
$F = 50$ raffinate solvent flow rate of secondary feed (kg/min)

solve for the steady–state composition (x_m) when (i) the raffinate solvent flow rate $b = 50$ kg/min, and (ii) the raffinate solvent flow rate $b = 100$ kg/min.

2.3. A company plans to make commercial alcohol by the process shown in Figure 2.8. The grain mash is fed through a heat exchanger, where it is heated to 77°C. The alcohol is removed at 60 percent (wt) alcohol from the first fractionating column; the bottoms contain no alcohol. The 60 percent alcohol is further fractionated to 90 percent alcohol and essentially pure water in the second column. Both stills operate at a 3-to-1 reflux ratio and heat is supplied to the bottom of the columns by steam.

Make complete material and energy balances on the process. Solve the linear equations via matrix techniques. The component material balances should be written in terms of the amount of that component in the stream rather than the product of the total flow rate times the mass fraction. (See Example 2.2.)

2.4. A methanol synthesis based on the reaction of hydrogen and carbon dioxide at high pressure is under development,

$$CO_2 + 3H_2 \rightarrow CH_3OH + H_2O$$

The H_2 and CO_2, which are reacted in stoichiometric ratios, are produced by reforming natural gas (stream 1), containing 0.5 percent by volume inerts. Sixty percent conversion through the reactor is obtained. Calculate the composition of each stream for this process when 60 kg mol/hr of methanol are to be produced. Forty percent of stream 5 is recycled.

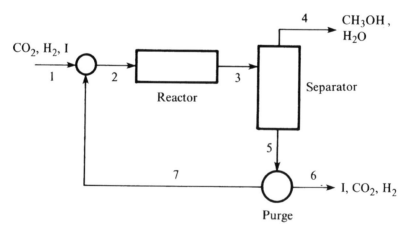

Figure 2.34: Problem 2.4.

2.5. Evaporation of a Sugar Solution in an Evaporator

An evaporator is being designed to evaporate a 10 percent (wt) sugar solution. The boiling-point rise of the solution over that of water can be estimated as BPR°C $= 1.78x + 6.22x^2$, when x is the weight fraction of sugar in solution. Saturated steam at 121.1°C is used to heat the solution. The feed rate $F = 22680$ kg/h at 26.7°C. The heat capacity of the liquid solution is $c_p = 4.19 - 2.35x$ k/kg·K. The heat transfer coefficient is $U = 3123$ W/m². The heat of vaporization of water is 2200 kJ/kg. The reference temperature for enthalpy is 0°C. (See Figure 2.35.)

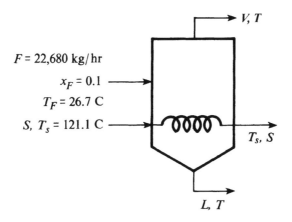

Figure 2.35: Problem 2.5.

Using a rearranged functionality matrix, specify the best design variable for this problem and the resulting computational sequence.

2.6. Perform a structural analysis on the functionality matrix in Figure 2.36.

	1	2	3	4	5	6	7	8	9	10
1			A	A			A			A
2	A						A	A		
3						A		A	A	A
4	A	A				A		A	A	
5	A		A		A					
6	A		A		A		A	A		

Figure 2.36: Problem 2.6.

2.7. Solvent Extraction Problem

Benzene, toluene, and other aromatic compounds can be recovered by solvent extraction with sulfur dioxide. As an example, a catalytic reformate stream containing 70 percent by weight benzene and 30 percent nonbenzene material is passed through the countercurrent extractive recovery scheme shown in Figure 2.37. One thousand kg of the reformate stream and 3000 kg of sulfur dioxide are fed to the system per hour. The benzene product stream contains .15 kg of sulfur dioxide per kg of benzene. The raffinate stream contains all the initially charged nonbenzene materials as well as .25 kg of benzene per kg of nonbenzene material. The remaining component in the raffinate stream is SO_2.

Eight hundred kg of benzene per hour with 0.25 kg of nonbenzene material per kg of benzene flow per hour through stream 10. Seven hundred kg of benzene per hour with 0.07 kg of nonbenzene material per kg of benzene flow through stream 7. Ninety percent of the SO_2 entering units 1 and 2 is returned to the system.

Write the material balances to determine the flow rates of each unknown stream. Perform a structural analysis on the linear equations that describe the system.

Solve the problem using the optimum solution strategy as well as all equations simultaneously.

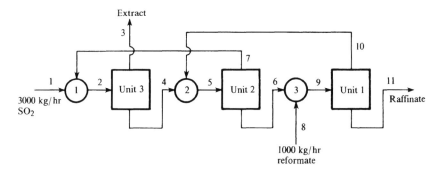

Figure 2.37: Problem 2.7.

2.8. Chemical Equilibrium Problem

The principal reactions in the production of synthesis gas by partial oxidation of methane with oxygen are

$$CH_4 + \frac{1}{2}O_2 \rightleftharpoons CO + 2\ H_2 \tag{1}$$

$$CH_4 + H_2O \rightleftharpoons CO + 3\ H_2 \tag{2}$$

$$H_2 + CO_2 \rightleftharpoons CO + H_2O \tag{3}$$

Determine the O_2/CH_4 reactant ratio that will produce an adiabatic equilibrium temperature of 1200°C at an operating pressure of 2 MPa, when the reactant gases are preheated to an entering temperature of 540°C.

Basic Data

Assuming that the gases behave ideally, so that the component activities are identical with component partial pressures, the equilibrium constants at 1200°C for the three reactions are respectively:

$$K_1 = \frac{p_{CO}\, p_{H_2}^2}{p_{CH_4}\, p_{O_2}^{1/2}} = 4.19 \times 10^9 \; MPa^{3/2} \tag{4}$$

$$K_2 = \frac{p_{CO}\, p_{H_2}^3}{p_{CH_4}\, p_{H_2O}} = 1831.3 \; MPa^2 \tag{5}$$

$$K_3 = \frac{p_{CO}\, p_{H_2O}}{p_{CO_2}\, p_{H_2}} = 2.6058 \tag{6}$$

Enthalpies of the various components at 540°C and 1200°C are presented in the following table (entries are in kJ/mol):

Component	540°C	1200°C
CH_4	−31317	19560
H_2O	−210168	−181541
CO_2	−359677	−322657
CO	−89428	−66934
H_2	23443	43932
O_2	24813	48351

Solution Strategy

Because of the magnitude of K_1, the equilibrium constant for the first reaction, that reaction is assumed to go to completion at 1200°C. That is, virtually no unreacted oxygen will remain in the product gases at equilibrium.

Use the following nomenclature:

- x_1 mole fraction of CO in the equilibrium mixture
- x_2 mole fraction of CO_2 in the equilibrium mixture
- x_3 mole fraction of H_2O in the equilibrium mixture
- x_4 mole fraction of H_2 in the equilibrium mixture
- x_5 mole fraction of CH_4 in the equilibrium mixture
- x_6 number of moles of O_2 per mole of CH_4 in the feed
- x_7 number of moles of product gases in the equilibrium mixture per mole of CH_4 in the feed

Then a system of seven simultaneous equations may be derived from three atom balances (oxygen, hydrogen, and carbon), an energy balance (enthalpy of reactants must equal enthalpy of products), a mole fraction constraint, and the two remaining equilibrium relations.

2.9. Select a best set of design variables for the heat exchanger problem in Figure 2.38. Discuss your solution strategy.

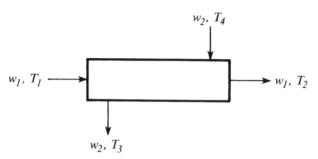

Figure 2.38: Problem 2.9.

The known variables are

w_1 = the process mass flow rate
U = the overall heat–transfer coefficient
T_1 = the inlet process temperature
T_4 = the inlet coolant temperature
c_p = the process and coolant heat capacities

The rate of heat transfer is to be modeled using a log–mean temperature difference.

2.10. For the reactor system shown in Figure 2.39, develop a best steady–state computational strategy. Discuss fully your computational options.

The species A undergoes the following reaction

$$A \rightarrow B$$

with the rate expression

$$r_A = -k\, C_A \quad \text{(mol/vol time)}$$

The following variables are known:

F_1 = inlet volumetric flow
F_6 = outlet volumetric flow
C_{A_1} = inlet concentration (mol/vol)
C_{A_3} = inlet concentration
C_{A_6} = outlet concentration
V_1, V_2 = reactor volumes
k = reaction rate constant

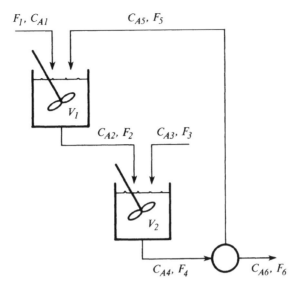

Figure 2.39: Problem 2.10.

2.11. Fluid Catalytic Cracking

The fluid catalytic cracking process has been widely employed for converting heavy hydrocarbons into more valuable lighter products, especially gasoline. A simplified schematic diagram of the process is shown in Figure 2.40.

Gas oil feed is introduced at the bottom of the riser, where it meets hot catalyst from the regenerator. The cracking reaction starts immediately as the mixture of catalyst and gas oil passes through the riser into the reactor. During this encounter, the catalyst is deactivated because of the carbonaceous material deposited on its surface. The product vapor is transferred to the fractionation section and the spent catalyst flows down through the reactor standpipe into the regenerator, where the carbonaceous deposits are burned off.

A simple steady–state mathematical model of the system is to be considered. There are 21 major variables, given in Table 1. The structure of the model is given in Table 2. Most of the equations are given except the carbon balance (equation 2), the energy balances (equation 8 and equation 12), and the coke balance (equation 11). You are to derive the appropriate equations.

Now perform a structural analysis to determine the best set of design variables. You can use a modified functionality matrix to make this analysis easier. Enter an A if the variable can be uniquely solved (linear, exponential, etc.) and a B if a multiple root exists or implicit solution is needed.

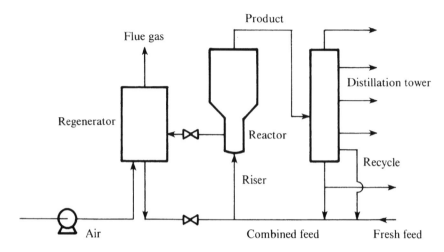

Figure 2.40: Problem 2.11.

How many of the following conventional design variables can be chosen as platform variables?

$$F_a \qquad H_{rx}$$
$$F_t H_{rg} \qquad T_a$$

Which other design variables do you recommend?

2.12 Rather than the Newton (tangent) method for root solving, $f(x) = 0$, a secant algorithm is often used. This method requires two initial guesses and generates the next guess by linear extrapolation. Develop a MATLAB M-file to implement this algorithm. Perform calculations on equation (2.3.2). Compare this algorithm to that of tangent.m..

2.13 Determine the condition number of A

$$A = \begin{bmatrix} 3.021 & 2.714 & 6.913 \\ 1.031 & -4.273 & 1.121 \\ 5.084 & -5.832 & 9.155 \end{bmatrix}$$

Determine the solution

$$Ax = b$$

where b is

$$b = \begin{pmatrix} 12.648 \\ -2.121 \\ 8.407 \end{pmatrix}$$

Now determine the solution when $a(2,2) = -4.275$.

What conclusions can you draw from this problem?

Steady–State Lumped Systems

Table 1: List of Major Variables

(1)	C_{rg}	=	Coke on regenerated catalyst, wt fraction
(2)	C_{sc}	=	Coke on spent catalyst, wt fraction
(3)	F_a	=	Air flow rate, lb/hr
(4)	F_c	=	Cat. circulation rate, lb/hr
(5)	F_t	=	Total feed rate, lb/hr
(6)	H_{rg}	=	Regenerator catalyst hold–up, lb
(7)	H_{rx}	=	Reactor catalyst hold–up (riser hold–up included), lb
(8)	K_0	=	Cracking reaction velocity constant, $\text{hr}^{-1}\ \text{conc}^{-1}$
(9)	PO_2	=	Oxygen concentration at the exit of reg. bed, mole fraction
(10)	R_{cb}	=	Coke burning rate, lb/hr
(11)	R_{cf}	=	Coke forming rate, lb/hr
(12)	S_v	=	Volume space velocity, vol/hr/vol
(13)	T_a	=	Air temperature, °F
(14)	T_f	=	Combined feed temperature, °F
(15)	T_{mix}	=	Mixed temperature in the riser, °F
(16)	T_r	=	Reference temperature for cracking, °F
(17)	T_{rg}	=	Regenerator temperature, °F
(18)	T_{rx}	=	Reactor temperature, °F
(19)	α	=	Decay velocity constant, hr^{-1}
(20)	β_v	=	Cat.–to–oil ratio, vol cat./vol oil
(21)	ϵ_w	=	Conversion, wt fraction

Table 2: Design Equations of FCC Model

Reactor

1. *Coke Make, lb/hr*

$$R_{cf} = a_{cc} \exp\left(-\frac{E_{cc}}{RT_r}\right)\left(\frac{H_{rx}}{F_c}\right)^n \cdot F_c + a_{shc} F_c + a_{ac} F_t$$

2. *Carbon Balance*

$$f(F_c, C_{rg}, C_{sc}, R_{cf}) = 0$$

3. *Cat.–to–Oil Ratio, vol cat./vol oil*

$$\beta_v = F_c \cdot \rho_0/(F_t \, \rho_{cat})$$

4. *Space Velocity, vol oil/hr/vol cat.*

$$S_v = F_t/\rho_0/V_r$$

5. *Cracking Reaction Velocity Constant, hr^{-1}, $conc^{-1}$*

$$K_0 = \frac{F\rho_0 k_0^1}{\rho_1 S_v \left(1 + C_{rg}^k\right)} e^{-\frac{Q_0}{RT_r}}$$

6. *Decay Velocity Constant, hr^{-1}*

$$\alpha = \alpha_0 \, e^{-\frac{Q_2}{RT_r}}$$

7. *Conversion wt fraction*

$$\epsilon_w = 1 - \frac{\alpha/\beta_v S_v}{\frac{\alpha}{\beta_v S_v} + \frac{K_0}{S_v}\left(1 - e^{-\alpha/\beta_v S_v}\right)}$$

8. *Heat Balance*

$$f(F_c, T_{rg}, T_{rx}, F_t, T_f, \epsilon_w) = 0$$

Model parameters include ΔH_{cr} = Heat of cracking (BTU/lb cracked)
λ_0 = Heat of vaporization of gas oil (BTU/lb)

(*continued*)

Table 2: Design Equations of FCC Model (continued)

Regenerator

9. *Oxygen Balance*

$$PO_2 = 0.21 \exp\left\{-\frac{a_{cb} \exp\left(-\frac{E_{cb}}{RT_{rg}}\right) C_{rg} M_g H_{rg} \alpha_{co}}{(0.21) F_a M_c}\right\}$$

10. *Coke Burning Rate, lb/hr*

$$R_{cb} = M_c F_a (0.21 - PO_2) / \alpha_{co} M_g$$

11. *Coke Balance*

$$f(F_c, C_{sc}, C_{rg}, R_{cb}) = 0$$

12. *Heat Balance*

$$F(F_c, T_{rx}, T_{rg}, F_a, T_a, R_{cb}) = 0$$

Model parameters include
Q_{cb}=Heat of coke combustion (Btu/lb coke)

Riser

13. *Mixed Temperature, °F*

$$T_{\text{mix}} = \frac{C_{pc} T_c T_{rg} + C_{pf} F_t T_f - \lambda_0 F_t}{C_{pc} F_c + C_{pf} F_t}$$

14. *Reference Temperature for Cracking, °F*

$$T_r = 0.6 T_{rx} + 0.4 T_{\text{mix}}$$

2.14 Find the solution to the following over determined system of equations using **pinv**.

$$x_1 + x_2 = 1.98$$
$$2.05 x_1 - x_2 = .95$$
$$3.06 x_1 + x_2 = 3.98$$
$$-1.02 x_1 + 2 x_2 = .92$$
$$4.08 x_1 - x_2 = 2.90$$

Plot these straight lines and the pseudo–inverse solution. What conclusions can you draw?

2.15 For the problem of example 2.2, check the savings obtained by using the sparse matrix capability of MATLAB.

2.16 Perform a structural analysis for a flash distillation unit (Figure 2.41) that is flashing a feed stream with three components. Assume that the pressure and temperture of the flash unit is known. Determine a best set of design variables for this problem. Next assume that you specify the input conditions (flow and compoisition) and determine a best computational scheme for this problem without design variable choice.

2.17 A chemical plant for making triethylamine (Figure 2.42) consists of the following units. Develop the steady state material balances which give the amount in each stream for each component. Ammonia is fed to the reactor in a 20

$$3\ CH_3-OH + NH_3 \rightarrow (CH_3)_3\ N + 3\ H_2O$$

- Perform a structural analysis for this problem.
- Solve for the component amounts in each stream.

REFERENCES

Aris, R. and Amundson, N. R., *Chem. Engr. Sci.* **7**, 121, 132, 148 (1958).

Ayyub, B. M. and McCuen, R. H., *Numerical Methods for Engineers*, Prentice-Hall (1996).

Book, N. L. and Ramirez, W. f., "The Selection of Design Variables in Systems of Algebraic Equations," *AIChE Journal* **22**, No. 1, 55 (1976).

Book, N. L. and Ramirez, W. F., "Structural Analysis and Solution of Systems of Algebraic Design Equations," *AIChE Journal*, **30**, No. 4, 609 (1984).

Brian, A. and Breiner, M. M. G., *MATLAB for Engineers*, Addison-Wesley (1995).

Carnahan, B., Luther, H. A., and Wilkes, J. O., *Applied Numerical Methods*, Wiley, New York (1969).

Christensen, J. H., "The Structuring of Process Optimization," *AIChE Journal* **16**, 177 (1970).

Christensen, J. H. and Rudd, D. F., "Structuring Design Computations," *AIChE Journal* **15**, No. 1, 94 (1969).

Edie, F. C. and Westerberg, A. W., "Decision Variable Selection to Avoid Hidden Singularities in Resulting Recycle Calculations for Process Design," *Chem. Engr. Journal* **2**, 114 (1971).

Etter, P. E., *Engineering Problem Solving with MATLAB*, Prentice-Hall (1993).

Friedman, F. and Ramirez, W. F., "Convergence Properties of Systems of Algebraic Equations—Explicit Loops," *AIChE Journal* **19**, 566 (1973).

Himmelblau, D. M., "Decomposition of Large–Scale Systems—II, Systems Containing Nonlinear Elements," *Chem. Engr. Sci.* **22**, 883 (1967).

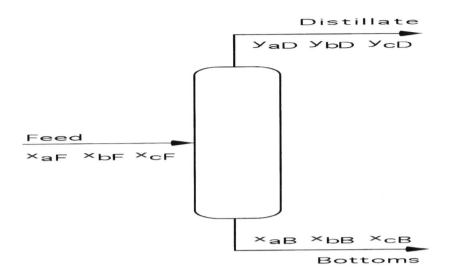

Figure 2.41: Problem 2.16 (Flash Distillation).

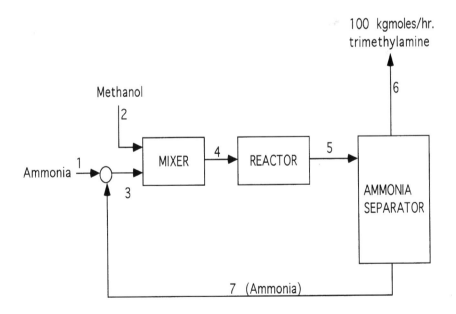

Figure 2.42: Problem 2.17 (Trimethylamine Production).

Koepcke, R. and Lapidus, L., *Chem. Engr. Sci.* **16**, 252 (1961).

Lapidus, L. et al., *AIChE Journal* **19**, 290 (1961).

Lee, W., Christensen, J. H., and Rudd, D. F., "Design Variable Selection to Simplify Process Calculations," *AIChE Journal* **12**, 1105 (1966).

Lee, W. and Rudd, D. F., "Ordering of Recycle Calculations," *AIChE Journal* **12**, No. 6, 1184 (1966).

Math Works, Inc., *The Student Edition of MATLAB Version 4 User's Guide*, Prentice-Hall (1995).

More, J., Garbow, B., and Hillstrom, K., *User Guide for MINPACK-1*, Arg. Nat. Lab. Report ANL–80–74, Argonne, Ill. (1980).

Motard, R. L., Shacham, M., and Rosen, E. M., "Design Variable Selection to Simplifying Process Calculations," *AIChE Journal* **21**, No. 3, 417 (1975).

Ramirez, W. F. and Vestal, C. R., "Algorithms for Structuring Design Calculations," *Chem. Engr. Sci.* **27**, 2243 (1972).

Rudd, D. F. and Watson, C. C., *Strategy of Process Engineering*, Wiley, New York (1968).

Sargent, R. W. H. and Westerberg, A. W., "SPEED–UP in Chemical Engineering Design," *Transactions of the Institute of Chemical Engineers* **42**, T 190 (1964).

Stadfherr, M. S., Gifford, W. A., and Scriven, L. E., "Efficient Solution of Sparse Sets of Design Equations," *Chem. Engr. Sci.* **29**, 1025 (1974).

Steward, D. V., "Partitioning and Tearing Systems of Equations," *J. SIAM Numerical Analysis Series B* **2**, 345 (1965).

Westerberg, A. W. and Edie, F. C., "Convergence and Tearing in Systems of Equations," *Chemical Engineering Journal* **2**, No. 9, 17 (1971).

Chapter 3

UNSTEADY-STATE LUMPED SYSTEMS

The systems to be studied in some detail here involve various aspects of a continuous–flow, well–agitated vessel, with interacting variables such as flow, pressure, mixing, and chemical reaction. The study of stirred tanks is important for two reasons. First, stirred tank vessels are an important and frequent piece of process equipment, and engineers must be capable of modeling this kind of system. Second, the stirred tank concept is an important idealized lumped unit which can be used effectively by an experienced engineer to simulate the behavior of more complicated process equipment.

This chapter deals only with initial–value problems, for which numerical methods are well developed.

3.1 SINGLE STEP ALGORITHMS FOR NUMERICAL INTEGRATION

3.1.1 Euler Method

When performing simulations of dynamic processes, the engineer must be concerned with the details of numerical integration. In numerical integration what is required is the projection of information known at a particular value of the independent variable, say time, forward to a new or future value of that independent variable. Consider the differential equation

$$dx/dt = f(x,t) \tag{3.1.1}$$

with the initial condition

$$x(0) = x_0 \tag{3.1.2}$$

If we expand x in a Taylor series about a known condition, we have

$$x(t + \Delta t) = x(t) + \frac{dx}{dt}(t)\,\Delta t + \frac{1}{2}\frac{d^2x}{dt^2}(t)\,\Delta t^2 + \cdots \tag{3.1.3}$$

Truncating after the first derivative term gives

$$x(t + \Delta t) = x(t) + \frac{dx}{dt}(t)\,\Delta t \qquad (3.1.4)$$

or

$$x(t + \Delta t) = x(t) + f(x,t)\Delta t \qquad (3.1.5)$$

Equation (3.1.5) can therefore be used to compute the value x at a new time, $t + \Delta t$, based upon a given value of the function $f(x,t)$ which is evaluated at the old time t. Since we know the initial condition at $t = 0$, we can start this forward–marching scheme and proceed out to any value of t in steps of Δt. This numerical algorithm of equation (3.1.5) is known as *Euler's method*. Unfortunately, the local truncation error for this algorithm is large and is of the order of Δt^2. To maintain accuracy with a truncation error of this size, the step size needs to be very small. In practice other methods of greater accuracy have been devised in order to allow for increased step sizes.

The Euler Method can be generalized to handle a set of differential equations

$$\frac{d\boldsymbol{x}}{dt} = \boldsymbol{f}(\boldsymbol{x},t) \qquad (3.1.6)$$

when $\boldsymbol{x}$ is an nth order vector. The Euler Method of equation (3.1.5) becomes

$$\boldsymbol{x}(t + \Delta t) = \boldsymbol{x}(t) + \boldsymbol{f}(\boldsymbol{x},t)\Delta t \qquad (3.1.7)$$

In a forward–marching method such as Euler's method, we are more interested in the **total error propagation** over multiple usage of the algorithm than in the local one–step error. If we let ϵ_i be the error between the approximate solution, x_i, and exact solution, $x(t_i)$ to the differential equation, then

$$\epsilon_i = x_i - x(t_i) \qquad (3.1.8)$$

The additional error in traversing the ith step is

$$\epsilon_{i+1} - \epsilon_i = (x_{i+1} - x_i) - \left(x(t_{i+1}) - x(t_i)\right) \qquad (3.1.9)$$

From Euler's method the difference between the approximate solutions is

$$x_{i+1} - x_i = \Delta t\, f(t_i,\, x_i) \qquad (3.1.10)$$

Also, from a Taylor series and use of the mean value theorem, the difference between the exact solutions can be expressed as (Carnahan et al., 1969)

$$x(t_{i+1}) - x(t_i) = \Delta t\, f(t_i,\, x(t_i)) + \frac{\Delta t^2}{2\,!} f'(\delta,\, x(\delta)) \qquad (3.1.11)$$

Here $f'(\delta, x(\delta))$ is the derivative of f with respect to the independent variable, t, evaluated at an appropriate value $t = \delta$.

Therefore,

$$\epsilon_{i+1} - \epsilon_i = \Delta t \left(f(t_i, x_i) - f(t_i, x(t_i)) \right) - \frac{\Delta t^2}{2} f'(\delta, x(\delta)) \quad (3.1.12)$$

Expanding $f(t_i, x_i)$ in a Taylor Series about $f(t_i, x(t_i))$ and using the mean value theorem again gives

$$f(t_i, x_i) - f(t_i, x(t_i)) = f'(\gamma, x(\gamma)) \left(x_i - x(t_i) \right) \quad (3.1.13)$$

Therefore, equation (3.1.12) is

$$\epsilon_{i+1} - \epsilon_i = \Delta t \, f'(\gamma, x(\gamma)) \left(x_i - x(t_i) \right) - \frac{\Delta t^2}{2} f'(\delta, x(\delta)) \quad (3.1.14)$$

Using the triangle inequality, $|z+w| \leq |z| + |w|$, gives

$$| \epsilon_{i+1} - \epsilon_i | \leq \Delta t \, K \, | x_i - x(t_i) | + \frac{\Delta t^2}{2} M \quad (3.1.15)$$

where $K = | f'(\gamma, x(\gamma)) |$ and $M = | -f'(\delta, x(\delta)) |$

or

$$| \epsilon_{i+1} | \leq (1 + \Delta t \, K) | \epsilon_i | + \frac{\Delta t^2}{2} M \quad (3.1.16)$$

Since the solution to the equation is known exactly at $t = 0$, then $\epsilon_0 = 0$. Equation (3.1.16) can be expressed by the following recursive relation:

$$| \epsilon_i | \leq \frac{M \Delta t}{2K} \left((1 + \Delta t \, K)^i - 1 \right) \quad (3.1.17)$$

Using the series expansion of $e^{\Delta t \, K}$, we also know that $(1 + \Delta t \, K) < e^{\Delta t \, K}$.

Therefore

$$| \epsilon_i | \leq \frac{M \Delta t}{2K} (e^{i \Delta t \, K} - 1) \quad (3.1.18)$$

or

$$| \epsilon_i | \leq \frac{M \Delta t}{2K} e^{i \Delta t \, K} \quad (3.1.19)$$

If we let the ith interval be the last interval N, then

$$i \, \Delta t = T \quad (3.1.20)$$

where T is the final time.

Therefore,

$$| \epsilon_N | \leq \frac{M \Delta t}{2K} e^{TK} \quad (3.1.21)$$

This equation shows that the overall multistep truncation error of Euler's method is of order Δt whereas the local truncation was of order Δt^2.

The **stability** of a numerical method is also important. Let us consider the equation

$$\frac{dx}{dt} = Kx \qquad (3.1.22)$$

Here $f(x,t) = Kx$ and Euler's Method for the problem is

$$x_{n+1} = x_n + \Delta t\, Kx_n \qquad (3.1.23)$$

or

$$x_{n+1} = (1 + \Delta t\, K)x_n \qquad (3.1.24)$$

We can also express x_{n+1} in terms of the initial condition, x_0

$$x_{n+1} = (1 + \Delta t\, K)^{n+1} x_0 \qquad (3.1.25)$$

To determine how errors propagate, let us assume that x_0 is perturbed by an error e_0 so that the approximate value for x is

$$x_0^a = x_0 - e_0 \qquad (3.1.26)$$

Using equation (3.1.25) we can get the approximate x at $n+1$ as

$$x_{n+1}^a = (1 + \Delta t\, K)^{n+1} x_0^a \qquad (3.1.27)$$

or

$$x_{n+1}^a = (1 + K\Delta t)^{n+1} x_0 - (1 + K\Delta t)^{n+1} e_0 \qquad (3.1.28)$$

which is

$$x_{n+1}^a = x_{n+1} - (1 + K\Delta t)^{n+1} e_0 \qquad (3.1.29)$$

In order for the method to be stable

$$|\,1 + K\Delta t\,| \leq 1 \qquad (3.1.30)$$

If this is the case, then after multiple steps the error will decrease and eventually disappear. Under these conditions the Euler Method for this problem is stable. Rewriting this singularity leads to the following condition for absolute stability

$$-2 < \Delta t\, K < 0 \qquad (3.1.31)$$

Note that the method is not absolutely stable for any positive values of K. If $K = -100$ then Euler's Method would be stable for

$$0 \leq \Delta t \leq .02 \qquad (3.1.32)$$

The Euler stability condition for

$$\frac{dx}{dt} = f(x,t) \qquad (3.1.33)$$

is

$$-2 < \Delta t\, \frac{\partial f}{\partial x} < 0 \qquad (3.1.34)$$

3.1.2 Runge–Kutta Methods

The fourth-order *Runge–Kutta* algorithm is one of the best all-around, single-step algorithms that has been developed and is probably the most widely used numerical integration scheme. It has a small local truncation error of order Δt^4 and a total truncation error of Δt^3.

A Runge–Kutta method is basically a slope-averaging procedure for computing the value of a vector $\boldsymbol{x}$ at a new time $t + \Delta t$ based upon information about the vector at time t. It is possible to develop a one-step method which involves only first-order derivative evaluations but which produces results equivalent in accuracy to higher-order Taylor series truncations. These single-step, higher-order-accuracy methods are called Runge–Kutta algorithms.

All Runge–Kutta algorithms can be expressed in the form

$$x_{i+1} = x_i + \Delta t \; \phi(t_i, \; x_i, \; \Delta t) \tag{3.1.35}$$

where ϕ is a suitably chosen approximation of $f(t, x)$ to have the accuracy of higher-order Taylor series expansions of $f(t, x)$. We will illustrate the procedure with the second-order Runge–Kutta method. Let ϕ be a weighted average of two derivatives k_1 and k_2 on the interval $t_i \le t \le t_{i+1}$,

$$\phi = a \; k_1 + b \; k_2 \tag{3.1.36}$$

Therefore

$$x_{i+1} = x_i + \Delta t \; (a \; k_1 + b \; k_2) \tag{3.1.37}$$

We let

$$k_1 = f(t_i, \; x_i) \tag{3.1.38}$$

and

$$k_2 = f(t_i + p\Delta t, \; x_i + q\Delta t k_1) \tag{3.1.39}$$

We now expand k_2 in a Taylor series:

$$k_2 = f(t_i, \; x_i) + p\Delta t \; f_t(t_i, \; x_i) + q\Delta t \; f(t_i, \; x_i) \; f_x(t_i, \; x_i) + O(\Delta t^2) \tag{3.1.40}$$

Here f_t denotes the partial of f with respect to t and f_x the partial with respect to x. Equation (3.1.37) now becomes

$$x_{i+1} = x_i + \Delta t \; \Big(a \; f(t_i, \; x_i) + b \; f(t_i, \; x_i) \Big) + \Delta t^2 \; \Big(bp \; f_t(t_i, \; x_i)$$

$$+ \; bq \; f(t_i, \; x_i) \; f_x(t_i, \; x_i) \Big) + O(\Delta t^3)$$

$$\tag{3.1.41}$$

We can now expand the exact solution to the differential equation in a Taylor series to obtain

$$x(t_{i+1}) = x(t_i) + \Delta t \; f(t_i, \; x(t_i)) + \frac{\Delta t^2}{2} \Big(f_t(t_i, \; x(t_i))$$

$$+ \; f_x(t_i, \; x(t_i)) \; f(t_i, \; x(t_i)) \Big) + O(\Delta t^3)$$

$$\tag{3.1.42}$$

If we equate like powers of Δt, then

$$\begin{aligned}
\Delta t^0 & \qquad x(t_i) = x_i \\
\Delta t^1 & \qquad f(t_i,\ x(t_i)) = (a+b)\ f(t_i,\ x_i) \\
\Delta t^2 & \qquad \tfrac{1}{2}\left(f_t(t_i,\ x(t_i)) + f_x(t_i,\ x(t_i))\ f(t_i,\ x(t_i))\right) \\
& \qquad = \left(bp\ f_t(t_i,\ x_i) + bq\ f_x(t_i,\ x_i)\ f(t_i,\ x_i)\right)
\end{aligned} \qquad (3.1.43)$$

Assuming that $x_i = x(t_i)$ and the same equality for all f functions, we have

$$\begin{aligned}
a + b &= 1 \\
bp &= 1/2 \\
bq &= 1/2
\end{aligned} \qquad (3.1.44)$$

This is a set of three equations in four unknowns. These equations are given in terms of one of the variables, such as

$$\begin{aligned}
a &= 1 - b \\
p &= 1/2b \\
q &= 1/2b
\end{aligned} \qquad (3.1.45)$$

Different choices of b yield different second–order Runge–Kutta methods. If $b = \frac{1}{2}$, the method is called the improved Euler's or Heun's method. If $b = 1$, the method is called the improved polygon or modified Euler's method. As demonstrated by this development, Runge–Kutta methods are not unique since they involve the choice of an arbitrary constant. All second–order methods involve the evaluation of two slopes k_1 and k_2 (equation (3.1.38) and equation (3.1.39)) and the value of y_{i+1} is a weighted average of these two slopes (equations (3.1.35) and (3.1.36)).

For the fourth–order Runge–Kutta algorithm, a four–step process is used to solve

$$\frac{dx}{dt} = f(x,t) \qquad (3.1.46)$$

A graphical interpretation of the process is given in Figure 3.1.

In the first step, the slope S_1 is computed from equation (3.1.32) based on known information at time t.

$$S_1 = f(x,t) \qquad (3.1.47)$$

In the second step, a new slope S_2 is calculated from equation (3.1.33) based on projecting the slope S_1 to a time $t + \Delta t/2$. The new value of the dependent variable is therefore $x + S_1(\Delta t/2)$.

$$S_2 = f\left(x + S_1(\Delta t/2),\ t + \Delta t/2\right) \qquad (3.1.48)$$

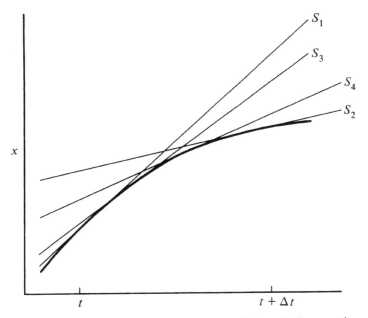

Figure 3.1: Runge–Kutta Algorithm Schematic.

The next step used to get a third slope S_3 is similar:

$$S_3 = f\left(x + S_2(\Delta t/2),\ t + \Delta t/2\right) \qquad (3.1.49)$$

In the last step, to obtain the slope S_4, a full–time increment is used based on the slope S_3.

$$S_4 = f(x + S_3 \Delta t,\ t + \Delta t) \qquad (3.1.50)$$

To get the new value of x at $t + \Delta t$, an average slope is used:

$$x(t + \Delta t) = x(t) + \frac{1}{6}(S_1 + 2S_2 + 2S_3 + S_4)\Delta t \qquad (3.1.51)$$

The most widely used single–step method is the Runge–Kutta modification due to Gill, called the Runge–Kutta–Gill algorithm. This algorithm for a vector set of differential equations is

$$\boldsymbol{x}(t+\Delta t) = \boldsymbol{x}(t) + \frac{1}{6}\left[\boldsymbol{s}_1 + 2\left(1 - 1/\sqrt{2}\right)\boldsymbol{s}_2 + 2\left(1 + 1/\sqrt{2}\right)\boldsymbol{s}_3 + \boldsymbol{s}_4\right]\Delta t \qquad (3.1.52)$$

where $s_1 = f(x, t)$

$$s_2 = f\left(x + s_1(\Delta t/2),\ t + \Delta t/2\right)$$

$$s_3 = f\left(x + s_1\left(-\frac{1}{2}+1/\sqrt{2}\right)\Delta t + s_2\left(1-1/\sqrt{2}\right)\Delta t, t+\Delta t/2\right)$$

$$s_4 = f\left(x - s_2\left(1/\sqrt{2}\right)\Delta t + s_3\left(1+1/\sqrt{2}\right)\Delta t,\ t + \Delta t\right)$$

All Runge–Kutta methods can be shown to be convergent, that is

$$\lim_{\Delta t \to 0} (x_i - x(t_i)) = 0 \qquad (3.1.53)$$

and the total truncation is one degree less than that of the local relative error. An estimate of the step sign Δt which will provide stability for fourth–order Runge–Kutta methods is (Polking, 1995)

$$-2.78 < \Delta t \frac{\partial f}{\partial x} < 0 \qquad (3.1.54)$$

3.1.3 MATLAB Runge–Kutta Routines

MATLAB has two implementations of Runge–Kutta algorithms, ode23 and ode45. Actually, there are two versions of each algorithm, an old original version and a new updated version. The command help funfun will tell you which version you have. If routines ode113, ode15s and ode235 are included, you have the new version. If you have the old version and would like access to the new version, it can be obtained by anonymous ftp from ftp.mathworks.com in the directory pub/mathworks/toolbox/mathlab/funfun. These programs can also be obtained on the world wide web under http://optimal.colorado.edu/~ramirez/chen4580.html.

Both versions of ode45 use variable step Runge–Kutta algorithms. Six slopes or derivatives are used to calculate an approximation of order five, and then another of order four. These are compared to develop an estimate of the error at the current step. It is required that the estimated error be less than a predetermined amount. Otherwise, the step is reduced. In the old version of ode45 and ode23, the error checking amount is computed as

$$|\text{error}| \leq \max(|\mathbf{x}^k|) * \text{tolerance} \qquad (3.1.55)$$

where x^k is the solution vector at the kth step. The default value of tolerance is 10^{-6}. This error checking method has several problems. Since the elements of x can be very different the |error| will be dominated by the largest elements of x. The error in calculating the smallest elements might be very large relative to the size of the smallest element,

Unsteady–State Lumped Systems

and still be small in comparison to errors in the larger elements. Often you need to reduce the value of `tolerance` to obtain accurate solutions using the old version of `ode45` and `ode23`.

The new version of `ode45` and `ode23` takes a more sophisticated approach to error checking. The solution x^k is computed at step k and each component of the solution is required to satisfy its own error condition by satisfying one of the following inequalities

$$\left|\text{error}_j^k\right| \leq \max\left(\left|x_j^k\right|, \left|x_j^{k-1}\right|\right) * \text{rtol}$$
$$\left|\text{error}_j^k\right| \leq \text{atol}_j$$
(3.1.56)

where `rtol` is the relative tolerance and the vector `atol` is an absolute tolerance vector with a value for each element of the solution vector. The default values as `rtol=`10^{-3} and each component of `atol=`10^{-6}.

The routine `ode23` uses four derivative evaluations to calculate an approximation of order three, and then another of order 2. These are compared to come up with an estimated error at the kth step. The same error checking methods as used of `ode45` are employed to find an appropriate step size. For most problems `ode45` is preferable to `ode23`. The syntax for using `ode45` and `ode23` are the same.

The syntax for the old version of `ode45` shown using `help ode45`

```
ODE45   Solve differential equations, higher order method.
        ODE45 integrates a system of ordinary differential
        equations using 4th and 5th order Runge-Kutta formulas.
        [T,Y] = ODE45('yprime', T0, Tfinal, Y0) integrates the
        system of ordinary differential equations described by
        the M-file YPRIME.M, over the interval T0 to Tfinal, with
        initial conditions Y0.
        [T, Y] = ODE45(F, T0, Tfinal, Y0, TOL, 1) uses tolerance
        TOL and displays status while the integration proceeds.

        INPUT:
        F       - String containing name of user-supplied problem
                  description.
                  Call: yprime = fun(t,y) where F = 'fun'.
                  t       - Time (scalar).
                  y       - Solution column-vector.
                  yprime - Returned derivative column-vector;
                           yprime(i) = dy(i)/dt.
        t0      - Initial value of t.
        tfinal- Final value of t.
        y0      - Initial value column-vector.
        tol     - The desired accuracy. (Default: tol = 1.e-6).
        trace   - If nonzero, each step is printed.
                  (Default: trace = 0).

        OUTPUT:
```

T   - Returned integration time points (column-vector).
Y   - Returned solution, one solution column-vector per tout-value.

The result can be displayed by: plot(tout, yout).

See also ODE23, ODEDEMO.

The syntax for the new version of ode45 is given below.

ODE45   Solve non-stiff differential equations, medium order method [T,Y] = ODE45('ydot',TSPAN,Y0) with TSPAN = [T0 TFINAL] integrates the system of first order differential equations y' = ydot(t,y) from time T0 to TFINAL with initial conditions Y0. Function ydot(t,y) must return a column vector. Each row in solution matrix Y corresponds to a time returned in column vector T. To obtain solutions at the specific times T0, T1, ..., TFINAL (all increasing or all decreasing), use TSPAN = [T0 T1 ... TFINAL].

[T,Y] = ODE45('ydot',TSPAN,Y0,OPTIONS) solves as above with default integration parameters replaced by values in OPTIONS, an argument created with the ODESET function. See ODESET for details. Commonly used options are scalar relative error tolerance 'rtol' (1e-3 by default) and vector of absolute error tolerances 'atol' (all components 1e-6 by default).

It is possible to specify tspan, y0 and options in ydot. If TSPAN or Y0 is empty, or if ODE45 is invoked as ODE45('ydot'), ODE45 calls [tspan,y0,options] = ydot([],[]) to obtain any values not supplied at the command line. TYPE R3BODY to see how this is coded.

As an example, the commands

   options = odeset('rtol',1e-4,'atol',[1e-4 1e-4 1e-5]);
   ode45('rigid',[0 12],[0 1 1],options);

solve the system y' = rigid(t,y) with relative error tolerance 1e-4 and absolute tolerances of 1e-4 for the first two components and 1e-5 for the third. When called with no output arguments, as in this example, ODE45 calls the default output function ODEPLOT to plot the solution as it is computed.

See also
   other ODE solvers: ODE23, ODE113, ODE15S, ODE23S
   options handling: ODESET, ODEGET

output functions: ODEPLOT, ODEPHAS2, ODEPHAS3
ydot examples: R3BODY, RIGID, TWOBODY, VDPNS

when computing over large intervals the new version of `ode45` and `ode23` allows you to change of value of `hmax`. The default value is `(tf-t0)/10` where `tf` is the final time and `t0` is the initial time. Reducing `hmax` sometimes assists and computation.

3.2 BASIC STIRRED TANK MODELING

We will use the `ode45` routines to solve for the dynamics associated with stirred tanks. Consider a simple tank with fluid flowing in at a known volumetric rate F_1, and a known output volumetric rate, F_2. Note that these rates can be functions of time and need not be constant. It is required to determine the level Z of the fluid in the tank at any time t (Figure 3.2).

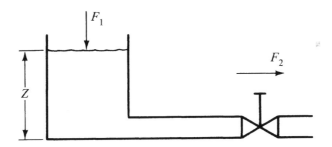

Figure 3.2: Continuous–Flow Tank.

Applying the conservation of mass to the system, we have

Rate of   Rate of   Rate of
Accumulation = Mass Into − Mass Out of
of Mass   System   System

$$\frac{d(\rho A Z)}{dt} = \rho F_1 - \rho F_2 \qquad (3.2.1)$$

where A = cross–sectional area of the tank.

Assuming that the density ρ and cross–sectional area A are constant gives

$$A \frac{dZ}{dt} = F_1 - F_2 \qquad (3.2.2)$$

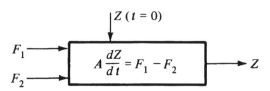

Figure 3.3: Information–Flow Diagram
for Continuous–Flow Tank.

Arranging in an information–flow diagram, we have Figure 3.3.

The describing relation is a first–order differential equation of the initial–value class of differential equations. That is, the initial condition $Z(t=0)$ is known and the height of the tank as a function of time $Z(t)$ is the desired result.

We will use both the old version of ode45 and the new version of ode45 to solve this problem. The new version is available from our web site as file ode45n.m. The values for F_1 and F_2, and the initial value for the level in the vessel Z_0 are supplied as knowns. Values chosen for this problem are an output flow $F_2 = .27$ m^3/min, $Z_0 = 30$ m, and a time varying input flow $F_1 = B \sin Wt + C$ where $B = .027$ m^3/min, $C = .27$ m^3/min, and $W = .1$ min^{-1}. The cross sectional area $A = .9$ m^2. The program using the old ode45 is given as file tank32o.m and is shown below.

```
% This file calls the built in ODE solver, ode45 (4th and 5th
% Runge-Kutta), to solve the dimensional basic tank model in
% section 3.2.
%
% Notice that the new version of ode45 in funfun can also be
% used, but provides better results with the same default
% tolerance.

% the default tolerance, 1.e-6, is not enough, since the graph
% looks coarse, so a tighter one, 1.e-10, is supplied.
%
[T, Z] = ode45('fun32', 0, 100.0, 30.0, 1.e-10, 1);

%
% F is the dimensional inflow rate.
%
A = 0.9;
B = 0.027;
F2 = 0.27;
W = 0.1;
C = 0.27;
F = B*sin(W*T) + C;

plot(T,Z,'-');
```

Unsteady–State Lumped Systems

```
title('Tank Height');
xlabel('Time, t (min)');
ylabel('Fluid level, Z (m)');
pause;
plot(T,F,'--');
title('Input Flow');
xlabel('Time, t (min)');
ylabel('Volumetric rate, F (m^3/min)');
```

Note that the default tolerance has been reduce to 10^{-10} in order to get results that graph well. The results of running the program are given in Figure 3.4 for tank height and the input flow as a function of time.

The file fun32.m defines the right hand side of the differential equation 3.2.2 which we are solving. This file is shown below.

```
function f = tank32(t,z)
% this file defines the dimesional basic tank model in section 3.2.
%

A = 0.9;
B = 0.027;
F2 = 0.27;
W = 0.1;
C = 0.27;
f = (B*sin(W*t) + C - F2)/A;
```

The program using the new version of ode45 is given as file tank32n.m and is shown below. Note that the default error checking parameters could be used directly. The results of running the program are the same as that shown in Figure 3.4.

```
% This file calls the new ODE solver which we all, ode45n (4th and
% 5th Runge-Kutta), to solve the dimensional basic tank model in
% section 3.2.
%
% The default tolerances are fine using the new ode45n routine for
% this problem
%
[T, Z] = ode45n('fun32',[ 0   100.0], 30.0);

%
% F is the dimensional inflow rate.
%
A = 0.9;
B = 0.027;
F2 = 0.27;
W = 0.1;
C = 0.27;
F = B*sin(W*T) + C;

plot(T,Z,'-');
title('Tank Height');
```

138 *Computational Methods for Process Simulation*

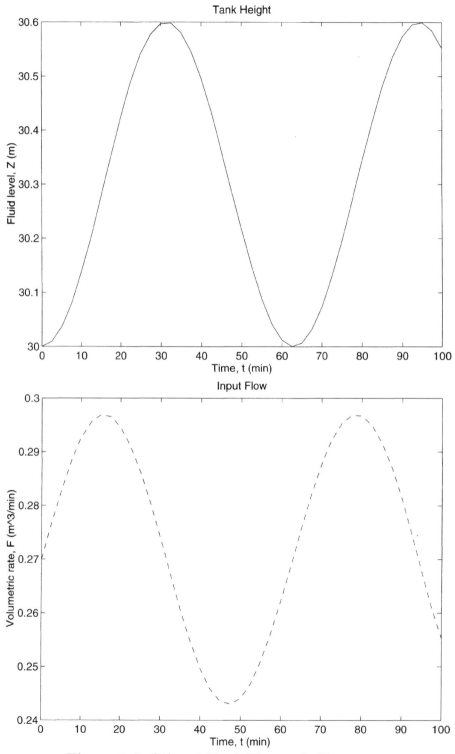

Figure 3.4: **Stirred Tank Dynamic Response**

```
xlabel('Time, t (min)');
ylabel('Fluid level, Z (m)');
pause;
plot(T,F,'--');
title('Input Flow');
xlabel('Time, t (min)');
ylabel('Volumetric rate, F (m^3/min)');
```

Even with this simple problem, we have some difficulty in working with the dimensional equation of (3.2.2) because it is difficult to specify the time in minutes when the program should be stopped. By using an order–of–magnitude approach, we can improve our understanding of the specific problem to be simulated. Order–of–magnitude scaling involves the defining of dimensionless independent and dependent variables that are of the order–of–magnitude of unity. We then can look at the order–of–magnitude of the resulting dimensionless coefficients to determine scaling arguments.

In order to perform order–of–magnitude scaling for this problem, we need a dimensionless distance, dimensionless time, and dimensionless volumetric flow rate, which we define as

$$Z^* = Z/Z_0 \qquad (3.2.3)$$

$$t^* = t/\tau \qquad (3.2.4)$$

$$F^* = F/F_2 \qquad (3.2.5)$$

where Z_0 is the initial height, F_2 the outlet flow rate, and τ the residence time of the tank based on the initial volume and output flow rate,

$$\tau = V_0/F_2 \qquad (3.2.6)$$

Each of the characteristic variables are chosen so that the respective dimensionless variable is order–of–magnitude one. Substituting equations (3.2.3) through (3.2.5) into the mass balance equation, (3.2.2), gives

$$\frac{dZ^*}{dt^*} = F_1^* - 1 \qquad (3.2.7)$$

In dimensionless form, the sinusoidal input flow is

$$F_1^* = \frac{B}{F_2}\sin(W\,\tau\,t^*) + \frac{C}{F_2} \qquad (3.2.8)$$

For this problem

$$\tau = 100 \text{ min} \qquad \frac{B}{F_2} = 0.1$$

$$W = 0.1 \text{ min}^{-1} \qquad \frac{C}{F_2} = 1$$

Equation (3.2.8) therefore becomes

$$F_1^* = 0.1 \sin(10t^*) + 1 \qquad (3.2.9)$$

From equations (3.2.7) and (3.2.9) we determine that all terms are of the same order–of–magnitude so that all are important to the problem. Also, we would like to observe the dynamic response for one– or two–time constants.

The MATLAB program dtank32n.m solves this problem using dfun32.m to define the dimensionless differential equation of (3.2.7). This program uses the new version of ode45. Program dtank32o.m solves the problem with the old version of ode45. The program dtank32n.m and dfun32.m follow.

Program dtank32n.m

```
% This file calls the new version of the ODE solver, ode45 (4th
% and 5th Runge-Kutta), to solve the dimensionless basic tank
% model in section 3.2.
%
[T, Z] = ode45n('dfun32',[ 0 2.0], 1.0);

%
% F is the dimensionless inflow rate.
%
F = 0.1*sin(10*T)+1;

plot(T,Z,'-', T,F,'--');
title('Dimensionless Tank Response (tank32)');
xlabel('Dimensionless time');
text(0.25, 1.07, 'Dimensionless inflow rate');
text(0.5, 0.99, 'Dimensionless tank height');
```

Program dfun32.m

```
function f = dtank32(t,z)
% this file defines the dimesionless basic tank model in section
% 3.2.
%
f = 0.1*sin(10*t) + 1 - 1;
```

The output from the program is shown in Figure 3.5. The results clearly show that the input cycles around .6 of a tank residence time as does the tank height. The relative size of both fluctuations is also evident.

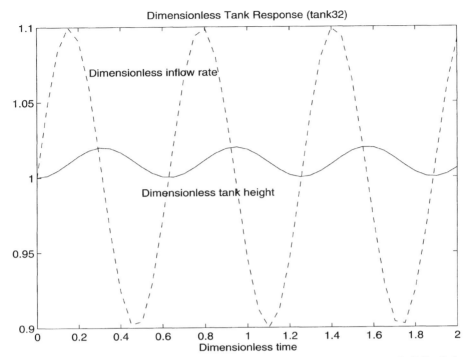

Figure 3.5: Dynamic Response of Dimensionless Tank Model

3.3 MULTISTEP METHODS

In order to solve the differential equation

$$\frac{dx}{dt} = f(x,t) \tag{3.3.1}$$

we have considered single–step methods, which can be written as

$$x_{i+1} = x_i + \int_{t_i}^{t_{i+1}} f(x,t)\, dt \tag{3.3.2}$$

However, we could also consider a multistep formulation:

$$x_{i+1} = x_{i-k} + \int_{t_{i-k}}^{t_{i+1}} f(x,t)\, dt \tag{3.3.3}$$

We can approximate the function $f(x,t)$ by an interpolating polynomial passing through points (t_j, f_j). This interpolating polynomial will be called $\phi_n(t)$ and is given by

$$\phi_n(t) = \sum_{j=1}^{n} a_j t^j \tag{3.3.4}$$

where n is the degree of the polynomial.

There are two types of interpolating polynomials that can be used. These are open formulas which are used to predict the x_{i+1} value based on known information up to x_i, and closed formulas which are used to correct the x_{i+1}. In both cases backward difference interpolating polynomials are used since we are using previous time information to determine current or future time behavior. In order to develop the backward difference formulas, we use Newton's fundamental formula for interpolating polynomials,

$$f(t) = \phi_n(t) + R_n(t) \tag{3.3.5}$$

where $f(t)$ is the continuous time function
$\phi_n(t)$ is the approximate polynomial of order n
$R_n(t)$ is the remainder or error associated with using the polynomial of order n

In order to generate $\phi_n(t)$ using backward differencing, we have

$$\phi_n(t - \alpha \Delta t) = f(t) - \alpha \nabla f(t) + \frac{\alpha(\alpha-1)}{2!}\nabla^2 f(t)$$

$$- \cdots \pm \frac{\alpha(\alpha-1)(\alpha-2)\cdots(\alpha-n-1)}{n!}\nabla^n f(t) \tag{3.3.6}$$

where the backward difference operator ∇, ∇^2, etc. is defined by,

$$\nabla f(t) = f(t) - f(t - \Delta t)$$

$$\nabla^2 f(t) = \nabla(\nabla f(t)) = \nabla\Big(f(t) - f(t - \Delta t)\Big)$$

$$= f(t) - 2f(t - \Delta t) + f(t - 2\Delta t) \tag{3.3.7}$$

$$\vdots$$

$$\nabla^n f(t) = \nabla\Big(\nabla^{n-1} f(t)\Big)$$

In order to avoid the sign change problem associated with equation (3.3.6), we let

$$\beta = -\alpha \tag{3.3.8}$$

therefore

$$\phi_n(t + \beta \Delta t) = f(t) + \beta \nabla f(t) + \frac{\beta(\beta+1)}{2!}\nabla^2 f(t)$$
$$+ \cdots + \frac{\beta(\beta+1)(\beta+2)\cdots(\beta+n+1)}{n!}\nabla^n f(t) \tag{3.3.9}$$

and

$$\nabla f(t) = f(t + \beta) - f(t)$$
$$\vdots \tag{3.3.10}$$
$$\nabla^n f(t) = \nabla^{n-1}\Big(f(t + \beta) - f(t)\Big)$$

In order to predict a value at t_{i+1}, we will use a formula that is based at t_i:

$$\phi_n(t) = f(t_i) + \beta \nabla f(t_i) + \cdots + \frac{\beta(\beta+1)(\beta+2)\cdots(\beta+n-1)}{n!} \nabla^n f(t_i) \quad (3.3.11)$$

therefore the integral

$$\int_{t_{i-k}}^{t_{i+1}} \phi_n(t)\,dt = \int_{t_{i-k}}^{t_{i+1}} \phi_n(t_i+\beta\Delta t)\,d(t_i+\beta\Delta t) = \Delta t \int_{-k}^{1} \phi_n(t_i+\beta\Delta t)\,d\beta \quad (3.3.12)$$

or

$$\int_{t_{i-k}}^{t_{i+1}} \phi_n(t)\,dt = \Delta t \left[\beta f(t_i) + \frac{\beta^2}{2}\nabla f(t_i) + \beta^2\left(\frac{\beta}{3}+\frac{1}{2}\right)\frac{\nabla^2 f(t_i)}{2!} + \cdots \right]_{\beta=-k}^{\beta=1} \quad (3.3.13)$$

For $k=3$, we are using an interpolating polynomial over the last four known data points $(t_i, t_{i-1}, t_{i-2}, t_{i-3})$, and are using a fourth-order interpolating polynomial. Therefore, using equations (3.3.13) and (3.3.3), and after algebraic manipulation, we have,

$$x_{i+1} = x_{i-3} + \Delta t \left(4f(t_i) - 4\nabla f(t_i) + \frac{8}{3}\nabla^2 f(t_i) + 0\nabla^3 f(t_i) \right. \\
\left. + \frac{14}{45}\nabla^4 f(t_i) \right) \quad (3.3.14)$$

If we are only interested in an estimate of x_{i+1} which is fourth-order correct, then we can truncate the last ∇^4 term and get

$$x_{i+1} = x_{i-3} + \frac{4}{3}\Delta t \left(2f(t_i) - f(t_{i-1}) + 2f(t_{i-2}) \right) \quad (3.3.15)$$

with a remainder error of $O(\Delta t^5)$.

A closed formula can be used to correct the x_{i+1} estimate.

The closed backward difference formula is based on t_{i+1} instead of t_i; therefore

$$\phi_n(t_i + \beta\Delta t) = f(t_{i+1}) + (\beta-1)\nabla f(t_{i+1}) + \frac{(\beta-1)}{2!}\beta\nabla^2 f(t_{i+1}) + \cdots$$

This gives, for the closed formula,

$$x_{i+1} = x_{i-k} + \left[\Delta t \left(\beta f(t_{i+1}) + \beta\left(\frac{\beta}{2}-1\right)\nabla f(t_{i+1}) \right) \right. \\
\left. + \left(\frac{\beta^2\left(\frac{\beta}{3}-\frac{1}{2}\right)}{2!}\nabla^2 f(t_{i+1}) + \cdots \right) \right]_{\beta=-k}^{\beta=1} \quad (3.3.16)$$

for $k=1$, which means that we will be placing an interpolating polynomial over the three points t_{i-1}, t_i, and t_{i+1}. We can show that for the fourth-order interpolating polynomial, the value of x_{i+1} is given by

$$x_{i+1} = x_{i-1} + \Delta t \left(2f(t_{i+1}) - 2\nabla f(t_{i+1}) + \tfrac{1}{3}\nabla^2 f(t_{i+1}) \right.$$
$$\left. + 0\nabla^3 f(t_{i+1}) - \frac{1}{90}\nabla^4 f(t_{i+1}) \right) \quad (3.3.17)$$

If we are only interested in an estimate x_{i+1}, which is fourth-order correct, then we can truncate the ∇^4 term and obtain

$$x_{i+1} = x_{i-1} + \frac{\Delta t}{3} \left(f(t_{i+1}) + 4f(t_i) + f(t_{i-1}) \right) \quad (3.3.18)$$

Note that this, as well as all corrector formulas, is implicit in x_{i+1} since $f(t_{i+1})$ involves x_{i+1}. This means that the implicit root solving methods of Section 2.3 must be used in order to solve equations of the type of (3.3.18).

There are a number of important predictor–corrector equations, including equations (3.3.15) and (3.3.18), which are the fourth-order Milne predictor–corrector formulas.

3.4 STIRRED TANKS WITH FLOW RATES A FUNCTION OF LEVEL

A more complex and realistic tank modeling case is to consider the situation in which both input and output flow rates are influenced by the level in the tank. In this case (Figure 3.6) the inflow F_1 passes through a fixed inlet valve from a pressure source, P_1. The pressure on the downstream side of the inlet valve is P_2, that is, the hydrostatic pressure in the tank at the level of the valve. In a similar fashion, the outflow passes through a fixed valve with the hydrostatic pressure P_2 on the upstream side discharging to pressure P_3. The flows F_1 and F_2 are influenced by the level Z and the pressures P_0 and P_3.

The equations that describe this system must be developed. Thus, applying the principle of conservation of mass to the system, we have

$$\begin{array}{c}\text{Rate of}\\ \text{Accumulation}\\ \text{of Mass}\end{array} = \begin{array}{c}\text{Rate of}\\ \text{Mass In}\end{array} - \begin{array}{c}\text{Rate of}\\ \text{Mass Out}\end{array}$$

$$A\frac{dZ}{dt} = F_1 - F_2 \quad (3.4.1)$$

The flows F_1 and F_2 can be related to the pressure drop across each valve by applying the mechanical energy balance across each valve. Here

Unsteady–State Lumped Systems

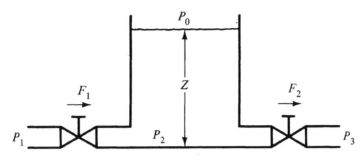

Figure 3.6: Tank in Which Flow Rates Are Influenced by Fluid Level.

we will assume that any hydraulic transients are very fast compared to the volume transients in the system, so that we can use the steady–state form of the mechanical energy balance equation, (1.6.18).

$$\begin{array}{c} \text{Rate of} \\ \text{Accumulation of} \\ \text{Mechanical Energy} \end{array} = \begin{array}{c} \text{Rate of} \\ \text{Mechanical} \\ \text{Energy In} \end{array} - \begin{array}{c} \text{Rate of} \\ \text{Mechanical} \\ \text{Energy Out} \end{array} + \begin{array}{c} \text{Rate of} \\ \text{Generation of} \\ \text{Mech. Energy} \end{array}$$

$$0 = \left(\frac{\langle v_1 \rangle^2}{2} + gZ_1 + \frac{P_1}{\rho} \right) - \left(\frac{\langle v_2 \rangle^2}{2} + gZ_2 + \frac{P_2}{\rho} \right) + \left(-h_L - \hat{W}_s \right) \quad (3.4.2)$$

The potential energy differences are negligible and there is no mechanical work, $\hat{W}_s$. Therefore, for the first valve we have

$$\frac{P_2 - P_1}{\rho} + \frac{\langle v_2 \rangle^2 - \langle v_1 \rangle^2}{2} + h_L = 0 \quad (3.4.3)$$

The steady–state mass balance across the valve gives

$$\begin{array}{c} \text{Rate of} \\ \text{Mass In} \end{array} = \begin{array}{c} \text{Rate of} \\ \text{Mass Out} \end{array}$$

$$\rho \langle v_1 \rangle S_1 = \rho \langle v_2 \rangle S_2 \quad (3.4.4)$$

or

$$\langle v_2 \rangle = \frac{\langle v_1 \rangle S_1}{S_2} \quad (3.4.5)$$

where S_1 = upstream pipe cross–section area and S_2 = the cross–section area at the throat of the valve. Combining equations (3.4.3) and (3.4.5) and solving for $\langle v_1 \rangle$ gives

$$\langle v_1 \rangle = \sqrt{\frac{2(-\Delta P/\rho - h_L)}{S_1^2/S_2^2 - 1}} \quad (3.4.6)$$

Letting the friction head be expressed as some fraction of the total pressure drop, we have

$$\frac{\Delta P}{\rho} + h_L = C_1^2 \frac{\Delta P}{\rho} \qquad (3.4.7)$$

Therefore, equation (3.4.6) becomes

$$S_1 \langle v_1 \rangle = F_1 = C_1 S_1 \sqrt{\frac{2}{(S_1^2/S_2^2 - 1)\rho}} \sqrt{(P_1 - P_2)} \qquad (3.4.8)$$

or

$$F_1 = C_{v_1} \sqrt{(P_1 - P_2)} \qquad (3.4.9)$$

where C_{v_1} is the valve constant for the inlet valve.

Finally, the hydrostatic pressure P_2 can be computed by application of the mechanical energy balance on the control volume of the tank itself. This gives

$$\frac{P_2 - P_0}{\rho} = gZ \qquad (3.4.10)$$

or

$$P_2 = P_0 + \rho g Z \qquad (3.4.11)$$

The three equations (3.4.1), (3.4.9), and (3.4.11) describe the dynamic behavior of the system and can be arranged in model form in at least three different ways, as shown in Figure 3.7. The difference between the models is the result of selecting a different equation to define each variable, and although each of these models is mathematically possible, *only the first model makes any sense from a physical viewpoint* (Franks, 1967). In this arrangement, each equation is used in its *natural form*: that is, the flows F_1 and F_2 are the result of exerting a pressure in a fluid across a valve and the variation in level Z is a result of variations in flows F_1 and F_2. There are two important reasons for strongly favoring the natural model over the other mathematically possible arrangements. The first is that the natural model, being based strictly on a cause–and–effect relationship, will provide insight to the engineer on the true mechanism of the system. Second, a non–natural model often leads to computational difficulties whereas a natural model is usually computationally stable. It should be noted that the natural model results in a solution strategy that uses the differential equations to compute the variables that appear as differentials. This leads to the process of numerical integration. The non–natural equation ordering results in having to use the process of numerical differentiation in order to obtain a solution to the equation set.

This problem can be easily solved on a digital computer using one of the MATLAB numerical integration routine for the solution of ordinary differential equations.

Again, an order–of–magnitude analysis can improve our modeling of this system. By the use of order–of–magnitude analysis, we can show

Unsteady–State Lumped Systems

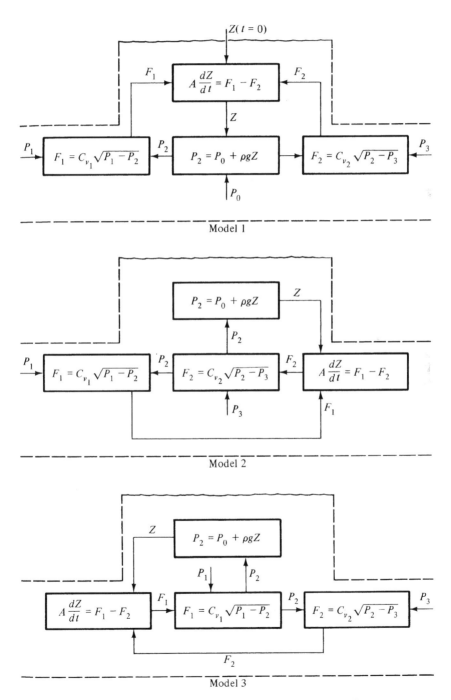

Figure 3.7: **Alternative Models for Tank System.**

that the quasi–steady–state analysis of using the steady–state mass and mechanical energy balance around the valve is justified. The dimensionless tank mass balance is

$$\frac{A Z_0}{\tau} \frac{dZ^*}{dt^*} = F_c(F_1^* - F_2^*) \qquad (3.4.12)$$

where

$$\tau = \frac{A Z_0}{F_c} \quad , \quad Z^* = \frac{Z}{Z_0} \quad , \quad t^* = \frac{t}{\tau} \qquad (3.4.13)$$

$$F_c = C_{v_1}\sqrt{(P_{10} - P_{20})} \quad , \quad F^* = \frac{F}{F_c} \qquad (3.4.14)$$

Therefore equation (3.4.12) becomes

$$\frac{dZ^*}{dt^*} = (F_1^* - F_2^*) \qquad (3.4.15)$$

The material balance on the valve is

$$\rho \frac{dV_v}{dt} = \rho(F_0 - F_1) \qquad (3.4.16)$$

Letting

$$V_v^* = \frac{V_v}{V_{v0}} \qquad (3.4.17)$$

gives

$$\frac{V_{v0}}{\tau} \frac{dV_v^*}{dt^*} = F_c(F_0^* - F_1^*) \qquad (3.4.18)$$

or

$$\frac{V_{v0}}{V_{t0}} \frac{dV_v^*}{dt^*} = (F_0^* - F_1^*) \qquad (3.4.19)$$

where

$$V_{t0} = A Z_0$$

Since the valve is assumed to always be filled with fluid, there is no time change of V_v^* with respect to time; therefore,

$$\frac{dV_v^*}{dt^*} = 0 \quad \text{or} \quad F_0^* = F_1^* \qquad (3.4.20)$$

Now let's consider the dynamic mechanical energy balance written around the valve,

$$\frac{d}{dt}\left[\rho\left(\hat{A} + \frac{v_2^2}{2}\right) V_v\right] = w\left[\frac{v_1^2}{2} - \frac{v_2^2}{2} + \frac{P_1 - P_2}{\rho}\right] - w\, h_L \qquad (3.4.21)$$

For an incompressible fluid

$$\hat{A} = -\int_{\hat{V}_R}^{\hat{V}} p \, d\hat{V} = 0 \qquad (3.4.22)$$

Equation (3.4.21) becomes

$$\rho V_v v_2 \frac{dv_2}{dt} = \frac{w}{2}(v_1^2 - v_2^2) + \frac{w}{\rho}(P_1 - P_2) - w \, h_L \qquad (3.4.23)$$

or assuming that the friction head can be expressed as a fraction of the total pressure drop, then

$$\rho V_v v_2 \frac{dv_2}{dt} = \frac{\rho F_1}{2}(v_1^2 - v_2^2) + \frac{\rho F_1}{\rho} C_1^2 (P_1 - P_2) \qquad (3.4.24)$$

Using the overall mass balance around the valve of equation (3.4.5) gives

$$\rho V_v \frac{S_1^2}{S_2^2} v_1 \frac{dv_1}{dt} = \frac{\rho F_1}{2}\left(1 - \frac{S_1^2}{S_2^2}\right) v_1^2 + F_1 C_1^2 (P_1 - P_2) \qquad (3.4.25)$$

Using the fact that

$$v_1 = F_1/S_1 \qquad (3.4.26)$$

gives

$$\rho \frac{V_v}{S_1^2} \left(\frac{S_1^2}{S_2^2}\right) F_1 \frac{dF_1}{dt} = \rho \frac{F_1}{2 S_1^2}\left(1 - \frac{S_1^2}{S_2^2}\right) F_1^2 + F_1 C_1^2 (P_1 - P_2) \qquad (3.4.27)$$

or

$$V_v \left(\frac{S_1^2}{S_2^2}\right) \frac{2}{(1 - S_1^2/S_2^2)} \frac{dF_1}{dt} = F_1^2 + \frac{2 C_1^2 S_1^2}{\rho(1 - S_1^2/S_2^2)}(P_1 - P_2) \qquad (3.4.28)$$

Now introducing order–of–one dimensionless variables

$$t^* = t/\tau$$

$$P^* = P/P_{10}$$

$$F^* = F/F_c$$

where the characteristic time and volumetric flow are

$$\tau = V_t/F_c$$

$$F_c = C_{v_1}\sqrt{P_{10} - P_{20}}$$

gives

$$\frac{V_v}{V_t}\left(\frac{S_1^2}{S_2^2}\right)\frac{2}{(1-S_1^2/S_2^2)}\frac{dF_1^*}{dt^*} = F_1^{*2} + \frac{2C_1^2 S_1^2}{\rho(1-S_1^2/S_2^2)}\frac{P_{10}}{F_c^2}(P_1^* - P_2^*) \quad (3.4.29)$$

or

$$\frac{V_v}{V_t}\left(\frac{S_1^2}{S_2^2}\right)\frac{2}{(1-S_1^2/S_2^2)}\frac{dF_1^*}{dt^*} = F_1^{*2} + \frac{2C_1^2 S_1^2}{\rho(1-S_1^2/S_2^2)}\frac{1}{C_{v_1}^2}\frac{P_{10}}{(P_{10}-P_{20})}(P_1^* - P_2^*) \quad (3.4.30)$$

which using the definition of C_{v_1} reduces to

$$\frac{V_v}{V_t}\left(\frac{S_1^2}{S_2^2}\right)\frac{2}{(1-S_1^2/S_2^2)}\frac{dF_1^*}{dt^*} = F_1^{*2} - \frac{P_{10}}{(P_{10}-P_{20})}(P_1^* - P_2^*) \quad (3.4.31)$$

Now the ratio V_v/V_t is normally 10^{-3} or smaller and the ratio S_1/S_2 is 10^2 or larger so that $(1 - S_1^2/S_2^2)$ is approximately $-S_1^2/S_2^2$ which makes the coefficient of equation (2.6.31)

$$\left(\frac{V_v}{V_t}\right)\left(\frac{S_1}{S_2}\right)^2 \frac{2}{(1-S_1^2/S_2^2)} \leq 10^{-3} \quad (3.4.32)$$

Therefore the valve dynamic term is small so that it can be neglected. This means that valve dynamics can be neglected and a quasi–steady–state assumption can be employed when considering tank dynamic residence times.

We now seek a numerical solution to the dimensionless model equations,

$$\frac{dZ^*}{dt^*} = F_1^* - F_2^* \quad (3.4.33)$$

$$F_1^* = \sqrt{\frac{P_{10}}{P_{10}-P_{20}}}\sqrt{P_1^* - P_2^*} \quad (3.4.34)$$

$$F_2^* = \left(\frac{C_{v_2}}{C_{v_1}}\right)\sqrt{\frac{P_{10}}{P_{10}-P_{20}}}\sqrt{P_2^* - P_3^*} \quad (3.4.35)$$

$$P_2^* = P_0^* + \left(\frac{\rho g Z_0}{P_{10}}\right) Z^* \quad (3.4.36)$$

We will use the following basic data:

$C_{v_2} = C_{v_1} = 1.2 \times 10^{-2}$ m^2/N$^{1/2}$ sec $\quad \rho = 10^3$ kg/m^3
$P_{10} = P_1 = 1.38 \times 10^5$ N/m^2 $\quad A = 0.465$ m^2
$P_{30} = P_3 = 1.08 \times 10^5$ N/m^2 $\quad Z_0 = 3.05$ m
$P_0 = 1.014 \times 10^5$ N/m^2 $\quad g = 9.8$ m/s^2

This gives

$$P_{20} = P_0 + \rho g Z_0 = 1.313 \times 10^5 \text{ N/m}^2$$

and the model equations (3.4.33)–(3.4.36) become

$$\frac{dZ^*}{dt^*} = F_1^* - F_2^* \qquad (3.4.37)$$

$$F_1^* = 4.541\sqrt{P_1^* - P_2^*} \qquad (3.4.38)$$

$$F_2^* = 4.541\sqrt{P_2^* - P_3^*} \qquad (3.4.39)$$

$$P_2^* = 0.735 + 0.2167\ Z^* \qquad (3.4.40)$$

with the initial conditions

$$Z_0^* = 1.0$$
$$P_1^* = 1.0 \qquad (3.4.41)$$
$$P_3^* = 0.783$$

We will use the predictor-correct routine available in the integration package ode113. This routine is a variable-order Adams–Bashford–Moulton method (Shampine and Gordon, 1975). The ode113 routine is usually the most efficient code in the MATLAB suite in terms of the number of function evaluations, but not necessarily the fastest. It is also relatively efficient in the presence of mild stiffness. It is an excellent all-round routine for integrating differential equations. This routine is available on the world wide web under http://optimal.colorado.edu/~ramirez/chen4580.html as ode113.m. The program tank34.m has been created to solve this problem. The differential equations are defined in program fun34.m. Both programs are given below and are available on the www.

Program tank34.m

```
% This file calls the new ODE solver, ode113, which is a multi
% step integration routine to solve the tank model of section
% 3.6.
%
[T, Z] = ode113('fun34',[ 0 3.0], 1.0);

%
% F1 is the dimensionless inflow rate;
% F2 is the dimensionless outflow rate;
% P2 is the dimensionless pressure at the tank bottom.
%
P2 = 0.735 + 0.2167*Z;
P1 = 1.;
P3 = .785;
F1 = 4.541*sqrt(P1-P2);
F2 = 4.541*sqrt(P2-P3);

plot(T,Z,'-', T,F1,'--', T,F2,'-.', T,P2,':');
```

```
title('Dimensionless Tank Response (tank34)');
xlabel('Dimensionless time');
text(0.5, 1.6, 'Dimensionless outflow rate');
text(0.5, 1.37, 'Dimensionless inflow rate');
text(0.5, 0.8, 'Dimensionless tank height');
text(0.5, 0.95, 'Dimensionless tank bottom pressure');
```

Program fun34.m

```
function f = tank34(t,z)
% this file defines the dimesionless model of the tank with flow
% rates a function of level in section 3.4.
%

%
% initial values:
%
P1=1.0; P3=0.785;

P2 = 0.735 + 0.2167*z;
F1 = 4.541*sqrt(P1-P2);
F2 = 4.541*sqrt(P2-P3);
f = F1 - F2;
```

Results are given in Figure 3.8. These results show that the system reaches steady-state in approximately 1.5 dimensionless time units or residence times.

3.5 ENCLOSED TANK VESSEL

This case is similar to the preceding except that the vessel is wholly enclosed (see Figure 3.9), and the pressure P_0 above the surface becomes a variable instead of a constant. Movement of the surface up and down will compress and expand the gas and cause the pressure to change. Assuming an ideal gas, we use the equation of state

$$P_0 V_G = nRT_G \tag{3.5.1}$$

where P_0 = gas pressure
 V_G = gas volume
 T_G = gas temperature
 n = number of moles of gas
 R = gas constant

Under the assumption of *adiabatic* conditions, the temperature of the system compression can be computed. Since no heat is lost from the system and all the work is converted into sensible heat, the energy balance on the closed system becomes

$$\Delta U + \Delta(KE) + \Delta(PE) = Q - \hat{W} \tag{3.5.2}$$

Unsteady–State Lumped Systems

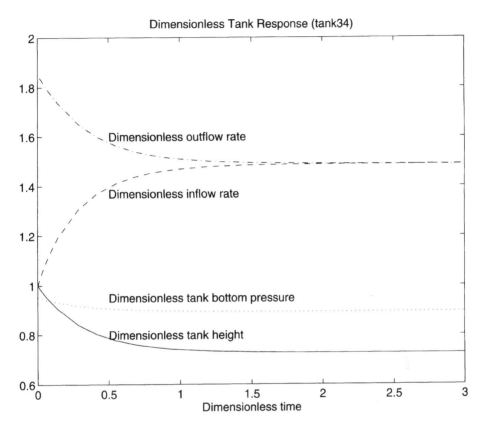

Figure 3.8: **Dynamic Response**

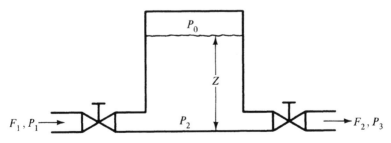

Figure 3.9: **Enclosed Vessel.**

where KE = kinetic energy
PE = potential energy

For adiabatic conditions $Q = 0$ and we will also assume that

$$\Delta(KE) = 0 \qquad \Delta(PE) = 0$$

for this system. Therefore, the energy balance becomes

$$\Delta U = -\hat{W} \qquad (3.5.3)$$

Applying this to a differential element in the gas phase, we have

$$dU = -d\hat{W} \qquad (3.5.4)$$

The internal energy can be related to the temperature of the system by

$$dU = c_v \, dT_G \qquad (3.5.5)$$

where c_v is the heat capacity at constant volume.

The only form of work in this system is PV work so that

$$d\hat{W} = P_0 \, d\hat{V} \qquad (3.5.6)$$

where $\hat{V}$ is the specific volume of the gas.

Combining equations (3.5.4), (3.5.5), and (3.5.6) gives

$$c_v \, dT_G = -\frac{P_0}{m} \, dV_G \qquad (3.5.7)$$

where m = the mass of the system. Using the ideal gas law and assuming c_v is constant, equation (3.5.7) can be integrated as

$$\frac{T_G}{T_{G_0}} = \left(\frac{V_{G_0}}{V_G}\right)^{R/C_v(MW)} \qquad (3.5.8)$$

The volume of the gas can be computed at any time from

$$V_G = V_0 - AZ \qquad (3.5.9)$$

where V_0 = total tank volume
A = cross-sectional area of the tank

The dimensionless equations using order-of-magnitude scaling become

$$\frac{dZ^*}{dt^*} = F_1^* - F_2^* \qquad (3.5.10)$$

Unsteady–State Lumped Systems

$$F_1^* = \sqrt{\frac{P_{10}}{P_{10} - P_{20}}} \sqrt{P_1^* - P_2^*} \tag{3.5.11}$$

$$F_2^* = \left(\frac{C_{v_2}}{C_{v_1}}\right) \sqrt{\frac{P_{10}}{P_{10} - P_{20}}} \sqrt{P_2^* - P_3^*} \tag{3.5.12}$$

$$P_2^* = P_G^* + \left(\frac{\rho g Z_0}{P_{10}}\right) Z^* \tag{3.5.13}$$

$$V_G^* = \frac{V_0}{V_{G_0}} - Z^* \tag{3.5.14}$$

$$T_G^* = \left(\frac{1}{V_G^*}\right)^{R/C_v(MW)} \tag{3.5.15}$$

$$P_G^* = \frac{P_{G_o}}{P_{10}} \frac{T_g^*}{V_g^*} \tag{3.5.16}$$

Here the dimensionless variables are defined as

$$Z^* = \frac{Z}{Z_0}, \quad F^* = \frac{F}{C_{v_1}\sqrt{P_{10} - P_{20}}}, \quad t^* = \frac{t\, C_{v_1} \sqrt{P_{10} - P_{20}}}{A Z_0}$$

$$V_G^* = \frac{V_G}{V_{G_0}}, \quad T_G^* = \frac{T_G}{T_{G_0}}, \quad P^* = \frac{P}{P_{10}}$$
$$\tag{3.5.17}$$

The information–flow diagram for this problem is given in Figure 3.10. Again, it is important to notice the natural arrangement of the equations.

If we use the same data as in Section 3.4 but add

$$V_0 = 2.83 \text{ m}^3 \qquad T_{G_0} = 338.6 \text{ K} \qquad V_{G_0} = 1.415 \text{ m}^3$$
$$c_v = 0.2 \text{ cal/g K} \qquad R = 1.987 \text{ cal/g mol K} \qquad MW = 29 \tag{3.5.18}$$

Then the scaled dimensionless equations for this case are

$$\frac{dZ^*}{dt^*} = F_1^* - F_2^* \tag{3.5.19}$$

$$F_1^* = 4.541 \sqrt{P_1^* - P_2^*} \tag{3.5.20}$$

$$F_2^* = 4.541 \sqrt{P_2^* - P_3^*} \tag{3.5.21}$$

$$P_2^* = P_G^* + 0.2167\, Z^* \tag{3.5.22}$$

$$V_G^* = 2 - Z^* \tag{3.5.23}$$

$$T_G^* = \left(\frac{1}{V_G^*}\right)^{0.343} \tag{3.5.24}$$

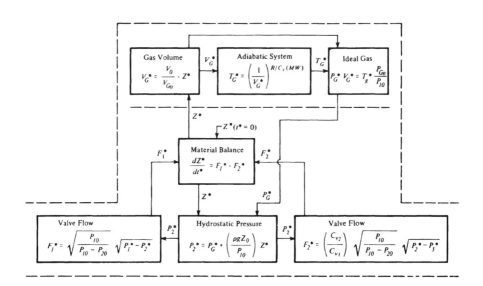

Figure 3.10: Information–Flow Diagram for Tank System Assuming Adiabatic Gas Volume.

$$P_G^* = .735\ T_G^*/V_G^* \qquad (3.5.25)$$

The results of running this problem are shown in Figure 3.11. The MATLAB routine `ode113` was used and the program is available on the www as `tank35.m` with the differential equations defined in `fun35.m`. Comparing the results of Sections 3.4 and 3.5 shows that the dynamic response of the closed system is significantly faster than that of the open system. This is because as the liquid tank volume increases, the gas volume decreases, causing an increase in the gas pressure P_G^* (equation (3.5.25)). This in turn increases the pressure P_2^*, which adjusts the flows F_1^* and F_2^*, so that an equilibrium state is approached more quickly than in the case with a constant gas phase pressure.

3.6 STIRRED TANK WITH HEATING JACKET

Figure 3.12 shows an agitated vessel with a steam jacket for heating the vessel. In order to describe the dynamic behavior of this system, we will have to apply both the conservation of mass and the conservation of energy for the system.

Unsteady–State Lumped Systems

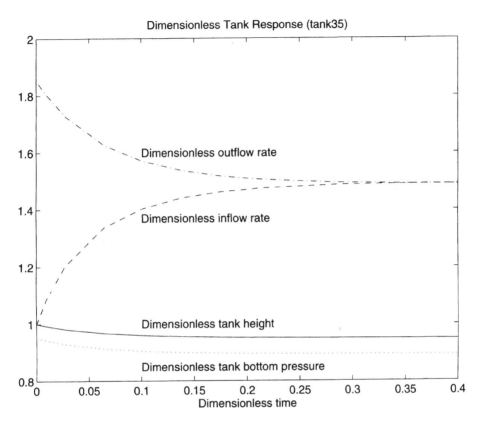

Figure 3.11: Dynamic Response of Enclosed Tank.

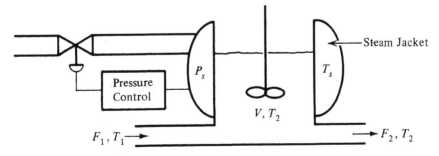

Figure 3.12: Steam–Jacketed Vessel.

Applying the equation for conservation of thermal energy to the system, we have

Rate of Accumulation of Thermal Energy = Rate of Thermal Energy In − Rate of Thermal Energy Out + Rate of Generation of Thermal Energy

$$\left[\frac{d}{dt}\int_{T_R}^{T_2} \rho V c_p \, dT\right] = \left[\int_{T_R}^{T_1} \rho c_p F_1 \, dT + UA(T_s - T_2)\right] - \left[\int_{T_r}^{T_2} \rho c_p F_2 \, dT\right] \quad (3.6.1)$$

where U = overall heat–transfer coefficient
A = cross–sectional area for heat transfer

Assuming that ρ, V, c_p, F_1, and F_2 are not functions of temperature gives

$$\rho c_p \frac{d}{dt} V \int_{T_r}^{T_2} dT = \rho c_p F_1 \int_{T_r}^{T_1} dT - \rho c_p F_2 \int_{T_r}^{T_2} dT + UA(T_s - T_2) \quad (3.6.2)$$

Evaluating the integrals, we have

$$\rho c_p \frac{d}{dt} V(T_2 - T_r) = \rho c_p F_1(T_1 - T_r) - \rho c_p F_2(T_2 - T_r) + UA(T_s - T_2) \quad (3.6.3)$$

The temperature in the jacket can be determined from the steam tables knowing the pressure in the jacket, P_s. Since the mechanism for heat transfer in the steam jacket is condensation, saturated steam properties can be used and

$$T_s = f(P_s) \quad (3.6.4)$$

where $f(P_s)$ is a known function for saturated steam (Keenan and Keyes, 1959). Equations (3.6.3) and (3.6.4) are combined with the mass balance

$$dV/dt = F_1 - F_2 \quad (3.6.5)$$

The information–flow diagram for the system is given in Figure 3.13.

Note that the energy balance is solved for the variable $y = V(T_2 - T_r)$ since it is the variable inside the derivative. This leads to ease in the use of the numerical integration routines for computer solution.

3.7 ENERGY BALANCES WITH VARIABLE PROPERTIES

This is the same problem discussed in the foregoing section, except now we will allow the heat capacity, c_p, to be a function of temperature.

$$c_p = a + bT + cT^2 \quad (3.7.1)$$

Unsteady-State Lumped Systems

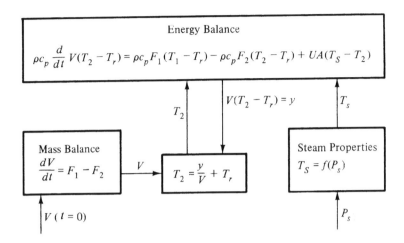

Figure 3.13: Information–Flow Diagram for Stirred Tank with Heating Jacket.

where a, b, and c are constants.

To handle this problem properly, we must go back to the definition of the heat capacity in terms of the total energy.

$$E = \int_{T_r}^{T} \rho V c_p \, dT \tag{3.7.2}$$

where E is total energy and T_r is the reference temperature.

The accumulation term in the energy balance is therefore

$$\frac{d}{dt} \int_{T_r}^{T} \rho V c_p \, dT \tag{3.7.3}$$

which must be evaluated. To evaluate a differential of an integral function, we must in general use *Leibniz's formula* for differentiating an integral, which can be stated as

$$\frac{d}{d\alpha} \int_{u_0(\alpha)}^{u_1(\alpha)} f(x, \alpha) \, dx = f(u_1, \alpha) \frac{du_1}{d\alpha} - f(u_0, \alpha) \frac{du_0}{d\alpha} + \int_{u_0(\alpha)}^{u_1(\alpha)} \frac{\partial f(x, \alpha)}{\partial \alpha} dx \tag{3.7.4}$$

which gives the following result when applied to the energy accumulation term with c_p a function of temperature.

$$\frac{d}{dt} \int_{T_r}^{T} \rho V c_p \, dT = \rho V c_p \frac{dT}{dt} - \rho_r V c_{p_r} \frac{dT_r}{dt} + \int_{T_r}^{T} \frac{\partial \rho V c_p}{\partial t} dT \tag{3.7.5}$$

Since $dT_r/dt = 0$ and ρ and c_p are not explicit functions of time, the accumulation term reduces to

$$\frac{d}{dt} \int_{T_r}^{T} \rho V c_p \, dT = \rho V c_p \frac{dT}{dt} + \int_{T_r}^{T} \rho c_p \frac{\partial V}{\partial t} dT \tag{3.7.6}$$

The term $\partial V/\partial t$ is obtained from the material balance to be $F_1 - F_2$. Therefore

$$\frac{d}{dt}\int_{T_r}^{T} \rho V c_p \, dT = \rho V c_p \frac{dT}{dt} + \int_{T_r}^{T} \rho c_p (F_1 - F_2) \, dT \qquad (3.7.7)$$

The total energy balance for the system is now

$$\rho V c_p \frac{dT_2}{dt} + \rho(F_1 - F_2)\int_{T_r}^{T_2} c_p \, dT =$$
$$\rho F_1 \int_{T_r}^{T_1} c_p \, dT + UA(T_s - T_2) - \rho F_2 \int_{T_r}^{T_2} c_p \, dT \qquad (3.7.8)$$

or

$$\rho V c_p \frac{dT_2}{dt} = \rho F_1 \int_{T_2}^{T_1} c_p \, dT + UA(T_s - T_2) \qquad (3.7.9)$$

If $c_p = a + bT + cT^2$, the energy balance becomes

$$\rho V(a + bT_2 + cT_2^2)\frac{dT_2}{dt} = -\rho F_1 \left(aT + \frac{bT^2}{2} + \frac{cT^3}{3}\right)\bigg|_{T_1}^{T_2} + UA(T_s - T_2) \qquad (3.7.10)$$

The information–flow diagram for the heated vessel with c_p a function of temperature is given in Figure 3.14.

3.8 TANKS WITH MULTICOMPONENT FEEDS

Finally, we consider a steam–jacketed vessel with an inlet stream that contains two components (A and B), and include the fact that the heat-transfer area is a function of vessel height (Figure 3.15).

In order to describe the dynamics of this system, we must include component material balances, an equation describing how the specific heat of the solution varies with the composition of the vessel, and the geometric relations which describe the heat-transfer area as a function of vessel volume. Since this is a binary mixture of components A and B, we can only write two material balances; we choose to write one component balance and the overall mass balance. We will assume that the total mass density ρ is a constant for this system. Thus, the component mass balance is

$$\frac{d}{dt}\rho_A V = \rho_{A_0} F_A - \rho_A F_2 \qquad (3.8.1)$$

The overall mass balance is

$$\frac{d}{dt}V = F_A + F_B - F_2 \qquad (3.8.2)$$

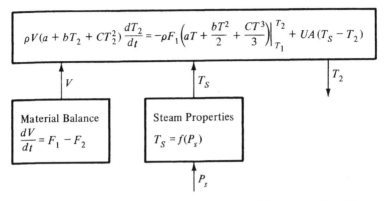

Figure 3.14: Information–Flow Diagram for the Steam–Jacketed Vessel with c_p a Function of Temperature.

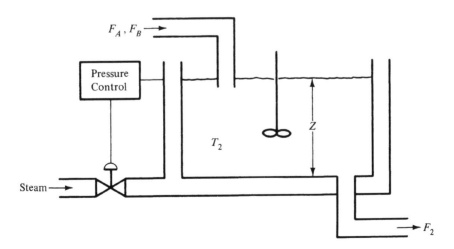

Figure 3.15: Steam–Jacketed Vessel with Mixing.

The specific heat of the vessel is

$$c_p = c_{pA}\frac{\rho_A}{\rho} + c_{pB}\frac{\rho_B}{\rho} \qquad (3.8.3)$$

where ρ is the total density, which is assumed constant.

The heat–transfer area is given by the sum of the transfer area on the bottom of the tank and the sides.

$$A = \pi D^2/4 + 4V/D \qquad (3.8.4)$$

The information–flow diagram for this problem is given in Figure 3.16.

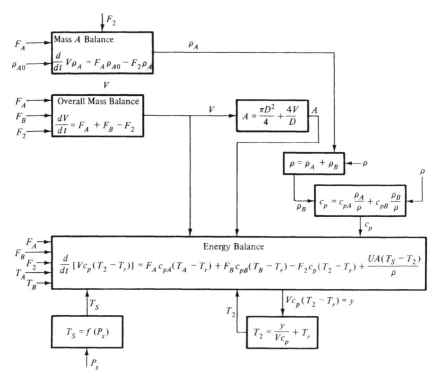

Figure 3.16: Information–Flow Diagram for Steam–Jacketed Vessel with Multicomponent Feeds.

3.9 STIFF DIFFERENTIAL EQUATIONS

Recently attention has been focused in the chemical engineering literature on stiff systems. These are sets of differential equations that contain a mixture of very fast dynamic equations and very slow dynamic equations. Single–step and multiple–step algorithms are known to have trouble with this type of equation system. To obtain completely accurate answers, the integration step size must be very small and this usually means prohibitively long computation times in order to observe the slow dynamic responses. On the other hand, if the integration step size is too big, the resulting inaccuracies in the fast dynamic parts can lead to general stability problems and inaccurate results.

However, often the real problem is not with the numerical algorithm but with the engineer developing the equation set. If one is interested in the slower dynamic parts of the problem, a quasi–steady–state assumption should be made for the fast parts of the problem. On the other hand, if one is interested in the fast parts of the problem, the value of the slower parts essentially remains constant over these very short time periods. Therefore, stiff systems of equations should not arise in most properly formulated simulations that use order–of–magnitude scaling in model formation.

However, even with proper order–of–magnitude scaling, stiff differential equation sets occasionally can arise in process modeling because of the appearance of fast dynamic modes during the running of a simulation. The work of Gear (1971) has shown that by adjusting both the order of a multistep integration method and the time step size an efficient algorithm can be developed for stiff differential equations.

MATLAB has two new stiff integration routines. These are ode15s and ode23s. The routine ode15s is a variable order (up to order 5) and a variable step size program that is based upon the Klopfenstein modification of classical backward difference formulas called numerical differential formulas (Klopfenstein, 1971). Standard backward difference formulas are also available as an option. In order to determine optimum step size and speed convergence of the implicit corrector formulas, the method depends upon the Jacobian, J, of the derivative function f in

$$\dot{x} = f(x, t) \tag{3.9.1}$$

where

$$J = \frac{\partial f}{\partial x} \tag{3.9.2}$$

The routine ode15s takes the approach of forming a new Jacobian only when the simplified Newton (chord) method is converging too slowly for the implicit corrector formulas. The code works surprisingly well when applied to a non-stiff problem because it forms few Jacobians and the extra linear algebra is performed efficiently. To learn more of the program, use the help ode15s command in MATLAB. The program is available on the world wide web under http://optimal.colorado.edu/~ramirez/chen4580.html in case your version of MATLAB does not yet have this routine. Also a detailed description of the integration formulas used are given in the file odesuite.ps.

A low order option is available as `ode23`. It is based upon a modified Rosenbrock single step formula similar to the Runge–Kutta method except it includes a Jacobian term. The method is therefore considered implicit since it requires the solution of systems of linear equations (Steihaug and Wolfbrandt, 1979).

It should be noted that these two programs `ode15s` and `ode23s` can handle sets of equations with a "Mass Matrix," M

$$M\dot{x} = f(x,t) \qquad (3.9.3)$$

3.10 CATALYTIC FLUIDIZED BEDS

Luss and Amundson (1968) have studied the dynamics of catalytic fluidized beds. The system is a good example of a stiff set of differential equations. Catalytic fluidized beds are utilized for a variety of reactions such as oxidation of naphthalene and ethylene and the production of alkyl chlorides. A batch fluidization reactor is usually built as a cylindrical shell with a support for the catalyst bed. The reactants enter from the bottom through a cone and cause the catalyst particles to be fluidized in the reactor. The reactants leave through a cyclone in which the entrained solids are separated and returned to the bed.

The catalyst particles are assumed to be small enough so that heat and mass transfer resistances can be lumped at the particle surfaces, and the reaction takes place in the porous volume of the catalyst. All particles are assumed to have the same temperature and partial pressure of the reactants. It is also assumed that one irreversible reaction $A \to B$ occurs in the bed. The dynamic model for the interstitial gas includes a mass balance:

$$V \frac{\epsilon \rho_g}{MP} \frac{dp}{dt} = \frac{q}{MP}(p_e - p) + a_v k_g V(p_p - p) \qquad (3.10.1)$$

where p = partial pressure of the reactant in the gas phase
p_e = partial pressure of the reactant at the entrance
p_p = partial pressure of the reactant in the particles
M = molecular weight of the reactant
P = total pressure
V = volume of the bed
ϵ = void fraction of the bed
ρ_g = density of the gas
q = gas mass flow rate
k_g = mass transfer coefficient

The left-hand-side is the rate of accumulation of reactant in the gas phase, the first term on the right-hand-side is the rate into the reactor minus the rate out of the reactor, and the second term on the right-hand-side is the mass transfer rate between the gas phase and the particles.

Unsteady–State Lumped Systems

The gas phase energy balance is

$$V \epsilon \rho_g c_g \frac{dT}{dt} = q c_g (T_e - T) + \frac{V 2\pi r h_w}{\pi r^2} (T_w - T) + a_v h_g V (T_p - T) \quad (3.10.2)$$

where T = gas phase temperature
T_p = particle temperature
T_e = inlet gas phase temperature
T_w = reactor wall temperature
c_g = heat capacity of the gas phase
h_w = heat transfer coefficient between the wall of the reactor and the gas
h_g = heat transfer coefficient between the gas and the catalyst
r = radius of the fluidized bed

The term on the left-hand-side is the rate of thermal accumulation. On the right-hand-side the first term is the sensible energy into the reactor minus the sensible energy out due to flow, the second term is heat transfer between the reactor wall and the gas phase and the third term is the heat transfer between the gas phase and the particles.

The model for the particles includes a mass balance:

$$\alpha v_p \frac{\rho_g}{MP} \frac{dp_p}{dt} = s_p k_g (p - p_p) - v_p \alpha k p_p \quad (3.10.3)$$

where α = void fraction of the particles
v_p = volume of each particle
s_p = area of a particle
k = reaction rate constant

The last term is the reaction rate mechanism. The energy balance gives,

$$v_p \rho_s c_s \frac{dT_p}{dt} = s_p h_g (T - T_p) + (-\Delta H) v_p \alpha k p_p \quad (3.10.4)$$

where c_s = heat capacity of the particles
$(-\Delta H)$ = heat of reaction

The model is simplified by introducing the following dimensionless groups:

$$A = \frac{\alpha v_p a_v}{\epsilon s_p} \qquad C = \frac{a_v c_s v_p \rho_s}{\epsilon s_p c_g \rho_g}$$

$$F = \frac{(-\Delta H) k_g}{h_g} \qquad H_g = \frac{a_v k_g MPV}{q}$$

$$H_T = \frac{a_v h_g V}{q c_g} \qquad H_w = \frac{2 h_w V}{r c_g q}$$

$$K = \frac{\alpha v_p}{s_p k_g} \qquad \tau = \frac{qt}{\epsilon \rho_g V}$$

and the fact that
$$(1-\epsilon)\frac{S_p}{V_p} = a_v$$
where a_v = interfacial area per unit volume.

The model of the fluidized bed therefore becomes

$$\frac{dp}{d\tau} = p_e - p + H_g(p_p - p) \tag{3.10.5}$$

$$\frac{dT}{d\tau} = T_e - T + H_T(T_p - T) + H_w(T_w - T) \tag{3.10.6}$$

$$A\frac{dp_p}{d\tau} = -H_g K k p_p + H_g(p - p_p) \tag{3.10.7}$$

$$C\frac{dT_p}{d\tau} = H_T F K k p_p + H_T(T - T_p) \tag{3.10.8}$$

The reaction rate constant obeys the Arrhenius temperature dependency

$$k = k_0 e^{-\frac{\Delta E}{RT_p}} \tag{3.10.9}$$

where k_0 = the pre/exponential factor
ΔE = the activation energy

Typical model parameters are
$A = .17141$ $H_g = 320$
$C = 205.74$ $H_T = 267$
$F = 8,000$ $H_w = 1.6$
$Kk = .0006 \exp\left(20.7 - \frac{15,000}{T}\right)$

As discussed by Luss and Amundson (1968), this problem has three possible steady-state conditions. The low temperature and high temperature conditions are stable while the intermediate condition is unstable. The equations are very stiff due to the presence of very rapid initial particle dynamic effects. Unless these are computed accurately, the calculations at longer times where the particle partial pressure and temperature nearly equal the gas phase values will be in error. We have developed the m-file ex310.m which solves this problem. The differential equations are given in file model310.m. Figure 3.17a gives the temperature dynamic approach to the low steady-state condition for the gas phase and particle temperatures. Notice that the time is plotted on a log scale in order to illustrate the early transient effects. There is a significant difference in the two temperatures at early times ($< 10^{-2}$ residence times), but after that the temperature differences are quite small. Figure 3.17b gives the partial pressure response curves. Again the differences between the gas and particle phases only exist at early times. It is also interesting to note the nonuniform approach to the steady-state value. Both ode15s and ode23s can be used successfully to compute these response curves with ode15s being slightly faster. The normal integration routines of ode45 and ode113 cannot handle this stiff problem.

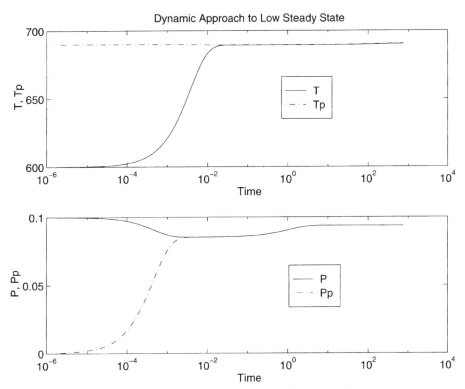

Figure 3.17: Temperature and Partial Pressure Response of Fluidized Bed

PROBLEMS

3.1. A large tank is connected to a smaller tank by means of a valve. The large tank contains nitrogen (assume it is an ideal gas) at 700 kPa, while the small tank is evacuated. The valve between the two tanks starts to leak and the rate of gas leakage is proportional to the pressure difference between the two tanks. Sometimes this relationship is given as

$$\text{Flow} = C_1\sqrt{P_1 - P_2} \tag{1}$$

while others think

$$\text{Flow} = C_1\sqrt{P(P_1 - P_2)} \tag{2}$$

(P = average absolute pressure (not gauge) across the valve) is better for a gas. Why?

The instantaneous initial flow rate is 2.5×10^{-5} kg mol/s.
Tank volumes are 28 m^3 and 14 m^3.
Assume constant temperature of 21°C in both tanks.

168   Computational Methods for Process Simulation

   a. Prepare a *neat* information–flow diagram for this problem. Show your work leading to the model given by this diagram.

   b. Determine how long it would take for the pressure in the small tank to be 0.25, 0.5, and 0.75 of its final value; first using equation (1) for the flow and then using equation (2). Write a MATLAB program to solve this problem.

   c. Are your results a linear function of time? Assuming the second flow equation is correct, how well would you say equation (1) represented the flow situation?

3.2. The dynamics of urea transfer and cerebral pressure: Hemodialysis patients sometimes experience headache, nausea, confusion, and even convulsions during treatment. These symptoms have been referred to as the disequilibrium syndrome, and are believed to be related to an increase in cerebrospinal (CSF) pressure which occurs during rapid dialysis. It has been suggested that this is due to the removal of urea from the blood plasma at a rate greater than it can be removed from the CSF.

The artificial kidney replaces the function of the natural kidney to eliminate waste metabolites from the body. The main waste metabolite is urea.

It is desired to develop a mathematical model of the urea transfer in the body, which includes a CSF compartment for simulating cerebral pressure transients. A three–compartment model can be used as shown in Figure 3.18.

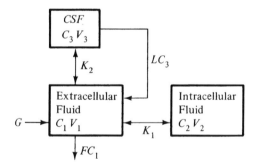

Figure 3.18: Urea Transfer Model.

In this model, there are mass transfer effects between the CSF and extracellular compartments (given by the mass transfer coefficient K_2) and between the extracellular compartment and the intracellular compartment (K_1). G is the natural rate of generation of the contaminant urea, L is the flow rate of CSF fluid that flows directly into the extracellular fluid, and F is the rate of urea removal due to dialysis with an artificial kidney.

Develop the dynamic response of the system and pay particular attention to the CSF pressure response for various removal rates, F. The CSF pressure can be expressed as

$$P = P_0(C_3 - C_1) + \pi$$

where π = normal CSF pressure and P_0 a proportionality constant.

Use the following data obtained from experiments with dogs.

K_1	=	0.15 W, ℓ/hr, where W = body weight of the dog in kg
K_2	=	1.1 x 10^{-4} W, ℓ/hr
P_0	=	84 mm H_2O ℓ/g
G	=	0.2 g/hr
V_1	=	extracellular fluid volume = 17.5 percent of total body weight ($\rho = 1$ g/cc)
V_2	=	intracellular fluid volume = 40.6 percent of total body weight
V_3	=	CSF volume = 0.4 ml/kg of body weight
L	=	Bulk flow term = CSF turnover of seven times daily
W	=	9 kg
$C_1(0)$	=	$C_2(0) = 5$ g/ℓ
$C_3(0)$	=	4.5 g/ℓ

and F is in the range of 0.5 ℓ/hr $< F <$ 5 ℓ/hr (could go even higher).

3.3. The dynamic response to concentration changes of a partly filled horizontal tank: A chemical plant often includes a surge tank between major process units. Such tanks are useful in smoothing out concentration and level changes, thus preventing excessive deviation in feed to subsequent process equipment.

The purpose of this study is to compare normalized concentration responses for models of the relative effects of inlet and outlet nozzles to the ideal case. The models shown in Figure 3.19 have been proposed.

Develop your comparison for a tank 0.43 m in diameter and a residence time of 300 sec. Step changes in inlet concentration are experienced. The initial concentration is zero.

3.4. Reactor Control with Cooling Jacket Dynamics

In the fermentation industry and in industry as a whole, it is often necessary to maintain process vessels at constant temperature. In order to do this, the vessel is equipped with some type of heating/cooling jacket by which the temperature can be controlled. A fully jacketed fermentation reactor is shown in Figure 3.20. Coolant is circulated through this cooling jacket to provide the cooling control for this reactor. This particular reactor was found to be difficult to control. It was suspected that since the cooling jacket capacity was large, it was possible that the control difficulties came from cooling jacket dynamics. Model

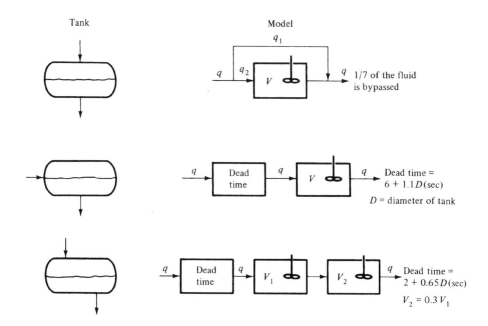

Figure 3.19: Alternative Surge–Tank Models.

the reactor and the jacket as stirred tanks. It can be assumed that the effective bulk jacket temperature is

$$T_c = \frac{T_{\text{in}} + T_{\text{out}}}{2}$$

You can use Newton's law of cooling to describe the heat–transfer process. The heat of fermentation can be measured and its dynamic response is shown in Figure 3.21 for a particular fermentation. The fermentation reactor is run as a batch reactor.

Model the reactor under conventional feedback control of

$$F_c = -K_1(T - T_{\text{set}}) - K_2 \int (T - T_{\text{set}})\, dt$$

Ramirez and Gee (1988) have proposed the following control policy:

$$F_c = -K_1(T - T_{\text{set}}) - K_2 \int (T - T_{\text{set}})\, dt - K_3(T_c - T_p) - K_4 \int (T_c - T_p)\, dt$$

with $K_3 = 0.7146$, $K_4 = 0.7052$, and $T_p = 10°\text{C}$.

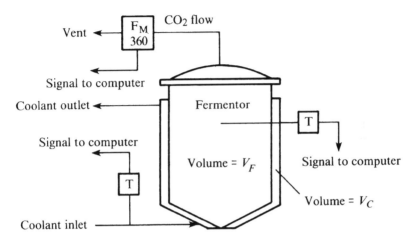

Figure 3.20: Fermentor.

Investigate the advantages of the revised control law if

$$0.1 < K_1 < 100$$
$$0.1 < K_2 < 10$$

Model Parameters

$V_F = 0.235$ m^3 $V_c = 0.032$ m^3
$\rho_F = 1040$ kg/m^3 $\rho_c = 1062$ kg/m^3
$C_{pf} = 4016$ J/kg °C $C_{pc} = 3768$ J/kg °C
$n = 879$ KJ m^2 °C $A_c = 1.746$ m^2
$0 \leq F_c \leq 168$ kg/hr

3.5. A laboratory binary distillation column, consisting of five trays, a total condenser and a reboiler, has been studied by Hu and Ramirez (1972). They have shown that a linearized version of the model describes the process dynamics reasonably well. The linear model is

$$\dot{x} = A\,x + B\,u + W\,d$$

where x represents the state variables of the seven methanol liquid phase deviation mole fractions, u represents the control variables of

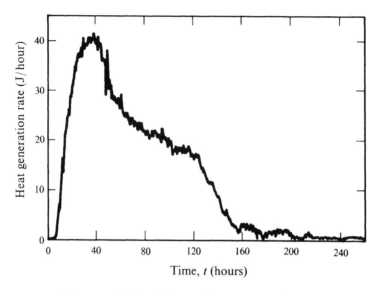

Figure 3.21: **Heat of Fermentation.**

normalized deviations in the reflux ratio and heat duty, and $\boldsymbol{d}$ represents the disturbance variables of the normalized deviations in feed flow (g mol $\min^{-1}$/g mol $\min^{-1}$), feed composition mole fraction, feed temperature deviations (°C/°C), and reflux temperature deviations (°C/°C). The matrix elements are

$$\boldsymbol{A} = \begin{bmatrix} -2.271 & 0.9535 & 0 & 0 \\ 1.449 & -2.428 & 0.9472 & -1.68 \times 10^{-4} \\ -1.37 \times 10^{-3} & 1.387 & -2.217 & 0.9259 \\ -1.85 \times 10^{-3} & -1.71 \times 10^{-2} & 1.1825 & -4.836 \\ -1.37 \times 10^{-4} & -1.17 \times 10^{-3} & 0 & 3.98 \\ -4.88 \times 10^{-4} & -4.52 \times 10^{-3} & 0 & -1.42 \times 10^{-4} \\ -1.89 \times 10^{-5} & -1.75 \times 10^{-4} & 0 & -5.48 \times 10^{-6} \end{bmatrix}$$

$$\begin{bmatrix} 0 & 0 & 0 \\ 2.36 \times 10^{-5} & 2.0 \times 10^{-4} & -1.99 \times 10^{-4} \\ 3.53 \times 10^{-5} & 2.99 \times 10^{-4} & -2.98 \times 10^{-4} \\ 0.9891 & 1.17 \times 10^{-3} & -6.61 \times 10^{-4} \\ -4.944 & 1.007 & 9.9 \times 20^{-6} \\ 3.822 & -4.739 & 1.203 \\ 0 & 4.09 \times 10^{-2} & -3.39 \times 10^{-2} \end{bmatrix}$$

Unsteady–State Lumped Systems

$$B = \begin{bmatrix} 0 & 0 \\ 3.412 \times 10^{-2} & 3.11 \times 10^{-5} \\ 5.105 \times 10^{-2} & 4.654 \times 10^{-5} \\ 6.895 \times 10^{-2} & 2.176 \times 10^{-1} \\ 4.715 \times 10^{-3} & 3.142 \times 10^{-2} \\ 1.817 \times 10^{-2} & 1.21 \times 10^{-1} \\ 7.04 \times 10^{-4} & 4.687 \times 10^{-3} \end{bmatrix}$$

$$W = \begin{bmatrix} 0 & 0 & 0 & 0 \\ 3.12 \times 10^{-5} & -3.85 \times 10^{-7} & -7.99 \times 10^{-7} & -4.19 \times 10^{-3} \\ 4.67 \times 10^{-5} & -5.75 \times 10^{-7} & -1.20 \times 10^{-6} & -6.27 \times 10^{-3} \\ -0.7362 & 2.593 & -2.43 \times 10^{-2} & -8.47 \times 10^{-3} \\ 3.15 \times 10^{-2} & -3.88 \times 10^{-4} & -8.07 \times 20^{-4} & -5.79 \times 10^{-4} \\ 0.1213 & -1.50 \times 10^{-3} & -3.11 \times 10^{-3} & -2.23 \times 10^{-3} \\ 4.70 \times 10^{-3} & -5.80 \times 10^{-5} & -1.21 \times 10^{-4} & -8.63 \times 10^{-5} \end{bmatrix}$$

Note that since the model is in terms of deviation variables, the steady-state condition is $x = u = d = 0$. Compute the dynamic response of the model to ± 10 percent load changes in each disturbance variable. Then compute the worst case 10 percent change in all disturbance variables.

3.6. Develop a MATLAB mfile that implements the Euler numerical integration method. Use your program to solve the enclosed tank problem of Section 3.5. Compare the computational time taken for your program to that of using **ode113**.

3.7. Modify m-file **ex310.m** to study the dynamic approach to the stable high steady state (start the particle temperature above 910°R) and then study the shift from the intermediate steady state to the low temperature steady state for the fluidized bed problem of section 3.10.

	Low Temp. Steady State	Intermediate Steady State	High Temp. Steady State
p	.0936	.0672	.0072
p_p	.0936	.0671	.0069
T	690.27	758.88	911.75
T_p	690.43	759.71	914.07

Discuss your results. Compare **ode15s** and **ode23s**.

3.8. A Parasitic Differential Equation

Study the use of **ode15s** and **ode23s** for the solution to the following deceptively easy differential equation

$$\frac{dx}{dt} = 5(x - t^2) \qquad x(0) = .08$$

Solve for x over the time interval $[0 \quad 5]$.

There is an analytical solution for this equation. Compare your numerical results to the analytic solution. You may have to adjust **rtol** and `atol` to get accurate solutions. Be sure to compute the cpu time needed to solve the equation. Why is this a difficult equation for numerical algorithms? Do you think `ode45` or `ode113` has a chance with this equation?

3.9 Simulation of Liquid Mixing in a Freezer–Crystallizer Vessel

In a secondary refrigerant freezing process for desalination, sea water is partially frozen by direct contact with a vaporizing refrigerant liquid in an agitated vessel. A slurry of ice crystals and brine is formed which is pumped continuously from the vessel to the ice–washing and melting parts of the process. Simulation of the freezing–crystallization step will be performed to assist in the design and scale–up of the process.

Assume that the freezing vessel consists of the following regions:
  a) A primary back–mix region (continuous stirred tank) which receives feed liquid directly from the vessel inlet and discharges exit liquid directly to the vessel outlet. Denote its volume by V_p.

  b) A secondary region which also behaves as an ideal mixer which exchanges liquid with the primary region only. Denote its volume by V_s.

  c) A fluid path, within the vessel, between the vessel inlet and outlet, causes some fluid to bypass the contents and to appear in the outlet immediately after injection.

  A. Develop a mathematical model which describes the dynamic response of the injection of a tracer into this vessel.

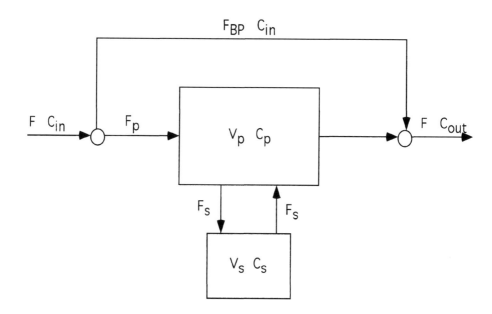

Unsteady–State Lumped Systems

Data: $\dfrac{V_p}{F} = \tau_p = 18.3$ min (primary tank time constant)

$\dfrac{V_s}{F} = \tau_S = 2.7$ min (secondary tank time constant)

$\dfrac{F_{BP}}{F} = f = 0.86$ (by–pass fraction)

$\dfrac{F_s}{F} = f_s = .17$ (fraction secondary flow)

$\dfrac{V_s}{V_p} = .15$

$C_{\text{in}} = 1$ g/l

Solve the problem using MATLAB.

Show the development of your equations.

Present the results graphically.

Discuss the results.

3.10 1) Experimental time versus temperature data has been taken from a jacketed vessel that is heated with steam. Determine a mathematical model for the heating of an initial charge of fluid in this vessel.

2) Develop a MATLAB program for determining the response of this batch system. Run the program for various UA values and compare with experimental data.

3) Can the equation be solved analytically? If yes, perform the calculations. If you have trouble integrating, use the **int** function in MATLAB which is part of the symbolic toolbox.

4) From experimental data (file **expdata.dat** available on the world–wide web), we want to determine the best value of the product of the overall heat transfer coefficient times the heat transfer area (UA). You need to use the **load** command which will give you a two–dimensional array with time and temperature called expdata. Is there a linear plot that allows for the use of linear regression? MATLAB has the function **polyfit** that can perform this fit. What is the optimal vaue of UA for this system? What do you think of the data? MATLAB also has a function **polyval** that allows you to get best predicted values.

5) Additional Data
$$T_{\text{steam}} = 110^\circ C$$
$$\rho V = 59.1 \text{ kg}$$
$$C_p = 1 \text{ cal}/^\circ C \text{ gram} = 4.2 \text{ kJ/kg } ^\circ C$$

REFERENCES

Carnahan, B., Luther, H. A., and Wilkes, J. O., *Applied Numerical Methods*, Wiley, New York (1969).

Franks, R. G., *Mathematical Modeling in Chemical Engineering*, Wiley, New York (1967).

Gear, G. W., *Numerical Initial Value Problems in Ordinary Differential Equations*, Prentice–Hall, Englewood Cliffs, N.J. (1971).

Hu, Y. C. and Ramirez, W. F., "Application of Modern Control Theory to Distillation Columns," *AIChE Journal* **18**, No. 3, 479 (1972).

Hull, T. E., Enright, W. H., and Jackson, K. R., "User's Guide for DVERK," TR No. 100, Dept. of Computer Science, University of Toronto (1976).

Keenan, J. H. and Keyes, F. G., *Thermodynamic Properties of Steam*, Wiley, New York (1959).

Klopfenstein, R. W., "Numerical Differentiation Formulas for Stiff Systems of Ordinary Differential Equations," *RCA Review* **32**, 447-462 (1971).

Luss, D. and Amundson, N. R., "Stability of Batch Catalytic Fluidized Beds," *AIChE Journal* **14**, 211-221 (1968).

Polking, J. C., *Ordinary Differential Equations Using MATLAB*, Prentice Hall (1995).

Ramirez, W. F. and Gee, D. A., "Optimal State Identification and Optimal Control of Batch Beer Fermentation," Proceedings of the ISMM International Symposium, Computer Applications in Design Simulation and Analysis, Honolulu, February (1988).

Shampine, L. F. and Gordon, M. K., *Computer Solutions of Ordinary Differential Equations: The Initial Value Problem*, W. H. Freeman, San Francisco (1975).

Steihaug, T. and Wolfbrandt, A., "An Attempt to Avoid Exact Jacobian and Non-linear Equations in the Numerical Solution of Stiff Differential Equations," *Math. Comp.* **33**, 521-534 (1979).

Chapter 4

REACTION-KINETIC SYSTEMS

In this chapter we consider the simulation of reaction-kinetic systems. The following examples illustrate the modeling of various reaction systems.

4.1 CHLORINATION OF BENZENE

The chlorination of benzene produces monochlorobenzene, dichlorobenzene, and trichlorobenzene through the successive reactions

$$C_6H_6 + Cl_2 \xrightarrow{k_1} C_6H_5Cl + HCl$$

$$C_6H_5Cl + Cl_2 \xrightarrow{k_2} C_6H_4Cl_2 + HCl$$

$$C_6H_4Cl_2 + Cl_2 \xrightarrow{k_2} C_6H_3Cl_3 + HCl$$

These reactions are carried out in a lead-lined or iron vessel, as shown in Figure 4.1. Ferric chloride ($FeCl_3$) is used as a catalyst. The vessel is fitted with cooling coils, since the reactions are exothermic. There is a reflux condenser, which returns vaporized chlorobenzenes to the system, while allowing the hydrogen chloride (HCl) and excess chlorine vapor to leave the system. In order to maintain the reacting mixture at a uniform temperature and to minimize mass transfer effects, the reacting mixture is well agitated. The amount of the chlorine gas which dissolves in the liquid phase is limited by the solubility of chlorine in the reacting mixture.

The following assumptions can be made:

1. There is no liquid or vapor hold–up in the reflux condenser, i.e., no dynamics are involved.
2. The system operates under isothermal and isobaric conditions.
3. Volume changes in the reacting mixture are negligible.

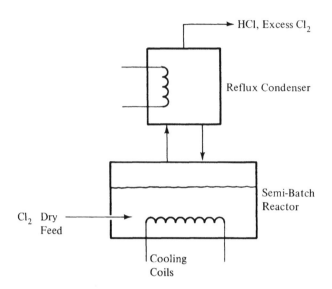

Figure 4.1: Reactor for Chlorination of Benzene.

4. Hydrogen chloride vaporizes and leaves the system.
5. There is negligible mass transfer resistance between the gaseous Cl_2 and the Cl_2 in solution, i.e., the Cl_2 goes immediately into solution up to the solubility limit.

Usually what is required is to know the time for (1) maximizing monochlorobenzene, (2) maximizing dichlorobenzene, and (3) maximizing trichlorobenzene.

We will assume that the feed rate of the dry chlorine is 1.4 kg mol of chlorine per hour per kg mol of initial benzene charge. The following rate constants are estimated values for the catalyst used at 55°C (assumed for this problem):

$$k_1 = 510 \text{ (kg mol/m}^3)^{-1} \text{ (hr)}^{-1}$$

$$k_2 = 64 \text{ (kg mol/m}^3)^{-1} \text{ (hr)}^{-1}$$

$$k_3 = 2.1 \text{ (kg mol/m}^3)^{-1} \text{ (hr)}^{-1}$$

There is negligible liquid or vapor hold–up in the reflux condenser. Volume changes in the reacting mixture are negligible, and the volume of liquid in the reactor remains constant at 1.46 m³/kg mol of initial benzene charge.

Hydrogen chloride has a negligible solubility in the liquid mixture. The chlorine gas fed to the system goes into the liquid solution immediately up to its solubility limit of 0.12 kg mol of chlorine per kg mol of original benzene and this value then remains constant. Each reaction is second–order as written.

The material balance for benzene is

$$\frac{dN_B}{dt} = \frac{-k_1 N_B N_C}{V} \qquad (4.1.1)$$

where N_B = moles of benzene
N_C = moles of chlorine
V = volume of reactor

The material balance for monochlorobenzene is

$$\frac{dN_M}{dt} = \frac{k_1 N_B N_C}{V} - \frac{k_2 N_M N_C}{V} \qquad (4.1.2)$$

where N_M = moles of monochlorobenzene
The material balance for dichlorobenzene is

$$\frac{dN_D}{dt} = \frac{k_2 N_M N_C}{V} - \frac{k_3 N_D N_C}{V} \qquad (4.1.3)$$

The material balance for trichlorobenzene is

$$\frac{dN_T}{dt} = \frac{k_3 N_D N_C}{V} \qquad (4.1.4)$$

The material balance for chlorine is

$$\frac{dN_C}{dt} = F - \frac{k_1 N_B N_C}{V} - \frac{k_2 N_M N_C}{V} - \frac{k_3 N_D N_C}{V} \qquad (4.1.5)$$

There is a maximum chlorine concentration of

$$N_{C,\text{max}} = 0.12 \, N_{B0} \qquad (4.1.6)$$

where N_{B0} = number of moles of benzene charged (which is 50 kg mol). Figure 4.2 presents the information–flow diagram for this system.

4.1.1 Order of Magnitude Analysis for Chlorination of Benzene

If the chlorination of benzene example is to be solved using an IMSL integration routine, then the describing equations should first be order-of-magnitude scaled. To do this we want to create dimensionless variables of order one. For this problem we can choose a characteristic concentration of N_{B0} and a characteristic time associated with the primary

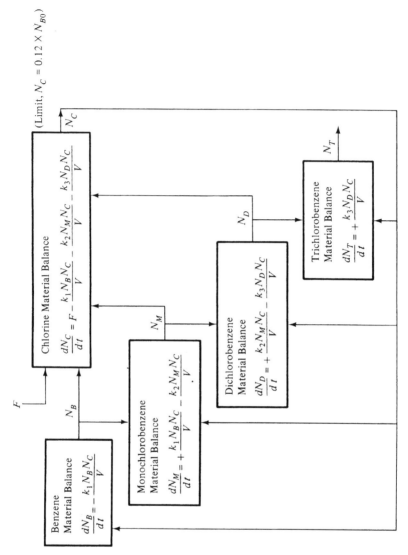

Figure 4.2: Information–Flow Diagram for Chlorination of Benzine.

Reaction–Kinetic Systems

reaction rate $\tau = \dfrac{V}{k_1}$. With these characteristic values the dimensionless variables of the problem are

$$N_B^* = \frac{N_B}{N_{B0}} \qquad N_T^* = \frac{N_T}{N_{B0}}$$

$$N_M^* = \frac{N_M}{N_{B0}} \qquad N_C^* = \frac{N_C}{N_{B0}}$$

$$N_D^* = \frac{N_D}{N_{B0}} \qquad t^* = \frac{t}{\tau}$$

Introducing these dimensionless variables, the describing differential equations become

$$\frac{dN_B^*}{dt^*} = -N_B^* \, N_C^* \tag{4.1.7}$$

$$\frac{dN_M^*}{dt^*} = N_B^* \, N_C^* - \frac{k_2}{k_1} N_M^* \, N_C^* \tag{4.1.8}$$

$$\frac{dN_D^*}{dt^*} = \frac{k_2}{k_1} N_M^* \, N_C^* - \frac{k_3}{k_1} N_D^* \, N_C^* \tag{4.1.9}$$

$$\frac{dN_T^*}{dt^*} = \frac{k_3}{k_1} N_D^* \, N_C^* \tag{4.1.10}$$

$$\frac{dN_C^*}{dt^*} = \frac{FV}{k_1} - N_B^* \, N_C^* - \frac{k_2}{k_1} N_M^* \, N_C^* - \frac{k_3}{k_1} N_D^* \, N_C^* \tag{4.1.11}$$

with

$$N_{C_{\max}}^* = .12 \frac{N_{B0}}{N_{B0}} \tag{4.1.12}$$

Substituting the given values into equations (4.1.7)–(4.1.12) gives the following:

Equation (4.1.7) stays the same, as

$$\frac{dN_B^*}{dt^*} = -N_B^* \, N_C^* \tag{4.1.13}$$

This implies that all terms in the benzene disappearance equation are of equal importance. Equation (4.1.8) becomes

$$\frac{dN_M^*}{dt^*} = N_B^* \, N_C^* - .1255 \, N_M^* \, N_C^* \tag{4.1.14}$$

This implies that the disappearance term due to the monochlorobenzene is approximately one order–of–magnitude less important than the generation term due to benzene. Both terms should be retained in the differential equation. Equation (4.1.9) becomes

$$\frac{dN_D^*}{dt^*} = 0.1255\ N_M^*\ N_C^* - 0.004118\ N_D^*\ N_C^* \qquad (4.1.15)$$

or

$$7.97\ \frac{dN_D^*}{dt^*} = N_M^*\ N_C^* - 0.0328\ N_D^*\ N_C^* \qquad (4.1.16)$$

Comparing equation (4.1.16) to (4.1.13) or (4.1.14) shows that the dichlorobenzene response is approximately one order-of-magnitude slower than that for the benzene consumption or monochlorobenzene generation response. Equation (4.1.10) becomes

$$\frac{dN_T^*}{dt^*} = 0.004118\ N_D^*\ N_C^* \qquad (4.1.17)$$

or

$$243\ \frac{dN_T^*}{dt^*} = N_D^*\ N_C^* \qquad (4.1.18)$$

This implies that the trichlorobenzene response is approximately two orders-of-magnitude slower than that for monochlorobenzene.

The problem is now ready for simulation. All terms should be retained. Depending upon the information desired, the problem should be run between 10 to 100 dimensionless time units.

A MATLAB program ex41.m has been created to solve this problem. The differential equations are defined in file model41.m. Results of running the problem are shown in Figure 4.3. The maximum amount of monochlorobenzene occurs at 0.78 hours and that for dichlorobenzene at 1.75 hours.

4.2 AUTOCATALYTIC REACTIONS

Autocatalysis is a term commonly used to describe the experimentally observable phenomenon of a homogeneous chemical reaction which shows a marked increase in rate with time, reaches its peak at about 50 percent conversion, and then drops off. The temperature has to remain constant and all ingredients must be mixed at the start for proper observation.

Selected for mathematical analysis is the catalytic thermal decomposition of a single compound A into two products B and C, of which B is the autocatalytic agent. Thus, A can decompose via two routes, a slow uncatalyzed one (k_1) and another catalyzed by $B(k_3)$. The three essential kinetic steps are

$$A \xrightarrow{k_1} B + C \quad \text{Start or background reaction} \qquad (4.2.1)$$

$$A + B \xrightarrow{k_2} AB \quad \text{Complex formation} \qquad (4.2.2)$$

$$AB \xrightarrow{k_3} 2B + C \quad \text{Autocatalytic step} \qquad (4.2.3)$$

Reaction–Kinetic Systems

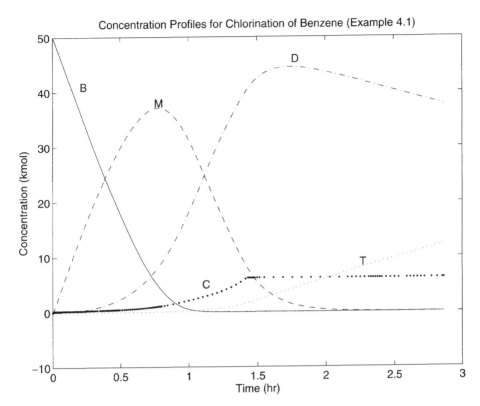

Figure 4.3: Results for Simulation of Chlorination of Benzene. Example 4.1.

These equations show that the autocatalytic agent B forms a complex, AB, with rate k_2. Next, the complex AB decomposes with rate k_3, thereby releasing B in addition to forming B and C.

Reactions k_2 and k_3 together form the path by which most of A decomposes. Reaction k_1 is the starter, but continues concurrently with k_2 and k_3 as long as there is any A.

The rates of formation of A, B, C, and AB, in accordance with the kinetic steps, are

$$r_1 = -k_1 C_A \tag{4.2.4}$$

$$r_2 = -k_2 C_A C_B \tag{4.2.5}$$

$$r_3 = -k_3 C_{AB} \tag{4.2.6}$$

where C_A, C_B, and C_{AB} are molar concentrations.

The basic material balances can be derived for scaled variables. The dimensionless variables are

$$C^* = \frac{C}{C_{A_0}}, \quad t^* = tk_1 \qquad (4.2.7)$$

and the material balance equations are

$$\frac{dC_A^*}{dt^*} = -C_A^* - \left(\frac{k_2}{k_1}C_{A_0}\right) C_A^* C_B^* \qquad (4.2.8)$$

$$\frac{dC_B^*}{dt^*} = C_A^* - \left(\frac{k_2 C_{A_0}}{k_1}\right) C_A^* C_B^* + \left(\frac{2k_3}{k_1}\right) C_{AB}^* \qquad (4.2.9)$$

$$\frac{dC_{AB}^*}{dt^*} = \left(\frac{k_2 C_{A_0}}{k_1}\right) C_A^* C_B^* - \left(\frac{k_3}{k_1}\right) C_{AB}^* \qquad (4.2.10)$$

$$\frac{dC_C^*}{dt^*} = C_A^* + \left(\frac{k_3}{k_1}\right) C_{AB}^* \qquad (4.2.11)$$

The information–flow diagram is given in Figure 4.4. It is expressed in terms of the two dimensionless variables α and β, defined as

$$\alpha = k_3/k_1 \qquad (4.2.12)$$
$$\beta = k_2 \, C_{A_0}/k_1 \qquad (4.2.13)$$

For $\beta = 10^4$, it has been observed that α, called the degree of autocatalysis, can be categorized as

$\alpha = 1$ to 10, the reaction is barely noticeably autocatalytic

$\alpha = 10$ to 100, the reaction is mildly autocatalytic

$\alpha = 100$ to 1000, the reaction is strongly autocatalytic

Typical results for $\alpha = 8$ are shown in Figure 4.5. This system is a mildly stiff set of differential equations and the use of ode23s results in a computational time that is approximately one order of magnitude faster than using the new version of ode45. The code is available using ode23s as ex42.m using the differential equations defined in model42.m.

4.3 TEMPERATURE EFFECTS IN STIRRED TANK REACTORS

The system to be considered is a continuous stirred tank reactor (CSTR), which is used for the decomposition of hydrogen peroxide. The reaction is first–order and homogeneously catalyzed. The general schematic is given in Figure 4.6. The reactant, hydrogen peroxide (H_2O_2), and the catalyst, sodium hydroxide, are fed into the reactor, while the product, water, and

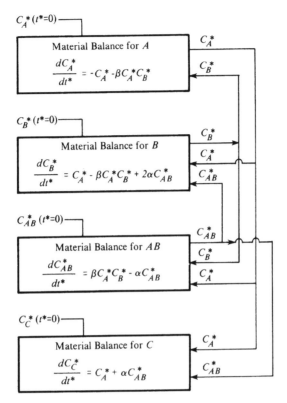

Figure 4.4: Information–Flow Diagram for Autocatalytic Reaction.

unreacted peroxide are continuously removed. The exothermic heat of reaction is continuously removed through a cooling water jacket which surrounds the reactor.

The following mass and energy balances describe the CSTR.

$$\rho V c_p \frac{dT}{dt} = HVkC - UA(T - T_A) - \rho F_0 c_p (T - T_0) \qquad (4.3.1)$$

$$V \frac{dC}{dt} = -VkC - F_0(C - C_0) \qquad (4.3.2)$$

where H = heat of reaction (cal/mol)
k = reaction rate constant
C = concentration of peroxide (mol/vol)
U = heat–transfer coefficient
A = heat–transfer area
T_A = average coolant temperature

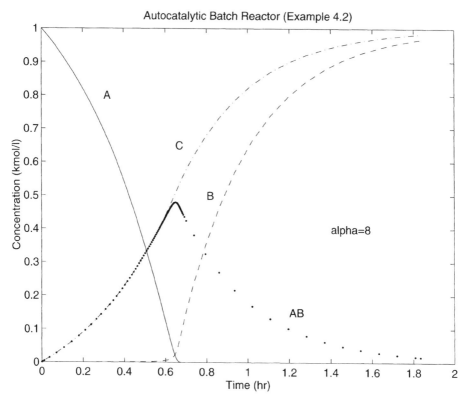

Figure 4.5: Autocatalytic Batch Reactor.

$F_0 =$ flow rate out of the reactor (vol/time)
$T_0 =$ inlet feed temperature
$C_0 =$ inlet peroxide concentration (mol/vol)

The heat removal driving force term involving $(T - T_A)$ is modeled by assuming that the cooling water temperature in the jacket is an arithmetic average of the input and output cooling temperatures. Employing this assumption, a steady-state energy balance across the cooling water jacket gives

$$(T - T_A) = \frac{T - T_{\text{in}}}{1 + F} \tag{4.3.3}$$

where
$\quad F \quad = UA/2Q_c\rho c_p$
$\quad T_{\text{in}} =$ inlet coolant temperature $\tag{4.3.4}$
$\quad Q_c \quad =$ coolant flow rate (mass/time)

For laminar flow in the coolant jacket, the overall heat-transfer coefficient

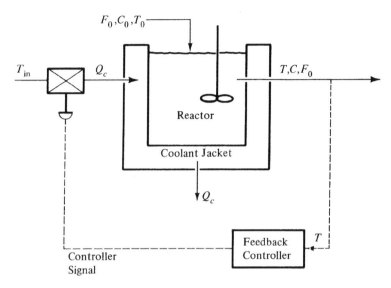

Figure 4.6: Reactor System Flow Sheet.

is related to the coolant mass flow rate according to (Perry, 1984)

$$U = C_1 Q_C^{1/3} \qquad (4.3.5)$$

where C_1 = constant.

The functional relationship between the mass flow rate of coolant and the controller output is usually given by the controller manufacturer, or can be determined experimentally. Figure 4.7 gives the control valve characteristic curve determined for this system. The fraction controller output is related to the proportional gain through

$$C' = \frac{G - G_{\min}}{G_{\max} - G_{\min}} \qquad (4.3.6)$$

$$G = K(T - T_{ss})C_2 + C_3 \qquad (4.3.7)$$

where G = the controller signal (ma)
$G_{\max}$ = maximum controller signal (ma)
$G_{\min}$ = minimum controller signal (ma)
K = proportional feedback controller gain
C_2 = unit conversion constant for controller, ma/°K
C_3 = steady–state controller output (ma)

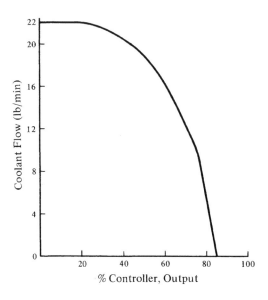

Figure 4.7: Control Valve Characteristic.

The reaction rate constant follows the Arrhenius temperature dependency of

$$k = k_0 e^{-E/T} \qquad (4.3.8)$$

where k_0 = the frequency factor, a constant
E = the activation energy

An approximate curve fit to the control valve characteristic curve of Figure 4.7 is

If $C' \geq 7.20$ then $Q_c = 22$ (lb/min)
If $C' < 20$ then $Q_c = -.00512\,(C' - 20)^2 + 22$ (lb/min) $\qquad (4.3.9)$

Also, the steady-state model for the cooling jacket is only valid for positive flows through the cooling jacket. Therefore the additional constraint

If $C' > 85.56$ then $Q_c = 0.01$ (lb/min) $\qquad (4.3.10)$

is added.

The file ex43.m solves this set of differential equations using ode23s. The model differential equations are defined in m-file model43.m. The program ex43a.m solves the problem using the stiff equation solver ode15.s.

Both stiff routines are much more efficient at solving this problem than the normal routines `ode23` or `ode45`.

The information–flow diagram for this system is given in Figure 4.8. Figures 4.9 and 4.10 show the phase plane plots for two values of the proportional gain K, for the system studied by Ramirez and Turner (1969). The phase plane representation involves plotting the two state variables, one as a function of the other with the independent variable, time, a parameter along the phase plane curves. Figures 4.9 and 4.10 show that the system without control is not stable, while the system with feedback control is stable (returns to the desired steady state).

A steady–state analysis can also predict the stability of a chemical reactor. Such an analysis is carried out by plotting the heat production and heat removal as a function of reactor temperature. Usually the heat production curve is sigmoid while that for heat removal is linear. Any intersection of the two curves results in a steady–state condition. Figure 4.11 shows that three intersections are possible (Van Heerden, 1953). If the slope of the heat production curve is greater than that of the heat removal curve, then the steady state is unstable. This is illustrated by point B in Figure 4.11. If the slope of the heat production curve is less than that of the heat removal curve, then the steady state is stable. This is illustrated by points A and C.

The heat production curve for this reactor is not sigmoid in shape as in the general theory because boiling occurs before the upper half of the curve can be completed. The heat removal curve intersects the heat production curve only at 30°C. Therefore, there was only one steady-state point for this system rather than the three which can be generally expected. Figure 4.12 shows that the steady state is stable for negative temperature perturbations and unstable for positive perturbations. This explains the run–away behavior of the reactor without control. The heat production and heat removal plot of Figure 4.12 was obtained using the MATLAB file `stable.m` wihch calls the file `qc.m`.

Other stability analysis techniques are discussed by Ramirez and Turner (1969). They include the eigenvalues of a linear analysis which give results that only apply to a small region about a steady-state value, and Liapunov's direct method via Kravoskii's theorem which accurately predicted regions of stability with and without feedback control for this nonlinear problem.

4.3.1 Mathematical Modeling of a Laboratory Stirred Tank Reactor

We desire to model a laboratory stirred tank reactor system shown in Figure 4.13 (Nyquist and Ramirez, 1971).

4.3.1.1 *Experimental*

The laboratory reactor is a Pyrex glass vessel to which an overflow spout has been added such that the liquid volume, with material overflowing,

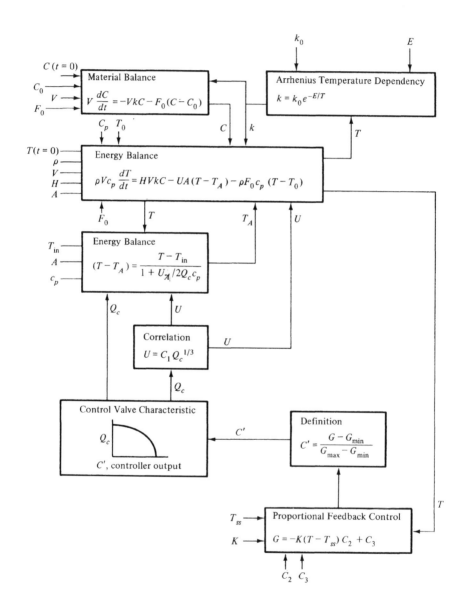

Figure 4.8: Information–Flow for Reactor with Temperature Effects.

Reaction–Kinetic Systems

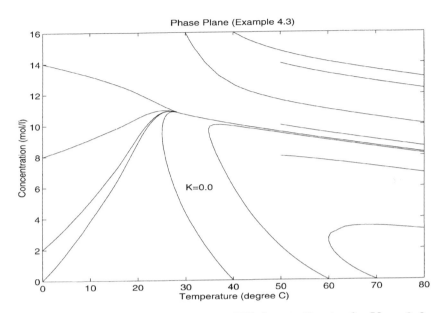

Figure 4.9: Phase Plane Without Control. $K = 0.0$.

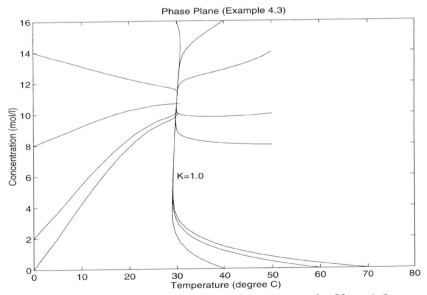

Figure 4.10: Phase Plane with Control. $K = 1.0$.

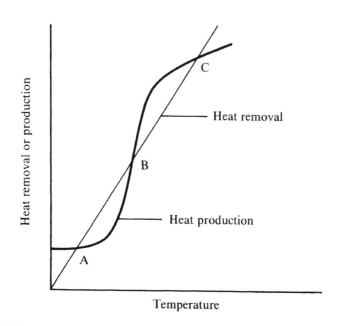

Figure 4.11: Steady–State Stability Analysis.

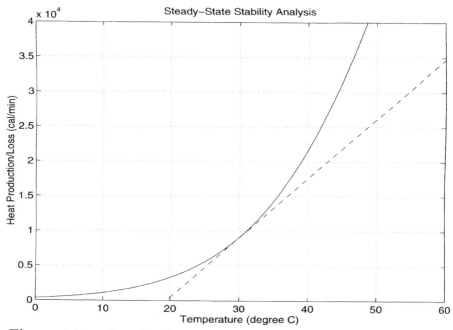

Figure 4.12: Steady–State Stability Analysis (Real System).

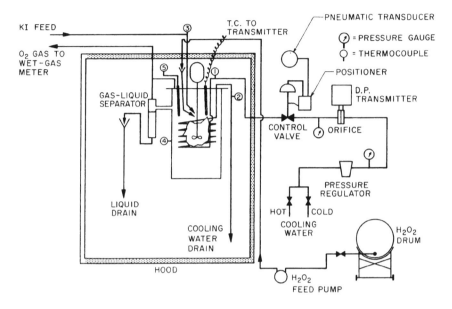

Figure 4.13: Laboratory Reactor System.

is approximately 16 ℓ. This vessel is hung from a gas-tight aluminum lid which is supported on a metal frame in the laboratory hood. A stirrer is mounted above the reactor and a gas-sealed shaft passes into the vessel to provide the required agitation. A stainless steel coil is supported by four polyethylene baffles inside the reactor. The connecting lines for this coil pass through the reactor lid. The sides of the reactor are wrapped with one half inch glass wool insulation. A liquid-gas separation chamber is attached to the overflow spout.

Both the H_2O_2 and the KI feeds enter the reactor through polyethylene lines terminating near the impeller. The H_2O_2 is metered directly from the shipping container with a positive displacement variable flow pump. It flows through a small funnel mounted above the reactor lid so that the actual flow and composition can be precisely sampled at the point of entry into the reactor. The catalyst is fed by a recycle feed system. Pressure for this feed system is provided by a small centrifugal pump which has a back-pressure valve on its outlet such that part of the material passes through a flow indicator on the control panel and on to the reactor while the remainder is recirculated to the feed tank.

The temperature and hydrogen peroxide concentration in the reactor are the state variables of interest. The temperature is measured by an iron constantan thermocouple, protected from corrosion by a thin polyethylene shield, mounted from the reactor lid. This m.V. signal is converted to m.A. signal with a potentiometer transmitter. The transmitter is adjusted for a temperature range of 20–40°C to produce a 4–20 m.A. signal at the output of the transmitter, which is then recorded. A voltage signal (0.5–2.5V, d.c.) is also available across the input terminals of the recorder.

The transmitter slowly drifts in its calibration but is constant during the period of a few hours. A second thermocouple (T.C. 5) is also mounted in the reactor and may be independently monitored on a laboratory potentiometer. Thus, the calibration of the temperature transmitter can be easily established at any time.

The hydrogen peroxide concentration in the reactor is not directly measured. Rather, the reaction rate is measured by continuously monitoring the oxygen gas evolution rate and converting this signal into the rate of hydrogen peroxide reacted per liter of reactor volume. The hydrogen peroxide concentration can thus be determined from the first–order reaction kinetic relation,

$$C = \frac{\text{Rate}}{k(T)} \qquad (4.3.11)$$

since the temperature is known.

Four thermocouples are available to measure other temperatures. These are the inlet cooling water (T.C. 1), the outlet cooling water (T.C. 2), the inlet KI feed stream (T.C. 3), and the outside reactor wall (T.C. 4) temperatures.

The cooling water flow rate is continuously indicated and can be recorded at the control panel. A differential pressure transmitter is mounted across an orifice in the water line feeding the control valve. This transmitter is adjusted to have a 4–20 m.A. output over the flow range. A 150 Ω resistor is placed across the output terminals to produce a corresponding voltage of 0.6–3.0 V.

The cooling water flow is controlled by a 1/2 in. air to close, pneumatic valve equipped with a valve positioner to overcome the inherent valve hysteresis. The pneumatic signal to the valve positioner is obtained from an electro–pneumatic transducer. A 4–20 m.A. signal is produced by placing a 5000 Ω resistor across a d.c. signal which can vary between 20 and 100 V.

Temperature control of the reactor is achieved by varying the analog signal to the valve controlling the cooling water flow. A simple proportional integral controller was used in the experimental model determination.

4.3.1.2 *Modeling*

A mathematical model of the experimental system is derived. The sys-

Reaction–Kinetic Systems

tem material balances and energy balance are presented. The experimental definitions of the control valve flow characteristics, heat–transfer coefficient, heat of stirring, and reaction rate follow.

The stoichiometry of the reaction for which the following balances are developed is given by

$$H_2O_2 \rightarrow H_2O(\ell) + \frac{1}{2} O_2(g) \tag{4.3.12}$$

The material balance for the hydrogen peroxide under the condition of constant volume is

$$\frac{dC}{dt} = \frac{F_p C_0}{V_x} - \frac{FC}{V_x} - R \tag{4.3.13}$$

The material balance for the iodide ion is

$$C_i = \frac{(F - F_p)}{F} C_{i0} \tag{4.3.14}$$

An oxygen balance is also required. The stoichiometry indicates 1/2 mole of oxygen results when one mole of hydrogen peroxide reacts. The reactor liquid is assumed saturated by oxygen so that the oxygen balance is the algebraic expression.

$$R = 2 F_{02}/V_x \tag{4.3.15}$$

Thus, the reaction rate can be directly monitored by measuring the rate of oxygen evolution.

The general energy balance over the reactor is

$$\frac{d}{dt}\left[V_H \rho \int_{T_0}^{T} C_p(T)dT\right] = (F-F_p)\rho_{KI} \int_{T_0}^{T_{KIi}} C_{pKI}dT + F_p \rho_p \int_{T_0}^{T_{pi}} C_{pP}dT$$

$$- F\rho \int_{T_0}^{T} C_p dT + Q_s + (-H)RV_x - Q_c \tag{4.3.16}$$

Here V_H includes an additional volume due to the mass of the reactor itself in addition to the liquid volume in the reactor. The heat capacity C_{pKI} is that of the catalyst, potassium iodide, and C_{pP} is that of the reactant, hydrogen peroxide.

Assuming that the heat of dilution is negligible, and that the heat capacity, the heat of reaction, and the density are constant over the 20°C range of the experiments and the inlet temperatures of KI and H_2O_2 are the same, the energy balance becomes

$$V_H \rho C_p \frac{dT}{dt} = F\rho C_p (T_i - T) + Q_s + (-H) RV_x - Q_c \tag{4.3.17}$$

The control valve flow characteristics and the functionality between the flow through the control valve and the valve signal voltage were determined experimentally. The voltage is applied across a resistor to the transducer, which produces a pneumatic signal to a valve positioner. The positioner regulates the pressure to the actuator to position the valve stem. Data for the valve flow characteristics as a function of the valve signal V_s was fit to the lowest–order polynomial giving a multiple correlation factor coefficient greater than 0.999. The resulting fourth–order polynomial is

$$F_c = 2.02 - 1.089 \times 10^{-2}\, V_s - 1.028 \times 10^{-3}\, V_s^2 \\ + 1.738 \times 10^{-5}\, V_s^3 - 8.03 \times 10^{-8}\, V_s^4 \tag{4.3.18}$$

The heat–transfer coefficient as a function of coolant flow (F_c) was also determined experimentally. A fifth–order polynomial with a multiple correlation coefficient of 0.9975 was fit to the data and is

$$UA = 4.905 \times 10^3\, F_c - 9.75 \times 10^3\, F_c^2 + 1.11 \times 10^4\, F_c^3 \\ - 6.316 \times 10^3\, F_c^4 + 1.415 \times 10^3\, F_c^5 \tag{4.3.19}$$

In order to define the heat transferred from the cooling coil, it is not convenient to use Newton's law of cooling directly since, because of the log mean temperature term, the exiting coolant temperature is implicit rather than explicit. By introducing an "effectiveness factor" concept, it is possible to directly calculate the energy transferred as

$$Q_c = \epsilon q C_{pc}(T - T_{ci}) \tag{4.3.20}$$

where ϵ is the effectiveness factor.

The derivation of the effectiveness factor expressions for various heat exchange situations can be found in Kays and London (1955) and the specific application to a CSTR in Aris (1969). For a coil in a stirred vessel

$$\epsilon = 1 - e^{-\text{NTU}} \tag{4.3.21}$$

where

$$\text{NTU} = \frac{UA}{qC_{pc}} \tag{4.3.22}$$

Since the heat–transfer coefficient is a function of coolant flow, then ϵ must also be a function of coolant flow. We then define a function β as

$$\beta = \epsilon q C_{pc} \tag{4.3.23}$$

so that equation (4.3.20) becomes

$$Q_c = \beta(T - T_{ci}) \tag{4.3.24}$$

Since the functionalities of UA with F_c and F_c with V_s have been determined, it is possible to compute β as a function of the valve signal, V_s, and then fit a simple polynomial to the resulting calculated points. The result is given as a third-order expression with a multiple correlation coefficient of 0.99898:

$$\beta = 1.51\times 10^3 - 27.26\ V_s + 0.1499\ V_s^2 - 2.38\times 10^{-4}\ V_s^3 \qquad (4.3.25)$$

The reactor is equipped with a relatively powerful stirrer so the conversion of mechanical energy to thermal energy is a necessary consideration. This conversion was established from a simple experiment suggested by the energy balance with only the heat of stirring important.

$$\frac{dT}{dt} = Q_s/V_h \rho C_p \qquad (4.3.26)$$

The reactor was cooled and then allowed to warm by the energy supplied from the stirrer. The slope of the temperature rise versus time curve was determined at various temperatures and was fit to the linear relation

$$\frac{dT}{dt} = 7.40\times 10^{-2} - 1.559\times 10^{-3}\ T \qquad (4.3.27)$$

with a root mean squared error of 0.000875.

For this calculation the equivalent volume for heat transfer also includes the equivalent volume of the coil in addition to the glass reactor and its contents. Combining equations (4.3.26) and (4.3.27) gives

$$Q_s = 1.736 \times 10^{+4} \left(7.42 \times 10^{-2} - 1.559 \times 10^{-3}\ T\right) \qquad (4.3.28)$$

Baxendale (1952), Liebhafsky (1932), and Liebhafsky and Mohammad (1933) have discussed and studied the kinetics of the reduction, in acid solution, of hydrogen peroxide by iodide ion. This decomposition is of the general form

$$\text{Reduced catalyst} + H_2O_2 \rightarrow \text{Oxidized catalyst} \qquad (4.3.29)$$

$$\text{Oxidized catalyst} + H_2O_2 \rightarrow \text{Reduced catalyst} + O_2 \qquad (4.3.30)$$

These two reactions add to the overall reaction

$$2H_2O_2 \rightarrow 2H_2O + O_2 \qquad (4.3.31)$$

The first of these reactions is thought to be represented by the two simultaneous and independent reactions,

$$I^- + H_2O_2 \rightarrow IO^- + H_2O \qquad (4.3.32)$$

$$I^- + H^+ + H_2O_2 \rightarrow HIO + H_2O \qquad (4.3.33)$$

The second reaction, equation (4.3.30), is

$$IO^- + H_2O_2 \rightarrow I^- + H_2O + O_2 \qquad (4.3.34)$$

Consideration must also be given to the equilibrium relation

$$2HIO \rightleftarrows 2I^- + H_2O + \frac{1}{2}O_2 \qquad (4.3.35)$$

When equations (4.3.32) and (4.3.33) are the rate controlling steps

$$R = 2k_0[I^-][H_2O_2] + 2k_1[H^+][I^-][H_2O_2] \qquad (4.3.36)$$

Liebhafsky and Mohammad have established that only the first term on the right of equation (4.3.36) is important under the condition of low hydrogen ion concentration. The hydrogen peroxide used in these experiments was DuPont's "Perone^{R}30," which is buffered to a pH range of 3.4–3.8, so the low hydrogen ion concentration situation applies. Therefore, the reaction rate is

$$R = kC_I C \qquad (4.3.37)$$

Batch runs were made at constant temperature and are presented in Figures 4.14 and 4.15 for two different drums of peroxide. An Arrhenius plot, Figure 4.16, gives the temperature dependency of the rate constant

$$k = 5.317 \times 10^9 \, \exp\left[-\frac{6.525 \times 10^3}{T + 273.16}\right] \qquad (4.3.38)$$

The information–flow diagram for the complete model is given in Figure 4.17. Three experimental runs are presented in Figures 4.18 to 4.20 along with the predicted response.

The first two consider step changes in reactor temperature set point for two different proportional gains. All changes were made from a steady–state operating condition. The final case presented is just stopping the feeds after steady state has been achieved.

The agreement between the model and these experimental runs is excellent, as is clear from the figures. Table 4.1 gives the parameters used for the simulation results presented in Figure 4.19. The poorest agreement is in Figure 4.18. It should be noted that both the temperature and the rate simulation are below the observed data. This indicates a consistency in the model and probably says the differences are the result of some error in determining the initial operating condition for this run. The model of Figure 4.17 is considered verified by these results.

The m-file `labreact.m` has been developed to solve for the dynamic response of this system. The differential equations are defined in file `labmodel.m`. The non-stiff equation solvers work well for this problem.

4.3.2 Dynamics of Batch Fermentation

It is common in industrial fermentations to operate at an experimentally determined optimal growth temperature for the yeast being used. Depending upon how the fermentation progresses, the temperature can be

Reaction–Kinetic Systems

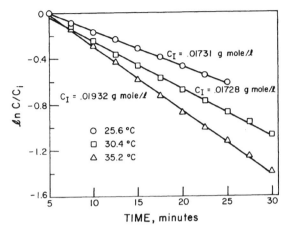

Figure 4.14: Batch Kinetic Data.

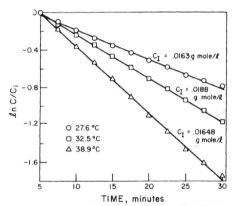

Figure 4.15: Batch Kinetic Data.

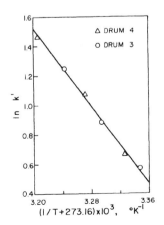

Figure 4.16: $\ln k'$ vs. $1/T + 273.16$.

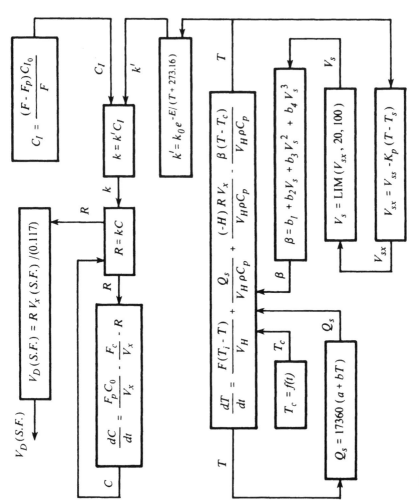

Figure 4.17: Information–Flow Diagram Complete System Model.

Reaction–Kinetic Systems

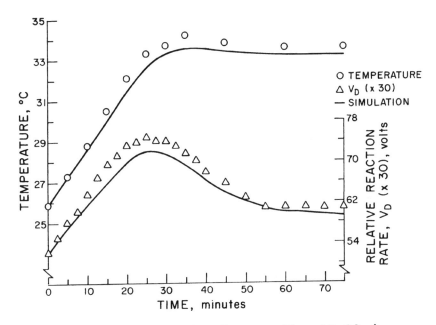

Figure 4.18: Reaction System. Run 11–02–A.

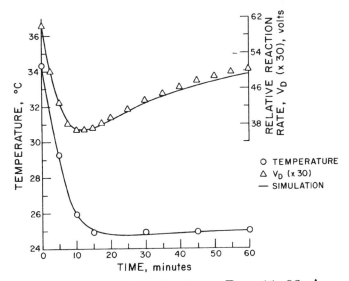

Figure 4.19: Reaction System. Run 11–03–A.

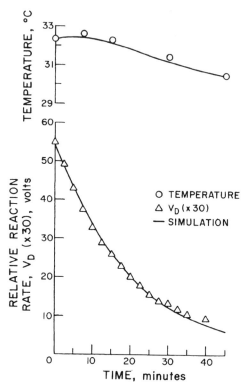

Figure 4.20: Reaction System (batch). Run 11–05–B.

raised (i.e., to speed up a slow fermentation) or lowered (i.e., to offset the formation of undesirable by-products) to keep the fermentation on schedule and to maintain the desired product quality.

In this section, we develop a simplistic model of beer fermentation. This model is based upon the work of Engasser (1981) and Gee and Ramirez (1987). The fermentation medium is assumed to contain three major sugars. These are glucose, maltose, and maltotriose and are the limiting nutrients. The fermentation is carried out in a batch reactor which can be considered to be ideally mixed. The material balance relations for each sugar are

Table 4.1: Parameters Used in Simulation of Run 11–03–A

Operating Parameters	
$V_H = 16.74 \; \ell$	$T_i = 24.5°$
$V_x = 15.78 \; \ell$	$T_s = 25.0°C$
$F = 0.1312 \; \ell/\text{min}$	$V_{ss} = 60 \; V$
$F_p = 0.032 \; \ell/\text{min}$	$K_p = 15 \; V/°C$
$-H = -22600 \; \text{cal/g mol} \; H_2O_2$	$T(t=0) = 34.33°C$
$C_0 = 8.53 \; \text{g mol}/\ell$	$C(t=0) = 0.310 \; \text{g mol}/\ell$

Inlet Water Temperature

t (min)	T_c (°C)
0	10.0
15	10.9
30	12.0
45	11.8
60	12.2

Glucose
$$\frac{dG}{dt} = -\mu_1 X \qquad (4.3.39)$$

Maltose
$$\frac{dM}{dt} = -\mu_2 X \qquad (4.3.40)$$

Maltotriose
$$\frac{dN}{dt} = -\mu_3 X \qquad (4.3.41)$$

where G = glucose concentration, g mol/m³
G = maltose concentration, g mol/m³
N = maltotriose concentration, g mol/m³
X = biomass (yeast) concentration, g mol/m³

The parameters μ_i are the specific reaction rates for the uptake of sugar i. These reactions are first–order in the biomass concentration. The specific rates must reflect the catalyzed enzymatic reactions that occur within yeast cells that convert sugar into alcohol. These rates are given by the Monod expression for glucose,

$$\mu_1 = \frac{V_G G}{K_G + G} \qquad (4.3.42)$$

a Monod form for maltose with inhibition due to the presence of glucose,

$$\mu_2 = \frac{V_M M}{K_M + M} \frac{K'_G}{K'_G + G} \qquad (4.3.43)$$

and the Monod form for maltotriose with inhibition due to the presence of both glucose and maltose

$$\mu_3 = \frac{V_N N}{K_N + N} \frac{K'_G}{K'_G + G} \frac{K'_M}{K'_M + M} \quad (4.3.44)$$

where V_i = maximum reaction velocity for the ith sugar, hr^{-1}
K_i = the Michaelis constant for the ith sugar, g mol/m^3
K'_1 = the inhibition constant for the ith sugar, g mol/m^3

The rates of biomass and ethanol are related to the consumption of the individual sugars by constant yield coefficients

$$\frac{dX}{dt} = -R_{XG}\frac{dG}{dt} - R_{XM}\frac{dM}{dt} - R_{XN}\frac{dN}{dt} \quad (4.3.45)$$

$$\frac{dE}{dt} = -R_{EG}\frac{dG}{dt} - R_{EM}\frac{dM}{dt} - R_{EN}\frac{dN}{dt} \quad (4.3.46)$$

where R_{Xi} = stoichiometric yield of biomass per mole of sugar reacted
R_{Ei} = stoichiometric yield of ethanol per mole of sugar reacted

These two equations, (4.3.43) and (4.3.44), can be analytically integrated to give

$$X(t) = X(t_0) + R_{XG}\Big(G(t_0) - G(t)\Big) + R_{XM}\Big(M(t_0) - M(t)\Big)$$
$$+ R_{XN}\Big(N(t_0) - N(t)\Big)$$
$$(4.3.47)$$

$$E(t) = R_{EG}\Big(G(t_0) - G(t)\Big) + R_{EM}\Big(M(t_0) - M(t)\Big)$$
$$+ R_{EN}\Big(N(t_0) - N(t)\Big)$$
$$(4.3.48)$$

An energy balance on the batch reactor gives

$$\frac{dT}{dt} = \frac{1}{\rho C_p}\left[\Delta H_{FG}\frac{dG}{dt} + \Delta H_{FM}\frac{dM}{dt} + \Delta H_{FN}\frac{dN}{dt} - \frac{UA}{V}(T - T_C)\right]$$
$$(4.3.49)$$

where ΔH_{Fi} = heat of fermentation of sugar i, J/mol
U = overall heat–transfer coefficient, J/hr m^2 °C
A = heat–transfer area, m^2
V = volume of fermenting mixture, m^3
T_c = coolant temperature, °C
C_p = mixture heat capacity, J/kg °C

The reaction rate constants follow the Arrhenius temperature dependency:

$$V_i = v_{i0} \exp\left(-E_{Vi}/R(T+273)\right) \quad (4.3.50)$$

$$K_i = k_{i0} \exp\left(-E_{Ki}/R(T+273)\right) \quad (4.3.51)$$

$$K'_i = k'_{i0} \exp\left(-E_{K'_i}/R(T+273)\right) \quad (4.3.52)$$

We will assume that we can control this fermenter using the feedback control law

$$u = K_p(T - T_{\text{set}}) \quad (4.3.53)$$

where

$$u = \frac{UA}{V} \quad (4.3.54)$$

We now perform a numerical simulation for the specific data given in Table 4.2. The MATLAB program ex432.m is used to solve this problem. The differential equations are defined in model432.m The old Runge-Kutta algorithm ode45 is used to solve this non-stiff problem. The error tolerance needs to be reduced to 10^{-9}. Results of the simulation include the sugar response curves of Figure 4.21. Glucose is consumed first, followed by maltose. The consummation of maltotriose is the slowest. Figure 4.22 shows the temperature response and Figure 4.23 the control action. While the temperature is below the set point temperature of 12°C, the coolant capacity remains at its minimum value of zero. This allows for the temperature of the reactor to increase due to the heat of fermentation. Once the set point temperature is reached, coolant begins to flow through the cooling jacket as prescribed by the cooling capacity response curve (Figure 4.23). This results in the fermentor temperature remaining at its desired value. Figure 4.24 shows the ethanol and biomass response curves. The transition in the ethanol production rate after glucose is consumed is obvious. Gee and Ramirez (1988) have used this model to investigate optimal control laws for batch fermentations, and Ramirez (1987) has discussed optimal state and parameter identification for this system.

Gee and Ramirez (1994) have extended this model and have also developed a complete flavor model for beer fermentation.

Table 4.2: Data

Physical Parameters

Sugar	R_{Ei}	R_{Xi}	ΔH_{Fi} (J/gmole)
Glucose (G)	1.92	0.134	-91.3×10^3
Maltose (M)	3.84	0.268	-226.3×10^3
Maltotriose (N)	5.76	0.402	-361.3×10^3

Arrhenius Constants

Parameter	Activation Energy (cal/g mol)	Natural Log of Frequency Factor
V_G (1/hr)	22.6×10^3	35.77
V_M (1/hr)	11.3×10^3	16.40
V_N (1/hr)	7.16×10^3	10.50
K_G (g mol/m^3)	-68.6×10^3	-121.30
K_M (g mol/m^3)	-14.4×10^3	-19.15
K_N (g mol/m^3)	-19.9×10^3	-26.78
K'_G (g mol/m^3)	10.2×10^3	23.33
K'_M (g mol/m^3)	26.3×10^3	55.61

Initial Conditions and Equipment Parameters

$G(t_0) = 70$ g mol/m^3
$M(t_0) = 220$ g mol/m^3
$N(t_0) = 40$ g mol/m^3

$X(t_0) = 175$ g mol/m^3
$T(t_0) = 8°C$

$A = 0.188$ m^2 (heat–transfer area)
$C_p = 4016$ J/kg °C
$\rho = 1040$ kg/m^3
$V = 0.1$ m^3

$K_p = 1 \times 10^6$ J/hr °C^2 m^3
$T_{\text{set}} = 12°C$
$u_{\min} = 0$ J/hr °C m^3
$u_{\max} = 40000$ J/hr °C m^3
$T_{\text{coolant}} = 0°C$

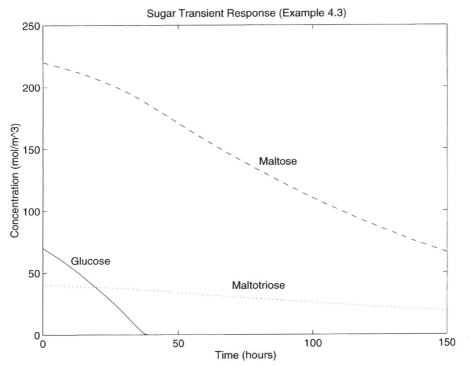

Figure 4.21: Sugar Transient Response.

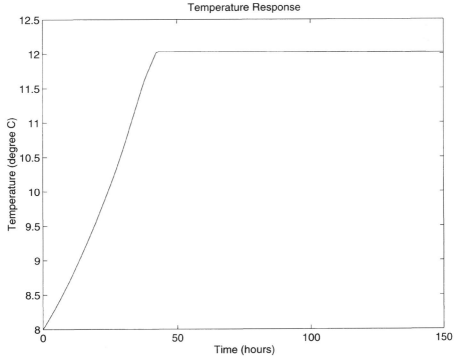

Figure 4.22: Temperature Response.

208   Computational Methods for Process Simulation

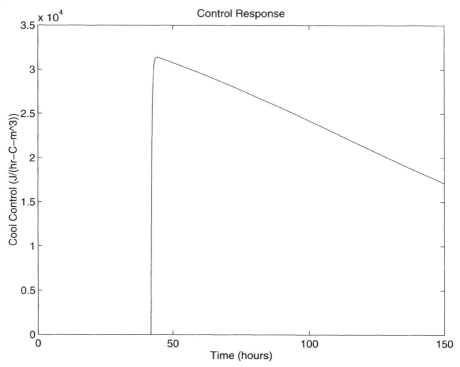

Figure 4.23: Control Response.

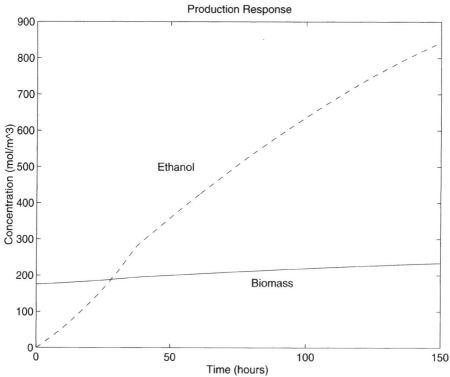

Figure 4.24: Production Response.

PROBLEMS

4.1. Develop the dynamic response for the batch autocatalytic reactions of Section 4.1 when
$$k_1 = 10^{-6}, \text{ sec}^{-1}$$
$$\beta = 10^4, \text{ } \ell/\text{mol}$$

and for the cases when

$$\alpha = 10 \qquad \alpha = 100 \qquad \alpha = 500$$

The initial concentration of A is 1 mol/ℓ.

4.2. Develop the phase planes for the exothermic reactor of Section 4.3 for the following set of numerical values:

A	=	3.18 ft^2	$G_{\max}$	=	20 ma
C_0	=	11.5 g mol/ℓ	H	=	22.6 K cal/mol
C_2	=	0.1068 ma/°K	T_0	=	22°C
C_1	=	62.5 (Btu/hr ft^2/°F)/ (lb/min)$^{1/3}$	T_{ss}	=	30°C
C_p	=	0.865 cal/°kg	T_{in}	=	10°C
F_0	=	0.5 ℓ/min	V	=	16.2 ℓ
$G_{\min}$	=	4 ma	ρ	=	1.1081 g/cc

These phase planes should be obtained for values of proportional feedback gain, K from 0 to 1. (*Note*: The steady–state value of the concentration and controller output must be determined before the dynamic simulation study is performed.)

4.3. In the last few years, it has been observed that sewage water has "self–purification" properties because there are small amounts of oxygen in solution with the water, which tend to oxidize and remove the contaminants. Research work done in this area has given industry some insight as to what the reaction mechanisms might be, but nobody has determined the limits of this method of purification and what might be done to improve it. You are to model this process, determine a few of its limitations, and tell how it can be changed to improve its purification potential.

Consider two mechanisms at this level of accuracy: (i) The reaction mechanism: The mass concentration of the oxygen demand of sewage particles S (mg/ℓ) as a function of time can be described as a first–order rate expression:

$$dS/dt = -K_1 S$$

where dS/dt is the rate of oxygen consumption. (ii) The oxygen mass transfer mechanism: The rate of reoxygenation of the water by aeration can be assumed to be linearly proportional to the oxygen deficit D (mg/ℓ). The oxygen deficit is defined as the mg/ℓ of oxygen needed to bring the inlet water up to its oxygen saturation point at a certain temperature. (Use K_2 as the proportionality constant.)

a. Write a computer program that will simulate the reaction reoxygenation process. You should be able to graph S and D as a function of time using the computer.

b. Using the following data, determine the maximum initial concentration of sewage (mg/ℓ) that will give a maximum oxygen deficit of 15 mg/ℓ. At what point in time does this maximum occur?

$$\text{Data: initial } \begin{aligned} D &= 3 \text{ mg}/\ell \\ K_1 &= 0.75/\text{day} \\ K_2 &= 0.35/\text{day} \end{aligned}$$

c. In some cases it might not be possible to lower the initial concentration of the sewage to the value you estimated in the previous part. What changes in equipment, conditions, or parameters could be made to overcome this restriction? (Hint: Among other things, consider changes in parameters or concentrations in your rate equation.)

4.4. An important industrial enzyme reaction is the isomerization of d–glucose to d–fructose catalyzed by glucose isomerase in solution. You are to model a batch reactor for carrying out this isomerization.

$$G \underset{k_2}{\overset{k_1}{\rightleftarrows}} F$$

The rate of reaction per unit volume is given by $r = r_0 a$, where a is the catalyst activity which decays with time as $da/dt = -k_0 a$ with an initial activity $a_0 = 1$. The rate parameter r_0 is described by

$$r_0 = (k_1 + k_2)G - k_2 G_0$$

where G is the glucose concentration in g mol/ℓ and G_0 is the initial glucose concentration.

a. Develop the dynamic equations for this system.
b. Develop an order–of magnitude set of equations for solution.
c. Develop the computer solution.

Basic Data

$$\begin{aligned} G_0 &= 2.8 \text{ g mol}/\ell & k_1 &= 7.11 \times 10^{-3} \text{ sec}^{-1} \\ k_0 &= 2.765 \times 10^{-2} \text{ sec}^{-1} & k_2 &= 5.06 \times 10^{-3} \text{ sec}^{-1} \end{aligned}$$

4.5. Develop an order-of-magnitude solution for the dispersion of atmospheric pollutants. The most commonly used models of atmospheric dispersion from continuous sources are the Gaussian plume models. For an infinite-line source such as might be used to simulate automotive emissions on a freeway, the model takes the form

$$X = \frac{q(A+B)}{\sqrt{2\pi}\,\sin\phi\,\sigma_z v}$$

$$\frac{dA}{dz} = -\frac{(z-H)A}{\sigma_z^2} \qquad A(z=0) = \exp\left(-\frac{H^2}{2\sigma_z^2}\right)$$

$$\frac{dB}{dz} = -\frac{(z+H)B}{\sigma_z^2} \qquad B(z=0) = \exp\left(-\frac{H^2}{2\sigma_z^2}\right)$$

where
X = concentration of pollutants (g/m^3)
σ_z = vertical dispersion coefficient (m)
q = rate of emission per unit length (g/m sec)
ϕ = angle between source line and wind direction
v = wind speed (m/sec)
z = vertical distance above ground (m)
H = effective height at which emissions occur (m)

Typical Values

σ_z	= 10 m	$\sin\phi$	= 0.94
v	= 4.44 m/sec	$z_{\max}$	= 50 m
H	= 1 m	q	= 10^{-4} g/m sec

4.6. Park and Ramirez (1989) have studied the dynamics of foreign protein secretion from yeast saccharomyces. In the development of genetically engineered organisms for the production of foreign proteins, it is useful to use the yeast secretion pathways. The production of secreted proteins offers advantages of initial protein separation and many naturally secreted proteins cannot be fully activated unless they are processed through the secretion pathway. A schematic picture of a yeast cell is shown in Figure 4.25. The secretion pathway starts with initial processing in the endoplasmid reticulum (ER) and then is continued in the golgi. From the golgi mature protein is secreted to the fermentation broth through the periplasma. Park and Ramirez have proposed the following mathematical model for the process of secretion dynamics:

$$\frac{dP_M}{dt} = \frac{\beta}{\alpha + \mu_x}\left[\mu_x(P_T - P_M) + \gamma f_p X\right]$$

$$\frac{dP_T}{dt} = f_p X$$

$$\frac{dX}{dt} = \mu_x X$$

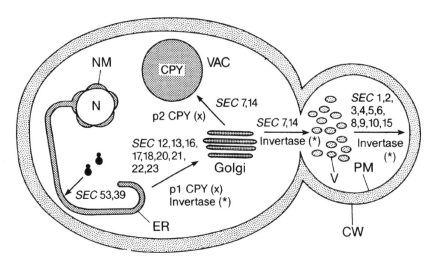

Figure 4.25: Yeast Cell.

where P_M = protein concentration in the broth medium
P_T = total protein
X = biomass concentration
μ_x = specific growth rate
f_p = synthesis rate
α, β, γ = constants

In order to test this model, experiments were performed that add cycloheximide to a fermentation to stop protein synthesis at the ribosomal level but not fully inhibiting other cellular activity.

Simplify the model under these conditions. Express your results in terms of the total internal protein concentration. Park and Ramirez obtained the experimental data given in Figure 4.26.

- • $\mu_x = 0$ ○ $\mu_x = .4$ △ $\mu_x = .1$ □ $\mu_x = .27$

Is the general model consistent with these data? What are specific reaction rate constants for this set of data?

4.7. For the decomposition of hydrogen peroxide reaction studied by Nyquist and Ramirez (1971), a second control variable is possible. This is the flow rate of potassium iodide catalyst solution to the reactor. Modify the process model to allow for a variable catalyst flow rate and therefore a variable catalyst concentration. The reaction rate is modeled as

$$R = k_0 e^{-E/RT} C_{KI} C$$

Solve for the steady–state condition when

Reaction–Kinetic Systems

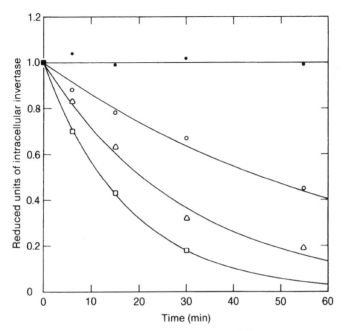

Figure 4.26: Experimental Data.

$$
\begin{aligned}
F_p &= \text{750 ml/min peroxide flow rate} \\
C_0 &= \text{2.25 g mol/}\ell \text{ inlet peroxide concentration} \\
V_X &= \text{reactor volume} = 15.78\ \ell \\
k_0 &= 9.775 \times 10^9 \\
E/R &= \text{6524 cal/g mol} \\
F_{KI} &= \text{100 ml/min catalyst flow rate} \\
V_H &= \text{energy reactor volume} = 17.36\ \ell \\
T_0 &= \text{coolant temperature} = 25°\text{C} \\
Q_s &= a + bT \quad a = 1284.6\ \text{cal/min} \\
    & \quad b = -23.61\ \text{cal/min}\ °\text{C} \\
\beta &= 122.1 \\
\rho &= 1\ \text{g/cc} \\
C_p &= 1\ \text{cal/g}\ °\text{C} \\
(-\Delta H) &= 22600\ \text{cal/gmol}
\end{aligned}
$$

Investigate the dynamic response of the system to ± 10 percent changes in the two control variables β and F_{KI}. What conclusions can you draw about these control variables?

4.8. We want to develop a model for a slurry reactor. This is a stirred tank reactor with suspended catalyst particles. The rate of reaction is first–order and obeys the Arrhenius temperature dependency. The reaction is endothermic. The thermal energy balance must consider both the catalyst and fluid phases; however, the heat transfer between the phases is rapid so that both the catalyst and fluid can be assumed to be at the same temperature. The catalyst activity decays as a first-order system

$$\tau \dot{\alpha} = -\alpha \qquad \tau = \text{time constant for catalyst decay}$$

The reaction rate expression is

$$\begin{aligned}
r &= \alpha k_0\, W_c C e^{-E/T} \\
k_0 &= \text{frequency factor} \\
W_c &= \text{weight of catalyst} \\
C &= \text{concentration} \\
E &= \text{activation energy}
\end{aligned}$$

Perform an order-of-magnitude analysis for this system and solve for the dynamic response for a 20 percent change in inlet flow, q. You are interested in the response over the time scale of one day at maximum. System parameters are

τ	Time constant of the activity (86400 sec)
α	Activity of catalyst (dimensionless, initially set to 1)
γ	Void fraction of catalyst bed (0.36)
C_{p_c}	Heat capacity of catalyst phase (0.2 cal/g°C)
ρ_c	Density of catalyst phase (2.4 g/cm^3)
ρ	Density of fluid phase (6.2 × 10^{-4} g/cm^3)
C_p	Heat capacity of fluid phase (0.57 cal/g°C)
V	Volume of reactor (1000 cm^3)
T	Temperature of reactor inlet T_0 600°C
q	Flow rate of fluid through reactor (1040 cm^3/hr)
ΔH	Heat of reaction (28080 cal/mol)
k_0	Arrhenius rate constant (0.368 hr^{-1})
W_c	Weight of catalyst bed (864 g)
C	Concentration (initial value C_0 2.92 × 10^{-6})
E	Activation energy of reaction (9323°K)

Now perform a simulation when the time scale is one year.

4.9. The fermentation problem of section 4.3.2 used the old ode45 routine to solve for the dynamic response with a very tight error tolerance. Investigate the use of the new ode45, ode113, ode23s, and ode15s for solving this problem. What error criteria are needed and which routine is most efficient for solving the problem?

4.10. We wish to determine the local stability of the CSTR system of section 4.3. To do this, we first linearize the nonlinear problem about the steady-state condition. You might want to use the symbolic toolbox of MATLAB to help in this linearization. This converts the problem to a linear set of differential equations

$$\dot{x} = Ax + b$$

In order to be stable, the eigenvalues of A must be negative. Compute the eigenvalues of the A matrix obtained by linearization. Is the steady state without feedback control locally stable?

Now redo the linearization adding feedback control. For a value of $K = 1$, is the system locally stable? What is the minimum value of K required for local stability? Determine if the nonlinear system is actually stable for this value of K by developing a phase-plane plot using various starting conditions.

4.11 Vleeschhouwer et al. (*AIChE Journal* **34**, No. 10, 1736 (1988)) discuss the transient behavior of a chemically reacting system in a CSTR. The chemical reaction they used to study transient behavior is the acid catalyzed hydration of oxiranemethanol to glycerol.

   a. Verify their mathematical model.

   b. Simulate the dynamic oscillatory response of the system approaching an arbitrarily stable steady state. Present results in a phase plane plot (Figure 4). Solve using a stiff differential equation algorithm.

4.12 Shanks and Bailey (*AIChE Journal* **33**, No. 12, 1971 (1987)) present a dynamic model for oxidation of CO over supported silver.

   a. Verify that for the reaction mechanisms proposed, the dynamic model of Equations 1–6 is valid.

   b. Solve the model for a step change of 1% CO and 10% O_2 to 10% CO and 10% O_2 using the best set of model parameters (Figure 1). Graphical presentation of the results and interpretation are necessary. In order to arrive at accurate starting steady–state values, run the problem using estimates of the steady state with input compositions of 1% CO and 10% O_2 until the initial steady state is obtained. Do your simulations agree with the authors'?

REFERENCES

Aris, R., *Elementary Reactor Analysis*, Second Ed., Prentice–Hall, Englewood Cliffs, N.J. (1969).

Baxendale, J. H., *Advances in Catalysis*, Vol. 4, Academic Press, New York (1952).

Engasser, J. M., Marc, I., Moll, M., and Duteurtre, B., *EBC Congress*, 579–583 (1981).

Gee, D. A. and Ramirez, W. F., "Optimal Temperature Control for Batch Beer Fermentation," *Biotech. Bioeng.* **31**, 224–234 (1988).

Gee, D. A. and Ramirez, W. F., "A Flavour Model for Beer Fermentation," *J. Inst. Brew* **100**, 321-329 (1994).

Kays and London, *Compact Heat Exchangers*, National Press (1955).

Liebhafsky, H. A., *J. Am. Chem. Soc.* **54**, 1792 (1932).

Liebhafsky, H. A. and Mohammad, A. J., *J. Am. Chem. Soc.* **55**, 3977 (1933).

Nyquist, J. K. and Ramirez, W. F., "Time Optimal Control of an Experimental Continuous Stirred Tank Reactor," *Chem. Engr. Sci.* **26**, No. 10, 1673 (1971).

Park, S. and Ramirez, W. F., "Dynamics of Heterologous Protein Secretion from *Saccharomyces cerevisiae*," *Biotech. Bioeng.* **33**, 272–281 (1989).

Perry, R., *Perry's Chemical Engineers' Handbook*, Sixth Ed., McGraw-Hill, New York (1984).

Ramirez, W. F., "Optimal State and Parameter Identification: An Application to Batch Fermentation," *Chem. Eng. Sci.* **42**, 2749–2756 (1987).

Ramirez, W. F. and Turner, B. A., "The Dynamic Modeling, Stability, and Control of a Continuous Stirred Tank Chemical Reactor," *AIChE Journal* **15**, No. 6, 853 (1969).

Van Heerden, C., *Ind. Eng. Chem.* **45**, 1242 (1953).

Chapter 5

VAPOR–LIQUID EQUILIBRIUM OPERATIONS

Chemical engineers have used the concept of vapor–liquid equilibrium for much of their treatment of separation processes such as distillation, absorption, and stripping. In this chapter, we examine the dynamic modeling and numerical solution of typical vapor–liquid equilibrium systems.

5.1 BOILING IN AN OPEN VESSEL

The phenomenon of the boiling of a pure fluid provides a situation in which the vapor pressure of the fluid tends to exceed the total pressure of the system. This gives rise to a stream of vapor rising from the liquid. It is important to realize that although heat is added to a boiling pure fluid, the temperature remains constant at the boiling temperature. The boiling point temperature is only a function of the system pressure. Heat supplied to the system goes into determining the boil–off rate.

Mechanistically, then, the system pressure establishes the temperature, and the heat flux to the system establishes the vapor boil–off rate.

Figure 5.1 shows an open system in which the fluid is boiling. The energy balance for this system is

$$Q = v\lambda \tag{5.1.1}$$

where Q = rate of heat supplied to the system
v = vapor boil–off rate, mass/time
λ = heat of vaporization

Also, the constraint on the system, namely that the fluid is boiling, states that

$$T = f(P) \tag{5.1.2}$$

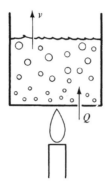

Figure 5.1: Boiling in an Open Vessel.

where P = system pressure

The information–flow diagram for this system is given in Figure 5.2. Note that the energy equation is used to compute the boil–off rate and the temperature is determined by the system pressure.

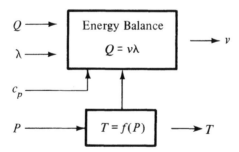

Figure 5.2: Information–Flow Diagram of Boiling in an Open System.

5.2 BOILING IN A JACKETED VESSEL (BOILER)

Most industrial boilers are enclosed vessels, as shown in Figure 5.3. In order to model this system, the following material and energy balances are used.

Vapor–Liquid Equilibrium Operations

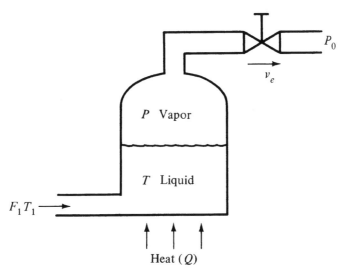

Figure 5.3: Boiling in a Jacketed Vessel.

Energy Balance (Liquid Phase):

$$Q = v\left(\lambda + c_{pL}(T - T_r)\right) - \rho F_1 c_{pL}(T_1 - T_r) \quad (5.2.1)$$

where Q = rate of heat supplied to boiler
F_1 = volumetric flow rate of liquid into boiler
λ = heat of vaporization at the system temperature
T_r = reference temperature
c_{pL} = liquid heat capacity

Mass Balance (Liquid Phase):

$$d\rho V_L/dt = \rho F_1 - v \quad (5.2.2)$$

Energy Balance (Vapor Phase; assume equilibrium):

$$T_L = T_G = T \quad (5.2.3)$$

Mass Balance (Vapor Phase):

$$dm/dt = v - v_e \quad (5.2.4)$$

where m = mass in vapor phase
v_e = mass flow of vapor through exit valve

Equation of State (Vapor Phase):

$$PV_G = \frac{mRT}{MW} \tag{5.2.5}$$

Size Relation:

$$V = V_L + V_G \tag{5.2.6}$$

where V = volume of the system.

Valve Equation:

A steady–state mechanical energy balance around the valve results in the following algebraic relation for flow of a gas across a valve (see Problem 3.1):

$$v_c = K_v\sqrt{P(P - P_o)} \tag{5.2.7}$$

The information–flow diagram of this system is given in Figure 5.4.

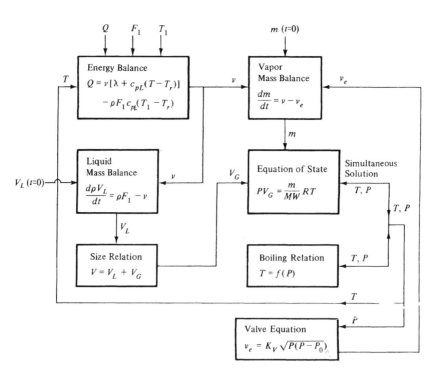

Figure 5.4: Information–Flow Diagram of a Boiler.

Vapor–Liquid Equilibrium Operations

It should be noted that the boiling relation and the equation of state both involve the temperature and pressure as unknowns. These two nonlinear algebraic equations must therefore be solved simultaneously.

EXAMPLE 5.1 Dynamics of a Steam–Power Boiler: Let us solve the following boiler dynamic response problem. The data for a steam–power boiler installation for a chemical plant are given in Table 5.1. The plant normally runs with a downstream pressure of 165 psia. Determine the unsteady–state response to an opening in the take–off valve that results in a downstream pressure of 150 psia. The boiling relation is given in the steam tables by Keenan and Keyes (1959) as

$$\log_{10} \frac{218 \text{atm}}{p \text{ atm}} = \frac{x}{499} \left[\frac{3.346 + 4.14 \; 10^{-2} \; x}{1 + 1.38 \; 10^{-2} \; x} \right] \quad (5.2.8)$$

where $x = 647.27 - T$ (°K).

Table 5.1: **Parameters for Boiler Example**

Variable	Value	Units
Q	20×10^3	Btu/hr
λ	853	Btu/lb
c_p	1	Btu/lb°F
T_r	32	°F
ρ	62.4	lb/ft^3
F_1	35.8	ft^3/hr
T_1	70	°F
$V_L(0)$	100	ft^3
V	200	ft^3
$m(0)$	20	lbs
MW	18	
R	10.73	psia ft^3/lb mol°R
K_v	6.03	lb$_m$/hr psia

MATLAB file `ex51.m` (Figure 5.5a) has been created to solve for the dynamic response of this system. The problem is somewhat stiff so the integration routine `ode15s` is used. Program `model51.m` (Figure 5.5b) defines the differential equations. It should be noted that all the algebra is calculated in this program including the nonlinear simultaneous solution for the system temperature and pressure using function `fsolve`. The nonlinear algebraic equations are defined in function `eq51.m` (Figure 5.5c).

Figure 5.5a: MATLAB program ex51.m to solve for boiler dynamics using ode15s.

```
% This file solves the boiler problem in section 5.2;
%
tt = cputime;
 global m Vl TK Patm;

m = 20; Vl = 100;        % initial conditions
TK = 380; Patm = 8;      % initial guess of T(K) and P(atm)

Q = 20*10^3;
lambda = 853;
Cpl = 1;
Tr = 32;
rho = 62.4;
F1 = 35.8;
Kv = 6.03;
P0 = 150;
T1 = 70;


        c0 = [m Vl]';
        [t, c] = ode15s('model51', [0, 0.4], c0);
time = cputime-tt

        N = length(c);

% prepare results for plotting

for I = 1:N,
   Vl = c(I,2);
   m = c(I,1);
   mplot(I) = m;
   x = fsolve('eq51',[TK,Patm]');
   TK = x(1);
   Patm = x(2);
   TF = (x(1)-273.15)*1.8+32;
   TFplot(I) = TF;
   Ppsia = x(2)*14.7;
   Ppsiaplot(I) = Ppsia;
   Vlplot(I) = Vl;
   v(I) = (Q + rho*F1*Cpl*(T1-Tr))/(lambda + Cpl*(TF-Tr));
   if Ppsia < P0
     ve(I) = 0;
    else
     ve(I) = Kv*sqrt(Ppsia*(Ppsia - P0));
    end
end


plot(t, Vlplot, '-', t, Ppsiaplot, '--', t, v, '*');
axis([0 0.4 80 160]);
```

Vapor–Liquid Equilibrium Operations

```
title('Boiler Dynamic Response (Example 5.2)');
xlabel('Time (hr)');
text(0.24, 150, 'P(psia)');
text(0.24, 115, 'Vl(ft^3)');
text(0.24, 95, 'v(lbs/hr)');
print -deps fig56.eps;
pause;
plot(t, mplot);
title('Vapor Mass Dynamic Response')
xlabel('Time (hr)');
ylabel('Vapor Mass, m (lbs)');
print -deps fig57.eps;
pause;
plot(t, TFplot)
title('Temperature Dynamic Response')
xlabel('Time (hr)');
ylabel('Temperature (F)');
print -deps fig58.eps;
pause;
plot(t, ve);
title('Exit Vapor Rate Dynamic Response');
xlabel('Time (hr)');
ylabel('Exit Vapor Rate (lbs/hr)');
print -deps fig59.eps;
```

Figure 5.5b: MATLAB program model51.m that defines the differential equations.

```
function cdot = ex51(t, c)
%
% this file defines the equation set in Section 5.2,
% note that all the nonlinear algebra is included in
% this program that computes the derivatives.
%
  global m Vl TK Patm;

Q = 20*10^3;              % Btu/hr
lambda = 853;             % Btu/lb
Cpl = 1;                  % Btu/(lb.F)
Tr = 32;                  % F
rho = 62.4;               % lb/ft^3
F1 = 35.8;                % ft^3/hr
T1 = 70;                  % F
V = 200;                  % ft^3
MW = 18;
R = 10.73;                % psia.ft^3/(lb.mol.R)
Kv = 6.03;                % lb/(hr.psia)
P0 = 150;                 % psia

m = c(1); Vl = c(2);
x = fsolve('eq51', [TK,Patm]');
```

```
TK = x(1);    % K
Patm = x(2);  % atm
TF = (x(1)-273.15)*1.8 +32;  % F
Ppsia = x(2) *14.7;  % psia

v = (Q + rho*F1*Cpl*(T1-Tr))/(lambda + Cpl*(TF-Tr));
if Ppsia < P0
    ve = 0;
else
    ve = Kv*sqrt(Ppsia*(Ppsia-P0));
end

cdot(1) = v - ve;
cdot(2) = F1 - v/rho;
cdot = cdot(:);
```

Figure 5.5c: MATLAB program eq51.m that defines the algebraic equations for the system temperature and pressure.

```
function s = equa(x)

% defines the Equation of State and the Boiling Relationship:
%
global m Vl;

V = 200;                % ft^3
MW = 18;
R = 10.73*1.8/14.7;     % atm.ft^3/(lb.mol.K)

Vg = V - Vl;

s(1) = x(2)*Vg - m*R*x(1)/MW;
s(2) =log10(218/x(2))-(647.27-x(1))/499*((3.346+4.14*10^(-2)*...
    (647.27-x(1)))/(1+1.38*10^(-2)*(647.27-x(1))));
```

The results of the simulation are shown in Figures 5.6–5.9. Figure 5.6 gives the response of the system pressure, P, the boilup gas flow rate, v and the liquid volume, Vl. Note that the liquid volume changes very slow whereas the response of the other two variables is quicker and have reached a new steady–state boiling condition. Figure 5.7 gives the dynamic response for the mass in the vapor phase. Its dynamic response is quicker than that of the other state variable of the liquid volume. This is one reason why the stiff differential equation solver is needed. The temperature dynamic response is given in Figure 5.8 and it shows that the system temperature has reached a new steady–state boiling condition The exit gas flow response is shown in Figure 5.9. Note that the zero constraint condition holds for times less than .11 hours and then there is a rapid response to its new steady–state operating condition. This very rapid dynamic response again require the stiff differential equation package for efficient and accurate numerical results.

Vapor–Liquid Equilibrium Operations 225

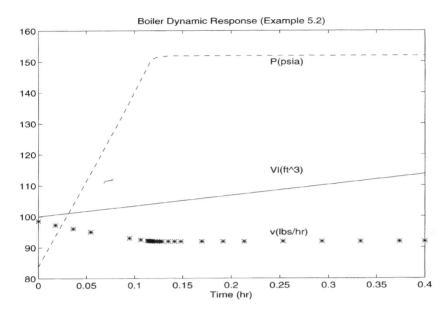

Figure 5.6: Results of Boiler Dynamics Response Problems. Example 5.1.

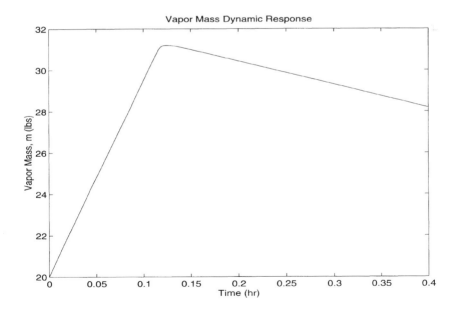

Figure 5.7: Vapor Mass Dynamic Response.

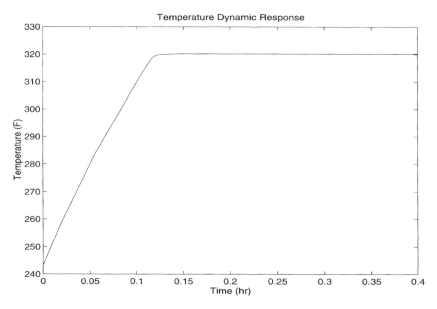

Figure 5.8: Temperature Dynamic Response.

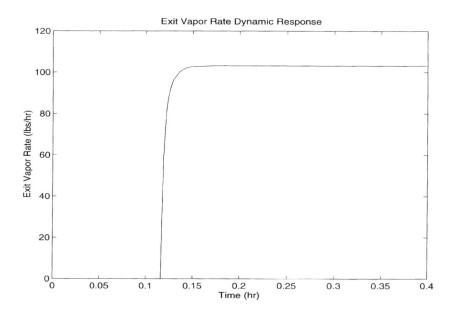

Figure 5.9: Exit Vapor Rate Dynamic Response.

5.3 MULTICOMPONENT BOILING— VAPOR–LIQUID EQUILIBRIUM

Here we consider a vessel in which a mixture of A and B are boiling (Figure 5.10). For mixtures, the boiling temperature is a function of both the system pressure and the composition of the mixture. As boiling proceeds in an open vessel, the composition of the fluid changes. This change causes the boiling temperature to vary. We are interested here in finding the boiling temperature and vapor composition for a known liquid composition.

Figure 5.10: Boiling of Binary Mixture.

The vapor composition in equilibrium with the liquid usually has a different composition than the liquid. For ideal gas and liquid mixtures the equilibrium relations are given by combining Dalton's and Raoult's laws.

$$\text{Dalton's Law}: \quad P_i = P y_i \tag{5.3.1}$$

where P_i = partial pressure of i
 P = system pressure
 y_i = mole fraction of i in vapor

$$\text{Raoult's Law}: \quad P_i = P_i^0 x_i \tag{5.3.2}$$

where P_i^0 = vapor pressure of i
 x_i = mole fraction of i in liquid phase

Combining these two laws gives

$$y_i = \frac{P_i^0}{P} x_i \tag{5.3.3}$$

For nonideal mixtures we define the fugacity f as

$$\tilde{F}_i = RT \ \ln f_i \tag{5.3.4}$$

where $\tilde{F}_i$ = partial molar Gibbs free energy $(\partial F/\partial N_i)_{T,P,N_j}$.

The condition of equilibrium is

$$\tilde{F}_i^L = \tilde{F}_i^V \tag{5.3.5}$$

or

$$RT \ln \left(x_i \, f_i^L \right) = RT \ln \left(y_i \, f_i^V \right) \tag{5.3.6}$$

where f_i^V = the vapor phase fugacity at the system pressure P
f_i^L = the liquid phase fugacity

The condition of equilibrium can therefore be written

$$y_i = \frac{f_i^L}{f_i^V} \, x_i \tag{5.3.7}$$

To compute the fugacity of the vapor phase for nonideal gases, we use the following procedure. At low pressures, the equation of state for an ideal gas is

$$PV = RT \tag{5.3.8}$$

multiplying both sides by dP gives

$$V \, dP = \frac{RT}{P} \, dP \tag{5.3.9}$$

or

$$V_{\text{ideal}} \, dP = RT \, d \, \ln P \tag{5.3.10}$$

since

$$dF = V \, dP + S \, dT \tag{5.3.11}$$

Considering a system at constant temperature and using equation (5.3.4) we have

$$V_{\text{actual}} \, dP = RT \, d \, \ln f_i^V \tag{5.3.12}$$

Combining equations (5.3.10) and (5.3.12) gives

$$\int_0^P (V_i - V_a) dP = \int_0^{\ln(P/f_i^V)} RT \, d \, \ln \left(\frac{P}{f_i^V} \right) \tag{5.3.13}$$

since

$$V_i = RT/P \tag{5.3.14}$$

and

$$V_a = Z_i RT/P \tag{5.3.15}$$

where Z_i = compressibility factor for component i. Here we have considered the nonideal equation of state to be adequately described by equation (5.3.15). Equation (5.3.13) therefore becomes

$$\int_0^P \frac{1 - Z_i}{P} \, dP = \ln \left(\frac{P}{f_i^V} \right) \tag{5.3.16}$$

For nonideal liquid mixtures,

$$f_i^L = P_i^0 \gamma_i \tag{5.3.17}$$

where γ_i = the activity coefficient for component i

Figure 5.11 gives the information–flow diagram for the solution to the problem of boiling of a nonideal binary mixture as illustrated in Figure 5.10. The information–flow diagram shows that an implicit relation is necessary for evaluation of the temperature. The solution of this type of implicit equation involves a root solving process. Root solving numerical methods were discussed in section 2.3

5.4 BATCH DISTILLATION

Figure 5.12 shows an apparatus for the batch distillation of a ternary mixture of A, B, and C. In this section, we develop the necessary equations to describe this problem.

Energy Balance:

$$c_p \frac{d}{dt} M T = q - vH \tag{5.4.1}$$

where M = total moles remaining in liquid phase, moles
 q = rate at which heat is applied to still, cal/min
 v = boil–up rate, moles/min
 H = total enthalpy, cal/moles
 H = $H_A y_A + H_B y_B + H_C y_C$
 H_i = $f_i(T)$

Normally the accumulation term $c_p \, dMT/dt$ is very small compared to the terms on the right side of equation (5.4.1) and is therefore neglected. This is essentially a quasi–steady–state assumption on the energy balance. The energy balance reduces to

$$v = q/H \tag{5.4.2}$$

Total Mass Balance:

$$\frac{d}{dt} M = -v \tag{5.4.3}$$

Component Balances:

$$\frac{d \, M x_i}{dt} = -v y_i \tag{5.4.4}$$

Equilibrium:

$$y_i = K_i(T) x_i \tag{5.4.5}$$

Figure 5.13 gives the information–flow diagram for batch distillation.

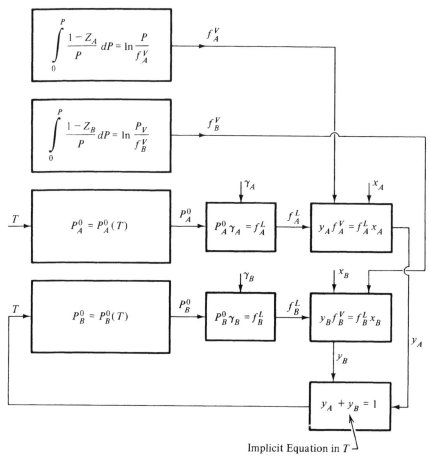

Figure 5.11: Information–Flow Diagram for Multicomponent Boiling.

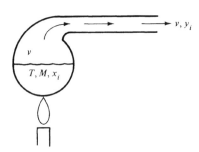

Figure 5.12: Batch Distillation.

Vapor–Liquid Equilibrium Operations

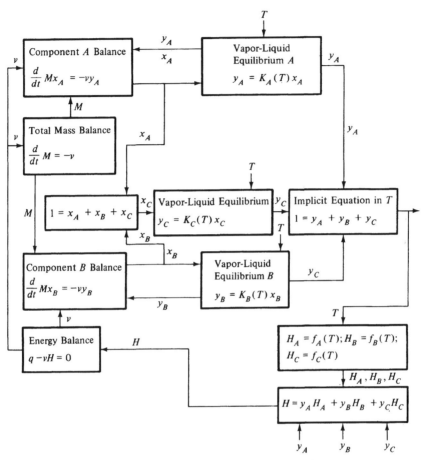

Figure 5.13: Information–Flow Diagram for Batch Distillation.

5.5 BINARY DISTILLATION COLUMNS

Here we are interested in modeling a typical tray in a binary distillation column, as shown in Figure 5.14. The major assumption we usually make for binary distillation is that of equal–molar overflow, namely that the molar vapor flow rate entering the tray is equal to the molar vapor flow rate leaving the tray, $V_{n+1} = V_n$.

The following material and energy balances describe the binary distillation column shown in Figure 5.15.

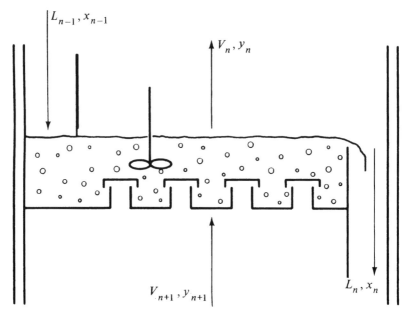

Figure 5.14: Binary Distillation Tray.

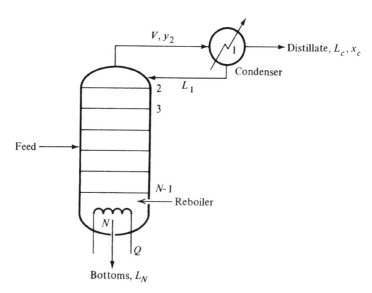

Figure 5.15: Binary Distillation Column.

Vapor–Liquid Equilibrium Operations

5.5.1 A Tray

The model for a typical tray is given below.

Equal Molar Flow:
$$V_{n+1} = V_n = V$$

Momentum Balance on Liquid Phase: This balance is written in order to consider the hydraulic delay that is usually experienced. The momentum balance written on a tray becomes

$$\frac{dL_n H_L}{dt} = h_d - h_w \tag{5.5.1}$$

where H_L = the liquid hold–up, the moles of liquid on the tray
h_d = the pressure head at the downcomer
h_w = the pressure head at the overflow weir

Since the liquid hold–up can be considered a constant, the momentum balance is the following proportionality

$$\frac{dL_n}{dt} \alpha \ (L_{n-1} - L_n) \tag{5.5.2}$$

By introducing the hydraulic time constant, τ, we have the following equality

$$\tau \frac{dL_n}{dt} = L_{n-1} - L_n \tag{5.5.3}$$

which describes the tray hydraulics.

Component Material Balance:

$$\frac{d}{dt}(H_v y_n + H_L x_n) = V y_{n+1} + L_{n-1} x_{n-1} - V y_n - L_n x_n \tag{5.5.4}$$

The vapor hold–up, H_v, is small compared to the liquid hold–up and is therefore neglected. The liquid hold–up is usually considered constant for a tray. The component material balance therefore becomes

$$H_L \frac{dx_n}{dt} = V y_{n+1} + L_{n-1} x_{n-1} - V y_n - L_n x_n \tag{5.5.5}$$

Vapor Liquid Equilibrium: The vapor–liquid equilibrium for a binary system can be expressed as

$$y_n = f(x_n) \tag{5.5.6}$$

This is just an algebraic statement of the usual x-y diagram for a binary system (McCabe and Smith (1954)).

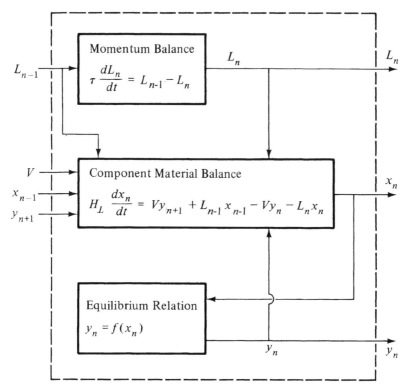

Figure 5.16: Information–Flow Diagram for a Typical Binary Distillation Tray.

The information–flow diagram for a typical tray is shown in Figure 5.16.

The main effect of the equal–molar overflow assumption is that the energy balance is not required in the solution model for a tray.

5.5.2 The Reboiler

For the reboiler, we have to include the energy balance to compute the vapor flow rate and overall material balance.

Energy Balance:

$$V = Q/\lambda \tag{5.5.7}$$

Vapor–Liquid Equilibrium Operations

where V = boil–off rate
Q = heat duty

Overall Material Balance:

$$dH_B/dt = L_{N-1} - L_N - V \tag{5.5.8}$$

Component Material Balance:

$$dH_B x_N/dt = L_{N-1}x_{N-1} - L_N x_N - V y_N \tag{5.5.9}$$

Equilibrium Relation:

$$y_N = f(x_N) \tag{5.5.10}$$

The information–flow diagram is shown in Figure 5.17.

5.5.3 The Condenser

For a total condenser, the following material balances can be written. The internal reflux ratio is L_c/V and Figure 5.18 gives the information–flow diagram.

Component Balance:

$$x_c = x_1 = y_2 \tag{5.5.11}$$

Overall Balance:

$$L_c + L_1 = V \tag{5.5.12}$$

5.6 MULTICOMPONENT DISTILLATION COLUMNS

In order to discuss the mathematical modeling of multicomponent distillation, a dynamic model of a ternary equilibrium column of five stages including a partial condenser and a reboiler will be discussed. In particular, the distillation of benzene, toluene, and xylene will be considered. The system is shown in Figure 5.19.

The modeling of the column is accomplished by individual analysis of each stage and then the lumping together of all the stages for an overall and simultaneous solution. These solution equations are integrated with respect to time for the dynamic response of the system.

Each stage of the column is idealized by the assumptions that (1) the vapor is in equilibrium with the liquid on the stage; (2) the composition of the liquid leaving the stage is the same as that on the stage; and (3) there are no gradients across the stage. These assumptions yield an ideal stage; that is, one with perfect mixing of the liquid, and equilibrium between the liquid and the vapor. Such an ideal stage is shown in Figure 5.20.

The equations for the various stages will depend upon the actual conditions, but they will all have the general form of the following equations. For any stage n, an overall mass balance, energy balance, and a

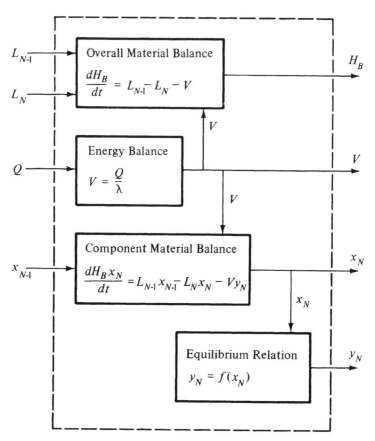

Figure 5.17: Information–Flow Diagram of the Reboiler in a Binary Distillation Column.

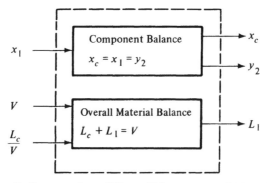

Figure 5.18: Information–Flow Diagram of Total Condenser in a Binary Distillation Column.

Vapor–Liquid Equilibrium Operations

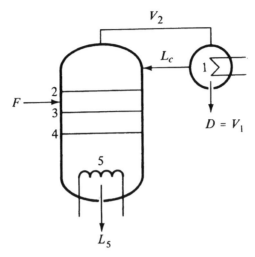

Figure 5.19: Multicomponent Distillation Column.

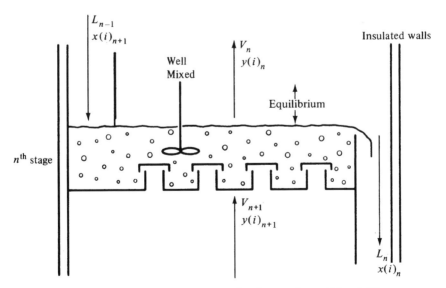

Figure 5.20: Pictorial Description of an Ideal Stage.

component balance for each component can be written. The total mole balance states that the change in hold–up is equal to the flow in minus the flow out.

$$dH_n/dt = \text{molar flow in} - \text{molar flow out} \tag{5.6.1}$$

Normally, the hold–up in the vapor phase is neglected and only the liquid hold–up is considered; this hold–up is usually constant. This gives the algebraic relation

$$\text{molar flow in} = \text{molar flow out} \tag{5.6.2}$$

The energy balance states that the change in energy content on a stage is equal to the flow of energy in minus the flow of energy out.

$$c_p \, dH_n T_n/dt = \text{rate of energy in} - \text{rate of energy out} \tag{5.6.3}$$

The change in heat content on any stage is usually small compared to heat flows and may be approximated as zero. This gives the algebraic relation

$$\text{rate of energy in} = \text{rate of energy out} \tag{5.6.4}$$

The component balances are best written in terms of mole fractions and are solved for the liquid phase composition, since there is no hold–up of the vapor. This states that the change in moles of component i in the liquid phase on the nth stage is equal to the flow in minus the flow out:

$$\frac{dH_n x_{i_n}}{dt} = \text{flow in} \cdot (x_{i,\text{in}}) - \text{flow out} \cdot (x_{i,\text{out}}) \tag{5.6.5}$$

Species material balances can be written for all components but it is most convenient to solve for the last component's mole fraction by subtracting the sum of all others from one.

$$x_N = 1 - \sum_{i=i}^{N-1} x_i \tag{5.6.6}$$

Since the liquid composition is known, the equilibrium vapor composition can be computed. A trial–and–error process is used to determine the tray temperature. The correct tray temperature results when the sum of the mole fractions of the vapor phase is unity:

$$\sum_{i=1}^{N} y_i = 1 \tag{5.6.7}$$

Equation (5.6.7) is implicit in the temperature, and root solving methods discussed in section 2.3 should be used to determine the temperature on each tray using this relationship.

Vapor–Liquid Equilibrium Operations

The information–flow diagrams for the various stages of this distillation column are given in Figures 5.21 to 5.25.

For the partial condenser, Figure 5.21, a reflux ratio specification is used instead of the energy balance, since the condenser duty is unknown and depends upon the state of the overhead product (vapor, liquid, or liquid below bubble point). The feed tray, Figure 5.23, naturally has the feed stream included as an input term, Otherwise it is similar to the other trays. The reboiler's heat balance, Figure 5.25, reflects the heat duty of the reboiler, Q^1.

Figure 5.26 shows how the various diagrams fit together. It is this general diagram that must be solved simultaneously for a given increment of time; and then all the differential equations must be incremented and a new solution found.

The computational strategy for solving the dynamic model is shown in Figure 5.27. It involves three basic sections that solve for the equilibrium tray temperatures, the simultaneous solutions of the algebraic equations giving the vapor and liquid flow rates, and the solution of the differential equations giving the liquid compositions.

Block 2 in the solution strategy initializes the problem by providing initial liquid phase compositions.

Block 4 determines the vapor compositions and tray temperatures that are in equilibrium with the known liquid compositions. This is a series of root solving solutions to implicit equations which compute the equilibrium temperatures on each tray subject to the physical constraint that the sum of the vapor phase mole fractions must equal unity for each equilibrium tray.

Block 6 solves the 5 overall tray material balances and 5 tray energy balances. These are linear coupled algebraic equations that can be conveniently solved simultaneously by matrix techniques.

Block 7 integrates the differential equations of the component material balances. A standard differential equation solver subroutine should be used. The output of this block is values for the liquid phase composition at a new time levels.

The MATLAB program `ex56.m` has been created to solve for the dynamic response of this multicomponent distillation column and is given in Figure 5.28. The differential equations are defined in program `model56.m` and given in Figure 5.29. All of the nonlinear root solving as well as the simultaneous linear algebra is included in this program so that every time that the dynamic process states are updated, the algebraic relations are satisfied. These calculations are repeated in the main program `ex56.m` at the output times in order that plots can be obtained for variables besides the dynamic states computed as outputs of the differential equation solver. The stiff differential equation solver `ode15s` is used. It is more than one order of magnitude more efficient at solving this equation set of differential equations than the nonstiff routine `ode45`.

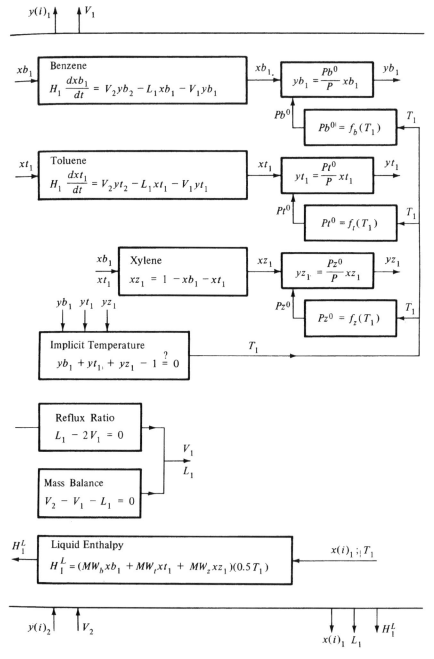

Figure 5.21: Information–Flow Diagram of Condenser.

Vapor–Liquid Equilibrium Operations

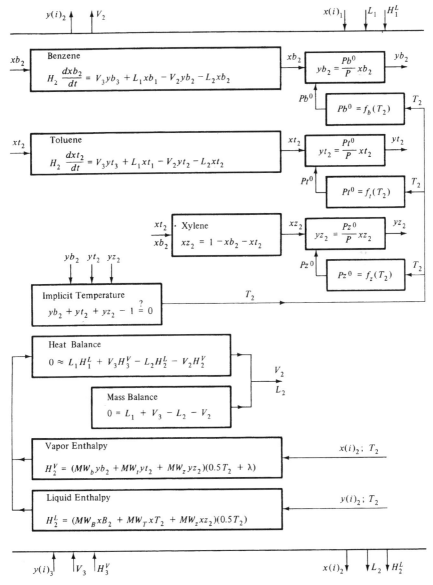

Figure 5.22: Information–Flow Diagram of First Tray.

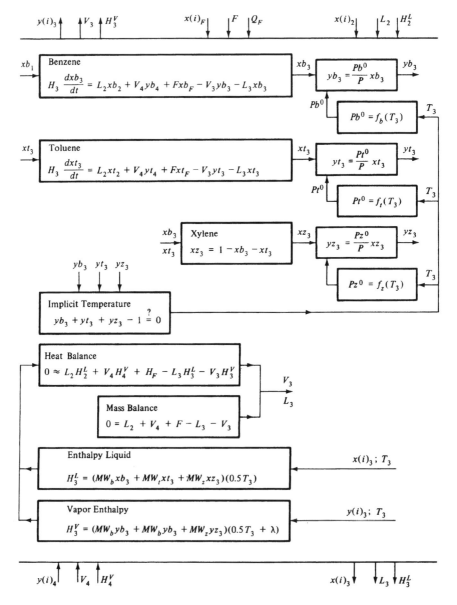

Figure 5.23: Information–Flow Diagram of Feed Tray.

Vapor–Liquid Equilibrium Operations

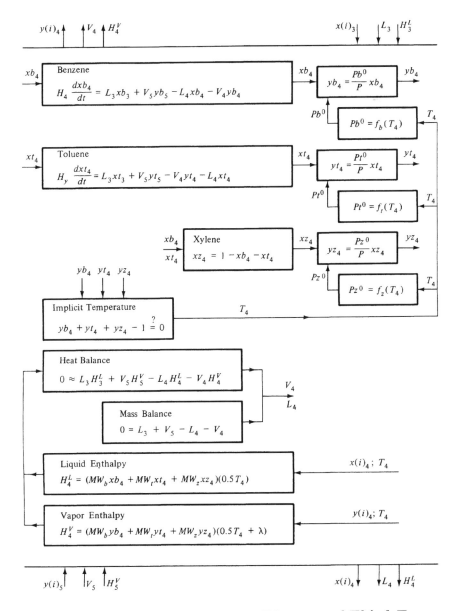

Figure 5.24: Information–Flow Diagram of Third Tray.

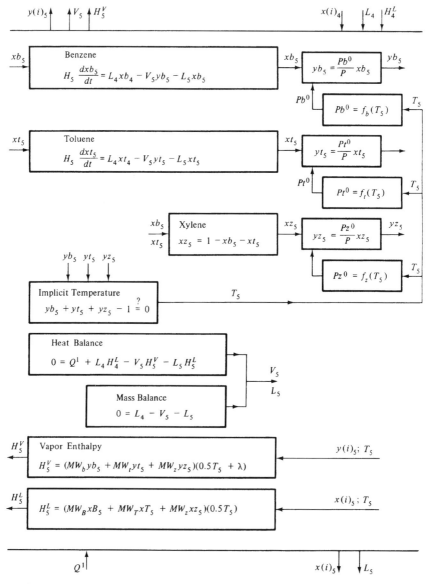

Figure 5.25: Information–Flow Diagram of Reboiler.

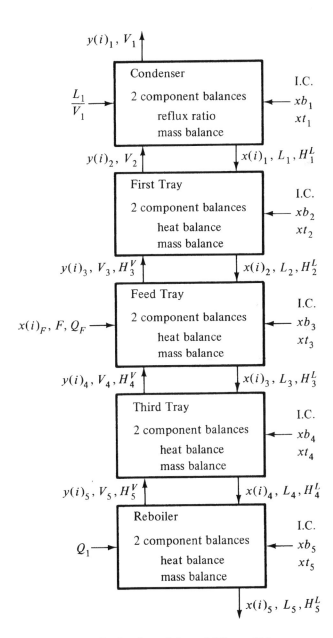

Figure 5.26: Relationship of Flow Diagrams. Figures 5.21 through 5.25.

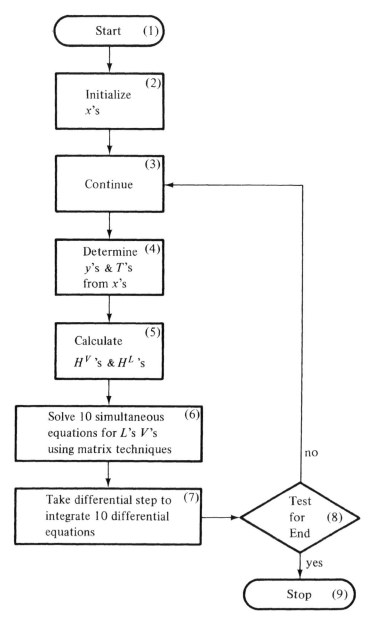

Figure 5.27: General Solution Diagram.

Figure 5.28: Main Program ex56.m to Compute Dynamic Response of Distillation Column.

```
% This file calls the stiff differential equation routine ode15s,
% to solve the distillation column problem in section 5.6;
%
  global coeff T xb xt xz JT right_hand;

tt=cputime;

% initial conditions
%
xb(1) = 0.3;      xt(1) = 0.4;
xb(2) = 0.25;     xt(2) = 0.35;
xb(3) = 0.25;     xt(3) = 0.35;
xb(4) = 0.25;     xt(4) = 0.35;
xb(5) = 0.2;      xt(5) = 0.3;


P = 1.0;          % atm
%
% parameters used to calculate HL and HV
%
MWb = 78.1; MWt = 92.1; MWz = 106.2;
lambda = 155;

% guess of initial T
%
for J = 1:5,
    T(J) = 200;
end

%
% matrix initialization
%
coeff = zeros(10);
right_hand = zeros(10, 1);

%
% parameters for matrix inversion
%
Q1 = 85000;
F = 2.5;
reflux = 2.0;
xtf = 0.30;
TF = 250;
xbf = (P - feval('Pz', TF) - xtf*(feval('Pt', TF)-feval('Pz', TF)))...
      /(feval('Pb', TF)-feval('Pz', TF));
HF = F*0.5*TF*(MWb*xbf + MWt*xtf + MWz*(1-xbf-xtf));

    for J = 1:5,
        xz(J) = 1 - xb(J) - xt(J);
```

```
        end
%
% matrix to solve for L and V
%
        coeff(1,1) = 1.0;
        coeff(1,2) = -reflux;
        coeff(2,1) = -1.0;
        coeff(2,2) = -1.0;
        coeff(2,4) = 1.0;
        coeff(4,1) = 1.0;
        coeff(4,3) = -1.0;
        coeff(4,4) = -1.0;
        coeff(4,6) = 1.0;
        coeff(6,3) = 1.0;
        coeff(6,5) = -1.0;
        coeff(6,6) = -1.0;
        coeff(6,8) = 1.0;
        coeff(8,5) = 1.0;
        coeff(8,7) = -1.0;
        coeff(8,8) = -1.0;
        coeff(8,10) = 1.0;
        coeff(10,7) = 1.0;
        coeff(10,9) = -1.0;
        coeff(10,10) = -1.0;

        right_hand(5) = -HF;
        right_hand(6) = -F;
        right_hand(9) = -Q1;

        c0 = [xb(1) xt(1) xb(2) xt(2) xb(3) xt(3) xb(4) xt(4) xb(5) xt(5)]';

        [t, c] = ode15s('model56', [0, 0.8], c0);
%
% set up plotting vectors
%
        N = length(t);
for I = 1:N,

        xb(1)=c(I,1); xt(1)=c(I,2);
        xb(2)=c(I,3); xt(2)=c(I,4);
        xb(3)=c(I,5); xt(3)=c(I,6);
        xb(4)=c(I,7); xt(4)=c(I,8);
        xb(5)=c(I,9); xt(5)=c(I,10);

        xbplot(I) = xb(5);
        xtplot(I) = xt(5);
        xzplot(I) = 1-xb(5)-xt(5);

    for J = 1:5,
        JT = J;
        xz(J) = 1-xb(J)-xt(J);
        T(J) = fzero('impT',T(J));
```

```
      yb(J) = feval('Pb',T(J))/P*xb(J);
      yt(J) = feval('Pt',T(J))/P*xt(J);
      yz(J) = feval('Pz',T(J))/P*xz(J);
   end
 ybplot(I) = yb(1);
 ytplot(I) = yt(1);
 yzplot(I) = yz(1);

   for J = 1:5,
      HL(J) = (MWb*xb(J)+MWt*xt(J)+MWz*xz(J))*(.5*T(J));
      HV(J) = (MWb*yb(J)+MWt*yt(J)+MWz*yz(J))*(.5*T(J)+lambda);
   end
 coeff(3,1) = HL(1);
 coeff(3,3) = -HL(2);
 coeff(3,4) = -HV(2);
 coeff(3,6) = HV(3);
 coeff(5,3) = HL(2);
 coeff(5,5) = -HL(3);
 coeff(5,6) = -HV(3);
 coeff(5,8) = HV(4);
 coeff(7,5) = HL(3);
 coeff(7,7) = -HL(4);
 coeff(7,8) = -HV(4);
 coeff(7,10) = HV(5);
 coeff(9,7) = HL(4);
 coeff(9,9) = -HL(5);
 coeff(9,10) = -HV(5);
 LV = coeff\right_hand;
 L(1)=LV(1); L(2)=LV(3); L(3)=LV(5); L(4)=LV(7); L(5)=LV(9);
 V(1)=LV(2); V(2)=LV(4); V(3)=LV(6); V(4)=LV(8); V(5)=LV(10);
 v1plot(I) = V(1);
 l5plot(I) = L(5);

end
time = cputime-tt

plot(t, v1plot);
title('Vapor Overhead Flow Rate (Example 5.6)');
xlabel('Time (hr)');
ylabel('Flow Rate (lb.mol/hr)');
%print -deps fig530.eps;
pause;
plot(t, l5plot);
title('Bottom Liquid Flow Rate')
xlabel('Time (hr)');
ylabel('Flow Rate (lb.mol/hr)');
%print -deps fig531.eps;
pause;
plot(t, ybplot, '-', t, ytplot, '--', t, yzplot, '*');
title('Overhead Vapor Composition');
xlabel('Time (hr)');
ylabel('Mole Fraction');
text(.04,.75,'benzene');
```

```
text(.32,.65,'toluene');
text(.4,.3,'xylene');
%print -deps fig532.eps;
pause;
plot(t, xbplot, '-', t, xtplot, '--', t, xzplot, '*');
title('Bottom Flow Composition');
xlabel('Time (hr)');
ylabel('Mole Fraction');
text(.07,.1,'benzene');
text(.2,.25,'toluene');
text(.3,.8,'xylene');
%print -deps fig533.eps;
```

Figure 5.29: Differential Equations for Dynamic Distillation Column. Program model56.m and other functions impT.m, Pb.m, Pt.m, Pz.m.

model56.m

```
function cdot = ex56(t, c)
%
% this file defines the equation set in Figure 5.4,

global coeff T xb xt xz JT right_hand

 xb(1)=c(1); xt(1)=c(2);
 xb(2)=c(3); xt(2)=c(4);
 xb(3)=c(5); xt(3)=c(6);
 xb(4)=c(7); xt(4)=c(8);
 xb(5)=c(9); xt(5)=c(10);
 for I = 1:5,
    xz(I) = 1-xb(I)-xt(I);
 end

holdup(1) = 0.05;
holdup(2) = 0.0125;
holdup(3) = 0.0125;
holdup(4) = 0.0125;
holdup(5) = 0.6;
 MWb = 78.1; MWt = 92.1; MWz = 106.2;
 lambda = 155;
 Q1 = 85000;
 F = 2.5;
 reflux = 2.0;
 xtf = .3;
 TF = 250;
 P = 1;
 xbf = (P -feval('Pz',TF)-xtf*(feval('Pt',TF)-feval('Pz',TF)))...
       /(feval('Pb',TF)-feval('Pz',TF));
 QF = F*0.5*TF*(MWb*xbf +MWt*xtf +MWz*(1-xbf-xtf));
%
```

```
% solve nonlinear equations to get y an T
%
for J = 1:5,
   JT = J;
   T(J) = fzero('impT',T(J));
   yb(J) = feval('Pb',T(J))/P*xb(J);
   yt(J) = feval('Pt',T(J))/P*xt(J);
   yz(J) = feval('Pz',T(J))/P*xz(J);
end
%
% calc HL and HV
%
for J = 1:5,
   HL(J) = (MWb*xb(J)+MWt*xt(J)+MWz*xz(J))*(.5*T(J));
   HV(J) = (MWb*yb(J)+MWt*yt(J)+MWz*yz(J))*(.5*T(J)+lambda);
end
%
% update matrix
%
 coeff(3,1) = HL(1);
 coeff(3,3) = -HL(2);
 coeff(3,4) = -HV(2);
 coeff(3,6) = HV(3);
 coeff(5,3) = HL(2);
 coeff(5,5) = -HL(3);
 coeff(5,6) = -HV(3);
 coeff(5,8) = HV(4);
 coeff(7,5) = HL(3);
 coeff(7,7) = -HL(4);
 coeff(7,8) = -HV(4);
 coeff(7,10) = HV(5);
 coeff(9,7) = HL(4);
 coeff(9,9) = -HL(5);
 coeff(9,10) = -HV(5);

LV = coeff\right_hand;
L(1)=LV(1); L(2)=LV(3); L(3)=LV(5); L(4)=LV(7); L(5)=LV(9);
V(1)=LV(2); V(2)=LV(4); V(3)=LV(6); V(4)=LV(8); V(5)=LV(10);
%
% derivatives
%
cdot(1) = (V(2)*yb(2) - L(1)*c(1) - V(1)*yb(1))/holdup(1);
cdot(2) = (V(2)*yt(2) - L(1)*c(2) - V(1)*yt(1))/holdup(1);
cdot(3) = (V(3)*yb(3) + L(1)*c(1) - V(2)*yb(2) - L(2)*c(3))...
          /holdup(2);
cdot(4) = (V(3)*yt(3) + L(1)*c(2) - V(2)*yt(2) - L(2)*c(4))...
          /holdup(2);
cdot(5) = (L(2)*c(3) + V(4)*yb(4) + F*xbf - V(3)*yb(3)...
           -L(3)*c(5))/holdup(3);
cdot(6) = (L(2)*c(4) + V(4)*yt(4) + F*xtf - V(3)*yt(3)...
           -L(3)*c(6))/holdup(3);
cdot(7) = (L(3)*c(5) + V(5)*yb(5) - L(4)*c(7) - V(4)*yb(4))...
          /holdup(4);
```

```
cdot(8) = (L(3)*c(6) + V(5)*yt(5) - L(4)*c(8) - V(4)*yt(4))...
          /holdup(4);
cdot(9) = (L(4)*c(7) - V(5)*yb(5) - L(5)*c(9))/holdup(5);
cdot(10) = (L(4)*c(8) - V(5)*yt(5) - L(5)*c(10))/holdup(5);
cdot = cdot(:);
```

impT.m

```
function f=impt(TT)
global coeff T xb xt xz JT
  P = 1;
  yb(JT) = feval('Pb',TT)/P*xb(JT);
  yt(JT) = feval('Pt',TT)/P*xt(JT);
  yz(JT) = feval('Pz',TT)/P*xz(JT);
  sum_y = yb(JT)+yt(JT)+yz(JT);
  f = sum_y - 1.0;
```

Pb.m

```
function f = pb(T)
f = 10^(-1686.8/((T-32)/1.8+273)+7.6546)/760;
```

Pt.m

```
function f = pt(T)
f = 10^(-3179.0/(T+459.69)+4.60);
```

Pz.m

```
function f = pz(T)
f = 10^(-3667.5/(T+459.69)+4.858);
```

Overhead and bottom dynamic response curves are shown in Figures 5.30 to 5.33 for a column as described in Table 5.2.

It is found that steady state is reached in the column in about 0.80 hr. At the time when steady state is reached, the rate of the overhead product is 1.72 lb mole/hr and there are 0.793 lb mole/hr being discharged from the reboiler. Since the feed stream has an input rate of 2.5 lb mole/hr, it can be seen that most of the feed goes out by the overhead. An important thing to note is also the fact that the bottoms are virtually benzene free, 0.00056 mole fraction. The feed mole fraction of benzene is 0.099 and the benzene has all gone into the overhead with a mole fraction of 0.145.

While the overhead has been enriched with benzene, it still has a large fraction of toluene (0.429) and xylene (0.426). While the bottoms flow is not great, 0.793 lb mole/hr, it has 0.97 mole fraction xylene, which is quite pure. The only other appreciable component is toluene. It should be noted that this column has not done a particularly good job in separating the components, since it has not greatly increased the composition

Vapor–Liquid Equilibrium Operations

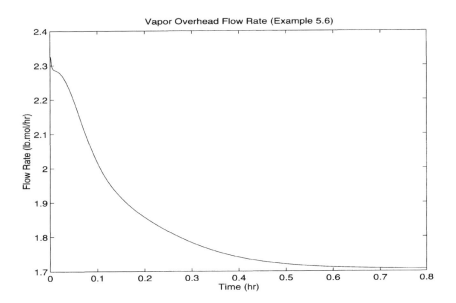

Figure 5.30: Overhead Flow vs. Time.

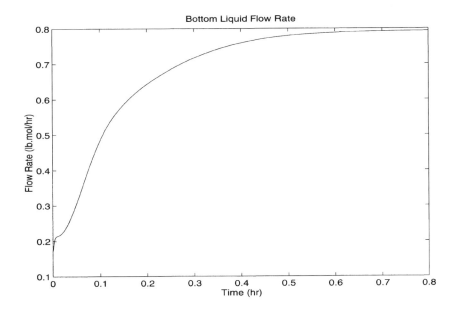

Figure 5.31: Bottoms Flow vs. Time.

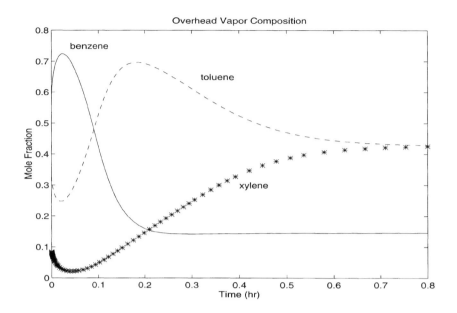

Figure 5.32: Overhead Composition vs. Time

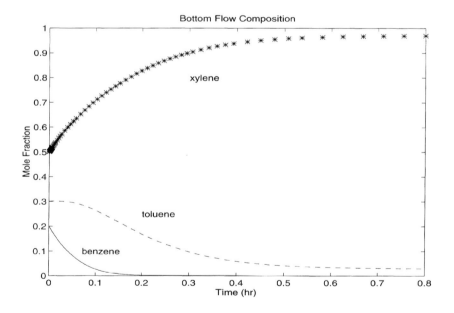

Figure 5.33: Bottoms Composition vs. Time.

Vapor-Liquid Equilibrium Operations

Table 5.2: Distillation Column Specifications

Feed Stream:	$F_F = 2.5$ lb mol/hr $x_T = 0.3$ Feed as a liquid at its bubble point $T_F = 250°F$
Condenser:	Reflux ratio = 2.0 Liquid hold–up = 0.05 lb mol
Trays:	Liquid hold–up = 0.0125 lb mol
Reboiler:	$Q^1 = 85000$ Btu/hr Liquid hold–up = 0.60 lb mol

of any one component in the products of the overhead, which has most of the flow. If the purpose of the column was to produce pure xylene from the bottoms, the flow is not large and it is very inefficient because of the large xylene mole fraction in the overhead of 0.426. The mathematical model can be used to investigate different operating conditions that might optimize production objectives. It can also be used to investigate the type of dynamic responses that can be expected to process changes which occur frequently in most industrial situations.

PROBLEMS

5.1. Modify Program ex56.m to study the dynamic response of the distillation column of section 5.6 under different heat duties and reflux ratios. What conclusions can you draw from this study? Now repeat this study when the feed composition of toluene is .2 mole fraction and .4 mole fraction.

Required Data

Components: Benzene, Toluene, Xylene
Vapor pressures: Assume Raoult's and Dalton's laws hold, $P_T = 1$ atm = constant.

$$\log_{10}(760\ P_B^0) = -\frac{1686.8}{T} + 7.6546 \qquad P^0 = \text{atm},\ T = °K$$

$$\log_{10} P_T^0 = -\frac{3179.0}{T} + 4.60 \qquad P^0 = \text{atm},\ T = °R$$

$$\log_{10} P_Z^0 = -\frac{3667.5}{T} + 4.858 \qquad P^0 = \text{atm},\ T = °R$$

Heat capacities:	Liquid streams = 0.5 Btu/lb°F Vapor streams = 0.3 Btu/lb°F
Tray hydraulics:	$\tau = 0$
Molecular weights:	Toluene = 92.1 lb/lb mol Benzene = 78.1 lb/lb mol Xylene = 106.2 lb/lb mol
Feed stream:	$F_F = 2.50$ lb mol/hr = constant $x_T = 0.30$ Feed is a liquid at its bubble point $T_F = 250°F$
Condenser:	Reflux ratio = L_c/V = 2.0, 5.0, and 10.0 Liquid hold–up = 0.05 lb mol Vapor hold–up = 0 $x_T = 0.40$ at $t = 0$ $x_B = 0.30$ at $t = 0$
Stages 1–3:	Liquid hold–up = 0.0125 lb mol Vapor hold–up = 0 $x_T = 0.35$ at $t = 0$ $x_B = 0.25$ at $t = 0$
Reboiler:	Q^1 = heat transferred to liquid = 85000, 50000, and 70000 Btu/hr λ = heat of vaporization = 155 Btu/lb Liquid hold–up = 0.60 lb mol Vapor hold–up = 0 $x_T = 0.30$ at $t = 0$ $x_B = 0.20$ at $t = 0$

5.2. Simulate the batch distillation of benzene, toluene, and xylene in a vessel holding 6 lb moles and with a heating rate of 85,000 Btu/hr.

5.3. Develop a model for the preheating of the boiler of section 5.2. The initial change starts at room temperature and the feed is not started until the system comes to the condition of boiling. How long is this heat-up period?

REFERENCES

Keenan, J. H. and Keyes, F. G., *Thermodynamic Properties of Steam*, Wiley, New York (1959).

McCabe, W. L. and Smith, J. C., *Unit Operations of Chemical Engineering*, McGraw–Hill, New York (1954).

Chapter 6

MICROSCOPIC BALANCES

In this chapter, we develop microscopic balances that are used for the simulation of distributed parameter systems. Distributed systems have spatial gradients as well as time changes. In order to properly describe these systems, the conservation laws must be applied to any point within the system rather than written over the entire macroscopic system. Not only will these point or microscopic conservation laws be developed in this chapter, but they will also be applied to several classical one–dimensional problems that have either analytical solutions or are initial-value problems that can be solved using computational techniques already introduced and discussed. An excellent treatment of microscopic balances is found in the book by Bird, Stewart, and Lightfoot (1960). Simplification of complex problems using order–of–magnitude analysis is also introduced.

6.1 CONSERVATION OF TOTAL MASS (EQUATION OF CONTINUITY)

The equation of continuity is derived by writing a mass balance over an arbitrary finite volume element $\Delta x \Delta y \Delta z$ through which a fluid is flowing, as is shown in Figure 6.1. We then take the limit as the volume element goes to zero to get the appropriate equation for any point in the spatial domain.

The rate of mass entering the volume element perpendicular to the x axis at x is $\rho v_x \Delta y \Delta z \,|_x$ and at the plane $x + \Delta x$ is $\rho v_x \Delta y \Delta z \,|_{x+\Delta x}$ Similar expressions can be written for the other planes. The rate of accumulation of mass within the volume element is $\partial(\rho \Delta x \Delta y \Delta z)/\partial t$. The mass balance therefore becomes

$$\begin{array}{c} \text{Rate of} \\ \text{Accumulation of Mass} \end{array} = \begin{array}{c} \text{Rate of} \\ \text{Mass In} \end{array} - \begin{array}{c} \text{Rate of} \\ \text{Mass Out} \end{array}$$

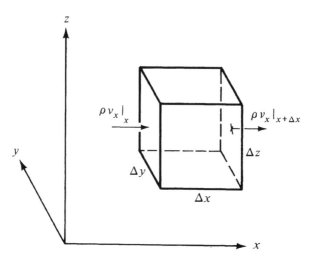

Figure 6.1: Differential Element for Equation of Continuity.

$$\frac{\partial(\rho \Delta x \Delta y \Delta z)}{\partial t} = \Delta y \Delta z(\rho v_x \mid_x - \rho v_x \mid_{x+\Delta x})$$
$$+ \Delta x \Delta z(\rho v_y \mid_y - \rho v_y \mid_{y+\Delta y})$$
$$+ \Delta x \Delta y(\rho v_z \mid_z - \rho v_z \mid_{z+\Delta z}) \qquad (6.1.1)$$

By dividing this equation by the volume element $\Delta x \Delta y \Delta z$ and taking the limit as the volume $(\Delta x \Delta y \Delta z)$ approaches zero, we get the point description

$$\frac{\partial \rho}{\partial t} = -\left(\frac{\partial}{\partial x}\rho v_x + \frac{\partial}{\partial y}\rho v_y + \frac{\partial}{\partial z}\rho v_z \right) \qquad (6.1.2)$$

This differential overall mass balance, which holds at any point within the volume of the system, is often called the *equation of continuity*. It describes the rate of change of the density of a fluid at any point in the system. The equation expressed in vector form is

$$\frac{\partial \rho}{\partial t} = -(\nabla \cdot \rho \boldsymbol{v}) \qquad (6.1.3)$$

The equation of continuity can be rearranged to become

$$\frac{\partial \rho}{\partial t} + v_x \frac{\partial \rho}{\partial x} + v_y \frac{\partial \rho}{\partial y} + v_z \frac{\partial \rho}{\partial z} = -\rho \left(\frac{\partial v_x}{\partial x} + \frac{\partial v_y}{\partial y} + \frac{\partial v_z}{\partial z} \right) \qquad (6.1.4)$$

The left side of equation (6.1.4) is the substantial derivative of the density, that is, the time derivative for a path following the fluid motion; and it can be written as

$$\frac{D\rho}{Dt} = \frac{\partial \rho}{\partial t} + v_x \frac{\partial \rho}{\partial x} + v_y \frac{\partial \rho}{\partial y} + v_z \frac{\partial \rho}{\partial z} \qquad (6.1.5)$$

Microscopic Balances

A very important special form of the equation of continuity is that for a fluid of constant density (incompressible fluid), for which the substantial derivative is zero. Equation (6.1.4) therefore becomes for an incompressible fluid,

$$\nabla \cdot \boldsymbol{v} = 0 \qquad (6.1.6)$$

6.2 CONSERVATION OF COMPONENT i

The mass balance for a species component i in a multicomponent mixture must also consider the rate of generation of the component by chemical reaction. The mass conservation law becomes

$$\begin{pmatrix}\text{Rate of}\\ \text{Accumulation}\\ \text{of Component } i\end{pmatrix} = \begin{pmatrix}\text{Rate In of}\\ \text{Component } i\end{pmatrix} - \begin{pmatrix}\text{Rate Out of}\\ \text{Component } i\end{pmatrix} + \begin{pmatrix}\text{Rate of}\\ \text{Generation of}\\ \text{Component } i\end{pmatrix}$$

$$\begin{aligned}\frac{\partial \rho_i \Delta x \Delta y \Delta z}{\partial t} &= \Delta y \Delta z (\rho_i v_{ix}|_x - \rho_i v_{ix}|_{x+\Delta x}) \\ &+ \Delta x \Delta z (\rho_i v_{iy}|_y - \rho_i v_{iy}|_{y+\Delta y}) \\ &+ \Delta x \Delta y (\rho_i v_{iz}|_z - \rho_i v_{iz}|_{z+\Delta z}) \\ &+ r_i (\Delta x \Delta y \Delta z) \end{aligned} \qquad (6.2.1)$$

where r_i is the rate of generation of i per unit volume.

By dividing this equation by the volume element $\Delta x \Delta y \Delta z$ and taking the limit as these dimensions go to zero, we have

$$\frac{\partial \rho_i}{\partial t} = -\left(\frac{\partial}{\partial x}\rho_i v_{ix} + \frac{\partial}{\partial y}\rho_i v_{iy} + \frac{\partial}{\partial z}\rho_i v_{iz}\right) + r_i \qquad (6.2.2)$$

Recognizing that $\rho_i v_{ix}$ is the mass flux of i in the x direction, n_{ix}, we can write equation (6.2.2) as

$$\frac{\partial \rho_i}{\partial t} = -\left(\frac{\partial n_{ix}}{\partial x} + \frac{\partial n_{iy}}{\partial y} + \frac{\partial n_{ix}}{\partial z}\right) + r_i \qquad (6.2.3)$$

In vector notation we have

$$\frac{\partial \rho_i}{\partial t} = -(\nabla \cdot \boldsymbol{n}_i) + r_i \qquad (6.2.4)$$

For systems in which molecular diffusion occurs, the mass flux is the sum of the bulk flow of species i and the diffusional flux of that species.

$$\boldsymbol{n}_i = \rho_i \boldsymbol{v} + \boldsymbol{j}_i = \omega_i \sum_{j=1}^{N} \boldsymbol{n}_j + \boldsymbol{j}_i \qquad (6.2.5)$$

where v = the mass average velocity of the fluid
ω_i = mass fraction
N = number of components
j_i = the diffusional flux

For *binary* diffusion we can relate the diffusional flux to the concentration gradient via Fick's first law.

$$j_A = -\rho \mathcal{D}_{AB} \nabla \omega_A \qquad (6.2.6)$$

where $\mathcal{D}_{AB}$ = the binary diffusion coefficient
ω_A = the mass fraction of A

To obtain the component balances in molar units, it is only necessary to divide the mass expressions by the molecular weight of the specific component:

Molar concentration = $C_i = \rho_i/M_i$
Molar flux = $N_i = n_i/M_i$
Molar rate of generation per unit volume = $R_i = r_i/M_i$

Therefore the component balance equation, (6.2.4), becomes

$$\frac{\partial C_i}{\partial t} + \nabla \cdot \boldsymbol{N}_i = R_i \qquad (6.2.7)$$

In terms of the bulk flow and diffusional flux we have

$$\boldsymbol{N}_i = C_i \boldsymbol{v}^* + \boldsymbol{J}_i^* = x_i \sum_{j=1}^{N} \boldsymbol{N}_j + \boldsymbol{J}_i^* , \qquad x_i = \text{mole fraction} \qquad (6.2.8)$$

where $\boldsymbol{v}^*$ is the molar average velocity and for binary systems the molar flux $\boldsymbol{J}_A^*$ is given by Fick's law to be

$$\boldsymbol{J}_A^* = -C\mathcal{D}_{AB} \nabla x_A \qquad (6.2.9)$$

Here x_A is the mole fraction of A.

A summary of the various component and overall mass balances is found in Tables 6.1, 6.2, and 6.3.

6.3 DISPERSION DESCRIPTION

For mass transport problems that do not involve molecular diffusion, such as turbulent-flow problems, or problems with highly complicated passages such as packed beds or porous media, a dispersion model is usually used. Here we assume that the controlling mass transport mechanisms are similar in character to the molecular case; and, therefore,

Table 6.1: Equation of Continuity in Various Coordinate Systems (Source: Bird et al., 1960)

Rectangular coordinates (x, y, z):

$$\frac{\partial \rho}{\partial t} + \frac{\partial}{\partial x}\rho v_x + \frac{\partial}{\partial y}\rho v_y + \frac{\partial}{\partial z}\rho v_z = 0$$

Cylindrical coordinates (r, θ, z):

$$\frac{\partial \rho}{\partial t} + \frac{1}{r}\frac{\partial}{\partial r}\rho r v_r + \frac{1}{r}\frac{\partial}{\partial \theta}\rho v_\theta + \frac{\partial}{\partial z}\rho v_z = 0$$

Spherical coordinates (r, θ, ϕ):

$$\frac{\partial \rho}{\partial t} + \frac{1}{r^2}\frac{\partial}{\partial r}\rho r^2 v_r + \frac{1}{r\sin\theta}\frac{\partial}{\partial \theta}(\rho v_\theta \sin\theta) +$$

$$\frac{1}{r\sin\theta}\frac{\partial}{\partial \phi}\rho v_\phi = 0$$

Table 6.2: Equation for the Conservation of Mass of Component A in Various Coordinate Systems (Source: Bird et al., 1960)

Rectangular coordinates:

$$\frac{\partial \rho_A}{\partial t} + \frac{\partial}{\partial x}n_{Ax} + \frac{\partial}{\partial y}n_{Ay} + \frac{\partial}{\partial z}n_{Az} = r_A$$

Cylindrical coordinates:

$$\frac{\partial \rho_A}{\partial t} + \frac{1}{r}\frac{\partial}{\partial r}r n_{Ar} + \frac{1}{r}\frac{\partial}{\partial \theta}n_{A\theta} + \frac{\partial}{\partial z}n_{Az} = r_A$$

Spherical coordinates:

$$\frac{\partial \rho_A}{\partial t} + \frac{1}{r^2}\frac{\partial}{\partial r}r^2 n_{Ar} + \frac{1}{r\sin\theta}\frac{\partial}{\partial \theta}(\sin\theta\, n_{A\theta}) +$$

$$\frac{1}{r\sin\theta}\frac{\partial}{\partial \phi}n_{A\phi} = r_A$$

Table 6.3: Equation of Continuity of A in a Dilute Binary System for Constant ρ and $\mathcal{D}_{AB}$ (Source: Bird et al., 1960)

Rectangular coordinates:

$$\frac{\partial \rho_A}{\partial t} + \left(v_x \frac{\partial \rho_A}{\partial x} + v_y \frac{\partial \rho_A}{\partial y} + v_z \frac{\partial \rho_A}{\partial z} \right) =$$

$$\mathcal{D}_{AB} \left(\frac{\partial^2 \rho_A}{\partial x^2} + \frac{\partial^2 \rho_A}{\partial y^2} + \frac{\partial^2 \rho_A}{\partial z^2} \right) + r_A$$

Cylindrical coordinates:

$$\frac{\partial \rho_A}{\partial t} + \left(v_r \frac{\partial \rho_A}{\partial r} + v_\theta \frac{1}{r} \frac{\partial \rho_A}{\partial \theta} + v_z \frac{\partial \rho_A}{\partial z} \right) =$$

$$\mathcal{D}_{AB} \left[\frac{1}{r} \frac{\partial}{\partial r} \left(r \frac{\partial \rho_A}{\partial r} \right) + \frac{1}{r^2} \frac{\partial^2 \rho_A}{\partial \theta^2} + \frac{\partial^2 \rho_A}{\partial z^2} \right] + r_A$$

Spherical coordinates:

$$\frac{\partial \rho_A}{\partial t} + \left(v_r \frac{\partial \rho_A}{\partial r} + v_\theta \frac{1}{r} \frac{\partial \rho_A}{\partial \theta} + v_\phi \frac{1}{r \sin \theta} \frac{\partial \rho_A}{\partial \phi} \right) =$$

$$\mathcal{D}_{AB} \left[\frac{1}{r^2} \frac{\partial}{\partial r} \left(r^2 \frac{\partial \rho_A}{\partial r} \right) + \frac{1}{r^2 \sin \theta} \frac{\partial}{\partial \theta} \left(\sin \theta \frac{\partial \rho_A}{\partial \theta} \right) + \frac{1}{r^2 \sin^2 \theta} \frac{\partial^2 \rho_A}{\partial \phi^2} \right] + r_A$$

we simply replace the molecular diffusion coefficients by new coefficients called dispersion coefficients, thus:

$$\frac{\partial \rho_A}{\partial t} + \left(v_x \frac{\partial \rho_A}{\partial x} + v_y \frac{\partial \rho_A}{\partial y} + v_z \frac{\partial \rho_A}{\partial z} \right) =$$

$$\tilde{D}_{Ax} \frac{\partial^2 \rho_A}{\partial x^2} + \tilde{D}_{Ay} \frac{\partial^2 \rho_A}{\partial y^2} + \tilde{D}_{Az} \frac{\partial^2 \rho_A}{\partial z^2} + r_A \quad (6.3.1)$$

Note that this is the same equation that appears in Table 6.3, except that we define dispersion coefficients for a species in a given direction. For example, $\tilde{D}_{Ax}$ is the dispersion coefficient for component A in the x direction. Actually the dispersion model can be derived rigorously by using the concepts of volume averaging developed by Slattery (1972), Whitaker (1962), and Friedman and Ramirez (1977).

Microscopic Balances

6.4 METHOD OF WORKING PROBLEMS

There are basically two attacks to setting up microscopic balance problems. These are to define the differential volume element for the problem at hand and then derive the describing equation, or to simplify the general equations. For problems with simple geometries, it is best to develop the equations for the individual case under study as the first step. In order to check the development, the general equations can then be simplified to make sure that the same describing equation set results. For problems with complicated geometries, it is probably best first to simplify the general equations.

6.5 STAGNANT FILM DIFFUSION

Let us consider the problem of determining the steady–state rate of evaporation into a stagnant nonmoving volume of gas over an open liquid surface, as shown in Figure 6.2.

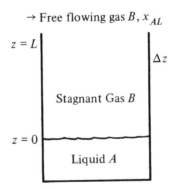

Figure 6.2: Stagnant Film Diffusion.

The mole fraction of vapor A in the free-flowing gas $(z = L)$ is x_{AL}, while that at the vapor-liquid surface $(z = 0)$ is x_{A0}. The vapor concentration at $z = 0$ can be computed from knowledge of the vapor pressure of liquid A. In order to solve this problem, we write a component material balance over a differential volume element $S\Delta z$. A differential element in only one dimension (Δz) is needed because the concentration is only a function of the one dimension, z.

$$\begin{array}{c} \text{Rate of} \\ \text{Accumulation} \\ \text{of Component} A \end{array} = \begin{array}{c} \text{Rate In} \\ \text{of Component } A \end{array} - \begin{array}{c} \text{Rate Out} \\ \text{of Component } A \end{array}$$

$$0 = N_A S \mid_z - N_A S \mid_{z+\Delta z} \tag{6.5.1}$$

Here S is the cross-sectional area of the container. Dividing by the differential element and taking the limit as $\Delta z \to 0$ gives

$$\frac{\partial N_A S}{\partial z} = 0 \qquad (6.5.2)$$

Since S is constant for this geometry,

$$\frac{\partial N_A}{\partial z} = 0 \qquad (6.5.3)$$

Upon integration, we have $N_A = $ constant. A material balance on component B, the stagnant gas, gives

$$\frac{\partial N_B}{\partial z} = 0 \qquad (6.5.4)$$

or $N_B = $ constant.

To evaluate the constant value for N_B, we use the boundary condition that gas B is not soluble in liquid A. This means that at $z = 0$, $N_B = 0$. Since N_B is a constant, that constant value must be zero, hence the term *stagnant* film diffusion.

To evaluate the constant flux for component A, we must also introduce the fact that the vapor A is diffusing in the gas B. Using Fick's first law, we have

$$N_A - x_A(N_A + N_B) = -C\mathcal{D}_{AB}\frac{\partial x_A}{\partial z} \qquad (6.5.5)$$

Using the fact that $N_A = $ constant and $N_B = 0$ gives

$$N_A(1 - x_A) = -C\mathcal{D}_{AB}\frac{dx_A}{dz} \qquad (6.5.6)$$

which can be integrated to give

$$\frac{N_A}{C\mathcal{D}_{AB}}L = \ln\left(\frac{1 - x_{AL}}{1 - x_{A0}}\right) \qquad (6.5.7)$$

The steady-state rate of evaporation is

$$N_A S = \frac{C\mathcal{D}_{AB} S}{L}\ln\left(\frac{1 - x_{AL}}{1 - x_{A0}}\right) \qquad (6.5.8)$$

6.6 CONSERVATION OF MOMENTUM (EQUATION OF MOTION)

For our differential volume element $\Delta x \Delta y \Delta z$ shown in Figure 6.1, we write a momentum balance that can be stated in words as

Rate of Momentum Accumulation = Rate In of Momentum − Rate Out of Momentum + Sum of Forces Acting on the System

Microscopic Balances

The generation term is stated in terms of forces acting on the system. This is a consequence of Newton's second law, which states that the rate of change of momentum is equal to a force, that is, $\mathbf{f} = d(m\mathbf{v})/dt$. It is important to realize that the momentum balance is a vector equation with components in each of three mutually orthogonal coordinate directions. We will consider in this derivation only the x component of the equation of motion in rectangular coordinates.

Momentum flows into and out of the volume element by two mechanisms: by *convection* (due to the bulk flow of the fluid) and by *molecular transfer* (due to the velocity gradients in the system).

Let us first consider the convective mechanism for the transport of the x component of momentum (ρv_x). This component can enter at plane x due to the flow v_x and leave at plane $x + \Delta x$ due also to the flow v_x. The x component of momentum can enter the volume element at plane y due to the flow v_y and exits due to this flow at plane $y + \Delta y$. The net convective x component of momentum is therefore

$$\Delta y \Delta z (\rho v_x v_x |_x - \rho v_x v_x |_{x+\Delta x}) + \Delta x \Delta z (\rho v_x v_y |_y - \rho v_x v_y |_{y+\Delta y})$$
$$+ \Delta x \Delta y (\rho v_x v_z |_z - \rho v_x v_z |_{z+\Delta z})$$

Now let's consider the molecular transport of momentum. The molecular mechanism is given by the stress tensor or molecular momentum flux tensor, $\boldsymbol{\tau}$. Each element τ_{ij} can be interpreted as the jth component of momentum flux transfer in the ith direction. We are therefore interested in the terms τ_{ix}. The rate at which the x component of momentum enters the volume element at face x is $\tau_{xx} \Delta y \Delta x |_x$, the rate at which it leaves at face $x + \Delta x$ is $\tau_{xx} \Delta y \Delta x |_{x+\Delta x}$, and the rate at which it enters at face y is $\tau_{yx} \Delta x \Delta z |_y$. The net molecular contribution is therefore

$$\Delta y \Delta z (\tau_{xx} |_x - \tau_{xx} |_{x+\Delta x}) + \Delta x \Delta z (\tau_{yx} |_y - \tau_{yx} |_{y+\Delta y})$$
$$+ \Delta x \Delta y (\tau_{zx} |_z - \tau_{zx} |_{z+\Delta z})$$

In most cases the only important forces acting on the system are those arising from the fluid pressure p and the gravitational force per unit mass g. The resultant force terms are therefore

$$\Delta y \Delta z (p |_x - p |_{x+\Delta x}) + \rho g_x \Delta x \Delta y \Delta z$$

The rate of accumulation of the x component of momentum is $(\Delta x \Delta y \Delta x)(\partial \rho v_x / \partial t)$.

Substituting these expressions into the conservation principle, dividing the resulting equation by volume element $\Delta x \Delta y \Delta z$, and taking the limit as Δx, Δy, and Δz approach zero gives

$$\frac{\partial}{\partial t} \rho v_x = -\left(\frac{\partial}{\partial x} \rho v_x v_x + \frac{\partial}{\partial y} \rho v_x v_y + \frac{\partial}{\partial z} \rho v_x v_z \right)$$
$$-\left(\frac{\partial}{\partial x} \tau_{xx} + \frac{\partial}{\partial y} \tau_{yx} + \frac{\partial}{\partial z} \tau_{zx} \right) - \frac{\partial p}{\partial x} + \rho g_x \quad (6.6.1)$$

The entire vector momentum balance equation is

$$\frac{\partial}{\partial t}\rho\boldsymbol{v} = -[\nabla \cdot \rho\boldsymbol{vv}] - \nabla p - [\nabla \cdot \boldsymbol{\tau}] + \rho\boldsymbol{g} \qquad (6.6.2)$$

which can be rearranged, with the help of the equation of continuity, to give

$$\rho\frac{D\boldsymbol{v}}{Dt} = -\nabla p - [\nabla \cdot \boldsymbol{\tau}] + \rho\boldsymbol{g} \qquad (6.6.3)$$

Tables 6.4, 6.5, and 6.6 show the equation of motion in several coordinate systems in terms of $\boldsymbol{\tau}$ and in terms of velocity gradients for a Newtonian fluid.

6.7 DISPERSION DESCRIPTION

For systems to be described with dispersion coefficients, we will assume that the dispersive transfer mechanisn is similar to that of the diffusion mechanism. We do not assume that all the dispersion coefficients are equal so that we have terms such as

$$\tau_{yx} = -\tilde{\mu}_{yx}\frac{dv_x}{dy} \qquad (6.7.1)$$

$$\tau_{zx} = -\tilde{\mu}_{zx}\frac{dv_x}{dz} \qquad (6.7.2)$$

The equations for dispersion look just like those for a Newtonian fluid except we substitute the dispersion coefficients for the viscosity.

6.8 PIPE FLOW OF A NEWTONIAN FLUID

Here, we will apply the momentum balance to determine the steady-state radial velocity profile for the flow in a pipe of a Newtonian fluid. Figure 6.3 describes the system.

Writing a momentum balance over the incremental volume $2\pi r \Delta r L$ gives the following:

$$\begin{array}{c}\text{Rate of Accumulation} \\ \text{of Component } z \\ \text{of Momentum}\end{array} = \begin{array}{c}\text{Rate In} \\ \text{of Component } z \\ \text{of Momentum}\end{array}$$

$$\begin{array}{c}\text{Rate Out of} \\ -\ \text{Component } z \\ \text{of Momentum}\end{array} + \begin{array}{c}\text{Sum of} \\ \text{Forces in the} \\ z\ \text{Direction}\end{array}$$

Table 6.4: Equation of Motion in Rectangular Coordinates (x, y, z) (Source: Bird et al., 1960)

In terms of $\boldsymbol{\tau}$:

x component
$$\rho\left(\frac{\partial v_x}{\partial t} + v_x \frac{\partial v_x}{\partial x} + v_y \frac{\partial v_x}{\partial y} + v_z \frac{\partial v_x}{\partial z}\right) =$$
$$-\frac{\partial p}{\partial x} - \left(\frac{\partial \tau_{xx}}{\partial x} + \frac{\partial \tau_{yx}}{\partial y} + \frac{\partial \tau_{zx}}{\partial z}\right) + \rho g_x$$

y component
$$\rho\left(\frac{\partial v_y}{\partial t} + v_x \frac{\partial v_y}{\partial x} + v_y \frac{\partial v_y}{\partial y} + v_z \frac{\partial v_y}{\partial z}\right) =$$
$$-\frac{\partial p}{\partial y} - \left(\frac{\partial \tau_{xy}}{\partial x} + \frac{\partial \tau_{yy}}{\partial y} + \frac{\partial \tau_{zy}}{\partial z}\right) + \rho g_y$$

z component
$$\rho\left(\frac{\partial v_z}{\partial t} + v_x \frac{\partial v_z}{\partial x} + v_y \frac{\partial v_z}{\partial y} + v_z \frac{\partial v_z}{\partial z}\right) =$$
$$-\frac{\partial p}{\partial z} - \left(\frac{\partial \tau_{xz}}{\partial x} + \frac{\partial \tau_{yz}}{\partial y} + \frac{\partial \tau_{zz}}{\partial z}\right) + \rho g_z$$

In terms of velocity gradients for a Newtonian fluid and constant ρ and μ:

x component
$$\rho\left(\frac{\partial v_x}{\partial t} + v_x \frac{\partial v_x}{\partial x} + v_y \frac{\partial v_x}{\partial y} + v_z \frac{\partial v_x}{\partial z}\right) =$$
$$-\frac{\partial p}{\partial x} + \mu\left(\frac{\partial^2 v_x}{\partial x^2} + \frac{\partial^2 v_x}{\partial y^2} + \frac{\partial^2 v_x}{\partial z^2}\right) + \rho g_x$$

y component
$$\rho\left(\frac{\partial v_y}{\partial t} + v_x \frac{\partial v_y}{\partial x} + v_y \frac{\partial v_y}{\partial y} + v_z \frac{\partial v_y}{\partial z}\right) =$$
$$-\frac{\partial p}{\partial y} + \mu\left(\frac{\partial^2 v_y}{\partial x^2} + \frac{\partial^2 v_y}{\partial y^2} + \frac{\partial^2 v_y}{\partial z^2}\right) + \rho g_y$$

z component
$$\rho\left(\frac{\partial v_z}{\partial t} + v_x \frac{\partial v_z}{\partial x} + v_y \frac{\partial v_z}{\partial y} + v_z \frac{\partial v_z}{\partial z}\right) =$$
$$-\frac{\partial p}{\partial z} + \mu\left(\frac{\partial^2 v_z}{\partial x^2} + \frac{\partial^2 v_z}{\partial y^2} + \frac{\partial^2 v_z}{\partial z^2}\right) + \rho g_z$$

Table 6.5: Equation of Motion in Cylindrical Coordinates (r, θ, z) (Source: Bird et al., 1960)

In terms of τ:

r component
$$\rho\left(\frac{\partial v_r}{\partial t} + v_r\frac{\partial v_r}{\partial r} + \frac{v_\theta}{r}\frac{\partial v_r}{\partial \theta} - \frac{v_\theta^2}{r} + v_z\frac{\partial v_r}{\partial z}\right) =$$
$$-\frac{\partial p}{\partial r} - \left[\frac{1}{r}\frac{\partial}{\partial r}r\tau_{rr} + \frac{1}{r}\frac{\partial \tau_{r\theta}}{\partial \theta} - \frac{\tau_{\theta\theta}}{r} + \frac{\partial \tau_{rz}}{\partial z}\right] + \rho g_r$$

θ component
$$\rho\left(\frac{\partial v_\theta}{\partial t} + v_r\frac{\partial v_\theta}{\partial r} + \frac{v_\theta}{r}\frac{\partial v_\theta}{\partial \theta} + \frac{v_r v_\theta}{r} + v_z\frac{\partial v_\theta}{\partial z}\right) =$$
$$-\frac{1}{r}\frac{\partial p}{\partial \theta} - \left[\frac{1}{r^2}\frac{\partial}{\partial r}r^2\tau_{r\theta} + \frac{1}{r}\frac{\partial \tau_{\theta\theta}}{\partial \theta} + \frac{\partial \tau_{\theta z}}{\partial z}\right] + \rho g_\theta$$

z component
$$\rho\left(\frac{\partial v_z}{\partial t} + v_r\frac{\partial v_z}{\partial r} + \frac{v_\theta}{r}\frac{\partial v_z}{\partial \theta} + v_z\frac{\partial v_z}{\partial z}\right) =$$
$$-\frac{\partial p}{\partial z} - \left[\frac{1}{r}\frac{\partial}{\partial r}r\tau_{rz} + \frac{1}{r}\frac{\partial \tau_{\theta z}}{\partial \theta} + \frac{\partial \tau_{zz}}{\partial z}\right] + \rho g_z$$

In terms of velocity gradients for a Newtonian fluid with constant ρ and μ:

r component
$$\rho\left(\frac{\partial v_r}{\partial t} + v_r\frac{\partial v_r}{\partial r} + \frac{v_\theta}{r}\frac{\partial v_r}{\partial \theta} - \frac{v_\theta^2}{r} + v_z\frac{\partial v_r}{\partial z}\right) =$$
$$-\frac{\partial p}{\partial r} + \mu\left[\frac{\partial}{\partial r}\left(\frac{1}{r}\frac{\partial}{\partial r}rv_r\right) + \frac{1}{r^2}\frac{\partial^2 v_r}{\partial \theta^2} - \frac{2}{r^2}\frac{\partial v_\theta}{\partial \theta} + \frac{\partial^2 v_r}{\partial z^2}\right] + \rho g_r$$

θ component
$$\rho\left(\frac{\partial v_\theta}{\partial t} + v_r\frac{\partial v_\theta}{\partial r} + \frac{v_\theta}{r}\frac{\partial v_\theta}{\partial \theta} + \frac{v_r v_\theta}{r} + v_z\frac{\partial v_\theta}{\partial z}\right) =$$
$$-\frac{1}{r}\frac{\partial p}{\partial \theta} + \mu\left[\frac{\partial}{\partial r}\left(\frac{1}{r}\frac{\partial}{\partial r}rv_\theta\right) + \frac{1}{r^2}\frac{\partial^2 v_\theta}{\partial \theta^2} + \frac{2}{r^2}\frac{\partial v_r}{\partial \theta} + \frac{\partial^2 v_\theta}{\partial z^2}\right] + \rho g_\theta$$

z component
$$\rho\left(\frac{\partial v_z}{\partial t} + v_r\frac{\partial v_z}{\partial r} + \frac{v_\theta}{r}\frac{\partial v_z}{\partial \theta} + v_z\frac{\partial v_z}{\partial z}\right) =$$
$$-\frac{\partial p}{\partial z} + \mu\left[\frac{1}{r}\frac{\partial}{\partial r}\left(r\frac{\partial v_z}{\partial r}\right) + \frac{1}{r^2}\frac{\partial^2 v_z}{\partial \theta^2} + \frac{\partial^2 v_z}{\partial z^2}\right] + \rho g_z$$

Table 6.6: Equation of Motion in Spherical Coordinates (r, θ, ϕ) (Source: Bird et al., 1960)

In terms of τ:

r component

$$\rho\left(\frac{\partial v_r}{\partial t} + v_r\frac{\partial v_r}{\partial r} + \frac{v_\theta}{r}\frac{\partial v_r}{\partial \theta} + \frac{v_\phi}{r\sin\theta}\frac{\partial v_r}{\partial \phi} - \frac{v_\theta^2 + v_\phi^2}{r}\right) =$$
$$-\frac{\partial p}{\partial r} - \left[\frac{1}{r^2}\frac{\partial}{\partial r}r^2\tau_{rr} + \frac{1}{r\sin\theta}\frac{\partial}{\partial \theta}(\tau_{r\theta}\sin\theta)\right.$$
$$\left. + \frac{1}{r\sin\theta}\frac{\partial \tau_{r\phi}}{\partial \phi} - \frac{\tau_{\theta\theta}+\tau_{\phi\phi}}{r}\right] + \rho g_r$$

θ component

$$\rho\left(\frac{\partial v_\theta}{\partial t} + v_r\frac{\partial v_\theta}{\partial r} + \frac{v_\theta}{r}\frac{\partial v_\theta}{\partial \theta} + \frac{v_\phi}{r\sin\theta}\frac{\partial v_\theta}{\partial \phi} + \frac{v_r v_\theta}{r} - \frac{v_\phi^2\cot\theta}{r}\right) =$$
$$-\frac{1}{r}\frac{\partial p}{\partial \theta} - \left[\frac{1}{r^2}\frac{\partial}{\partial r}r^2\tau_{r\theta} + \frac{1}{r\sin\theta}\frac{\partial}{\partial \theta}(\tau_{\theta\theta}\sin\theta)\right.$$
$$\left. + \frac{1}{r\sin\theta}\frac{\partial \tau_{\theta\phi}}{\partial \phi} + \frac{\tau_{r\theta}}{r} - \frac{\cot\theta}{r}\tau_{\phi\phi}\right] + \rho g_\theta$$

ϕ component

$$\rho\left(\frac{\partial v_\phi}{\partial t} + v_r\frac{\partial v_\phi}{\partial r} + \frac{v_\theta}{r}\frac{\partial v_\phi}{\partial \theta} + \frac{v_\phi}{r\sin\theta}\frac{\partial v_\phi}{\partial \phi} + \frac{v_\phi v_r}{r} + \frac{v_\theta v_\phi}{r}\cot\theta\right) =$$
$$-\frac{1}{r\sin\theta}\frac{\partial p}{\partial \phi} - \left[\frac{1}{r^2}\frac{\partial}{\partial r}(r^2\tau_{r\phi}) + \frac{1}{r}\frac{\partial \tau_{\theta\phi}}{\partial \theta} + \frac{1}{r\sin\theta}\frac{\partial \tau_{\phi\phi}}{\partial \phi}\right.$$
$$\left. + \frac{\tau_{r\phi}}{r} + \frac{2\cot\theta}{r}\tau_{\theta\phi}\right] + \rho g_\phi$$

Table 6.6 (continued): Equation of Motion in Spherical Coordinates (r, θ, ϕ) (Source: Bird et al., 1960)

In terms of velocity gradients for a Newtonian fluid with contstant ρ and μ :[a]

r component

$$\rho\left(\frac{\partial v_r}{\partial t} + v_r\frac{\partial v_r}{\partial r} + \frac{v_\theta}{r}\frac{\partial v_r}{\partial \theta} + \frac{v_\phi}{r\sin\theta}\frac{\partial v_r}{\partial \phi} - \frac{v_\theta^2 + v_\phi^2}{r}\right) =$$

$$-\frac{\partial p}{\partial r} + \mu\left(\nabla^2 v_r - \frac{2}{r^2}v_r - \frac{2}{r^2}\frac{\partial v_\theta}{\partial \theta} - \frac{2}{r^2}v_\theta\cot\theta - \frac{2}{r^2\sin\theta}\frac{\partial v_\phi}{\partial \phi}\right) + \rho g_r$$

θ component

$$\rho\left(\frac{\partial v_\theta}{\partial t} + v_r\frac{\partial v_\theta}{\partial r} + \frac{v_\theta}{r}\frac{\partial v_\theta}{\partial \theta} + \frac{v_\phi}{r\sin\theta}\frac{\partial v_\theta}{\partial \phi} + \frac{v_r v_\theta}{r} - \frac{v_\phi^2 \cot\theta}{r}\right) =$$

$$-\frac{1}{r}\frac{\partial p}{\partial \theta} + \mu\left(\nabla^2 v_\theta + \frac{2}{r^2}\frac{\partial v_r}{\partial \theta} - \frac{v_\theta}{r^2\sin^2\theta} - \frac{2\cos\theta}{r^2\sin^2\theta}\frac{\partial v_\phi}{\partial \phi}\right) + \rho g_\theta$$

ϕ component

$$\rho\left(\frac{\partial v_\phi}{\partial t} + v_r\frac{\partial v_\phi}{\partial r} + \frac{v_\theta}{r}\frac{\partial v_\phi}{\partial \theta} + \frac{v_\phi}{r\sin\theta}\frac{\partial v_\phi}{\partial \phi} + \frac{v_\phi v_r}{r} + \frac{v_\theta v_\phi}{r}\cos\theta\right) =$$

$$-\frac{1}{r\sin\theta}\frac{\partial p}{\partial \phi} + \mu\left(\nabla^2 v_\phi - \frac{v_\phi}{r^2\sin^2\theta} + \frac{2}{r^2\sin\theta}\frac{\partial v_r}{\partial \phi} + \frac{2\cos\theta}{r^2\sin^2\theta}\frac{\partial v_\theta}{\partial \phi}\right) + \rho g_\phi$$

[a]In these equations

$$\nabla^2 = \frac{1}{r^2}\frac{\partial}{\partial r}\left(r^2\frac{\partial}{\partial r}\right) + \frac{1}{r^2\sin\theta}\frac{\partial}{\partial \phi}\left(\sin\theta\frac{\partial}{\partial \theta}\right) + \frac{1}{r^2\sin^2\theta}\left(\frac{\partial^2}{\partial \phi^2}\right)$$

Microscopic Balances

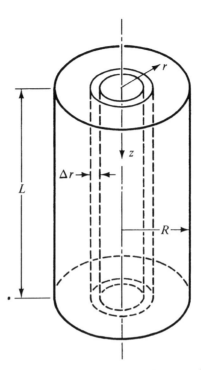

Figure 6.3: Pipe Flow of a Newtonian Fluid.

Steady state : $0 =$
Convective flow : $\rho v_z v_z (2\pi r \Delta r)|_{z=0} - \rho v_z v_z (2\pi r \Delta r)|_{z=L}$
Molecular transport : $+\tau_{rz}(2\pi r L)|_r - \tau_{rz}(2\pi r L)|_{r+\Delta r}$
Gravity force : $+\rho g_z (2\pi r L \Delta r)$
Pressure force : $+P_0(2\pi r \Delta r) - P_L(2\pi r \Delta r)$ (6.8.1)

Dividing by Δr and taking the limit as $\Delta r \to 0$ gives

$$0 = -2\pi L \frac{d\tau_{rz} r}{dr} + \rho g_z 2\pi r L + 2\pi r (P_0 - P_L) \tag{6.8.2}$$

or

$$\frac{d}{dr}(r\tau_{rz}) = \left(\frac{P_0 - P_L}{L} + \rho g\right) r \tag{6.8.3}$$

Integrating (6.8.3) gives

$$r\tau_{rz} = \left(\frac{P_0 - P_L}{L} + \rho g\right)\frac{r^2}{2} + c_1 \tag{6.8.4}$$

where c_1 is a constant of integration. Using the definition of a Newtonian fluid,

$$\tau_{rz} = -\mu \frac{dv_z}{dr} \qquad (6.8.5)$$

gives

$$\frac{dv_z}{dr} = -\left(\frac{P_0 - P_L}{L} + \rho g\right) \frac{r}{2\mu} + \frac{c_1}{\mu r} \qquad (6.8.6)$$

The boundary conditions for this problem are
    B.C.1   at $r = 0$ the velocity is finite
    B.C.2   at $r = R$ the velocity is zero ($v_z\,|_R = 0$)
(This is the no-slip-at-the-wall condition.)

Using B.C.1 we find that the constant c_1 in equation (6.8.6) must be zero. This gives

$$\frac{dv_z}{dr} = -\left(\frac{P_0 - P_L}{L} + \rho g\right) \frac{r}{2\mu} \qquad (6.8.7)$$

which upon integrating becomes

$$v_z = -\left(\frac{P_0 - P_L}{L} + \rho g\right) \frac{r^2}{4\mu} + c_2 \qquad (6.8.8)$$

Using B.C.2 we can evaluate c_2 and the final expression for the velocity profile v_z is

$$v_z = \left(\frac{P_0 - P_L}{L} + \rho g\right) \frac{R^2}{4\mu} \left[1 - \left(\frac{r}{R}\right)^2\right] \qquad (6.8.9)$$

This is the well-known parabolic velocity profile for laminar pipe flow.

6.9 DEVELOPMENT OF MICROSCOPIC MECHANICAL ENERGY EQUATION AND ITS APPLICATION

For isothermal problems of fluid dynamics, it is usually advisable to use a form of the energy balance that involves only mechanical energy terms. By taking the scalar product of the equation of motion, equation (6.6.3), with the velocity vector (**v**) we arrive at the mechanical energy balance (Bird et al., 1960).

$$\rho \frac{D}{Dt}\left(\frac{v^2}{2}\right) = -(\boldsymbol{v} \cdot \nabla p) - (\boldsymbol{v} \cdot [\nabla \cdot \boldsymbol{\tau}]) + \rho(\boldsymbol{v} \cdot \boldsymbol{g}) \qquad (6.9.1)$$

This scalar equation describes the rate of change of kinetic energy per unit mass for an element of fluid moving downstream.

Microscopic Balances

As discussed by Bird et al. (1960), we can rewrite this equation with the help of the equation of continuity as

Rate of Increase   Net Rate of Input of
in Kinetic Energy = Kinetic Energy by
Per Unit Volume   Virtue of Bulk Flow

$$\frac{\partial}{\partial t}\left(\frac{\rho v^2}{2}\right) = \left[-\left(\nabla \cdot \frac{\rho v^2 \boldsymbol{v}}{2}\right)\right]$$

Rate of Work Done by
− Pressure of Surroundings
on Volume Element

$$-[\nabla \cdot p\boldsymbol{v}]$$

Rate of Reversible   Rate of Work Done
− Conversion to   − by Viscous Forces
Internal Energy   on Volume Elements

$$-[p(-\nabla \cdot \boldsymbol{v})] - [(\nabla \cdot [\boldsymbol{\tau} \cdot \boldsymbol{v}])]$$

Rate of Irreversible   Rate of Work Done
− Conversion to   + by Gravity Force
Internal Energy   on Volume Element

$$-[(-\boldsymbol{\tau} : \nabla \boldsymbol{v})] + [\rho(\boldsymbol{v} \cdot \boldsymbol{g})] \tag{6.9.2}$$

Equation (6.9.2) is most often used for inviscous flow problems when $\tau = 0$. For one-dimensional problems, it is often more convenient to apply the macroscopic mechanical energy balance equation, (1.6.11), to a differential volume.

$$S\frac{\partial}{\partial t}(KE + PE + A) + \frac{\partial}{\partial z}w\left(\frac{1}{2}\frac{<v^3>}{<v>} + gZ\right)$$

$$+\frac{\partial}{\partial z}\frac{w}{\rho}p + \frac{\partial}{\partial z}E_v + \frac{\partial}{\partial z}\dot{W}_s = 0 \tag{6.9.3}$$

where KE = the kinetic energy per unit volume = $\rho v^2/2$
            PE = the potential energy per unit volume = $\rho g Z$
            A = the Helmholtz free energy per unit volume = $\rho(\hat{U} - T\hat{S})$
            $\hat{S}$ = the entropy per unit mass
            $\hat{U}$ = the internal energy per unit mass
            E_v = rate of mechanical energy irreversibly converted to thermal energy = wh_L
            h_L = head loss due to friction
            $\dot{W}_s$ = rate at which the system performs mechanical work

$$= w\hat{W}_s$$
$\hat{W}_s$ = work performed per unit mass
S = cross-sectional area

6.10 PIPELINE GAS FLOW

Here we want to consider the flow of a compressible gas down a pipe at steady state.

The equation of continuity is

$$w = \text{constant} \tag{6.10.1}$$

and the mechanical energy balance is

$$\frac{d}{dz}\frac{w}{\rho}p + \frac{d}{dz}E_v = 0 \tag{6.10.2}$$

Assuming ρ does not vary with distance to the extent that pressure does gives

$$\frac{w}{\rho}\frac{dp}{dz} + w\frac{d}{dz}\left(\frac{1}{2}\frac{v^2 zf}{4D}\right) = 0 \tag{6.10.3}$$

where f is the friction factor. For turbulent flow f is constant. Also v does not vary significantly with z; therefore,

$$\frac{dp}{dz} = -\frac{v^2 \rho f}{8D} \tag{6.10.4}$$

and introducing the volumetric flow rate F, we have

$$F = \frac{v\pi D^2}{4} \tag{6.10.5}$$

The mechanical energy balance equation, (6.9.3), then becomes

$$\frac{dp}{dz} = -f\frac{2}{\pi}\frac{F^2 \rho}{D^5} \tag{6.10.6}$$

The equation of state for an ideal gas is

$$\rho = \frac{p(mw)}{RT} \tag{6.10.7}$$

The information-flow diagram for solving this problem is shown in Figure 6.4.

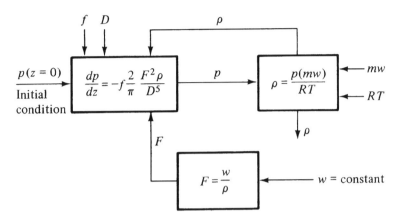

Figure 6.4: Pipeline Gas Flow.

6.11 DEVELOPMENT OF MICROSCOPIC THERMAL ENERGY BALANCE AND ITS APPLICATION

The conservation law for total energy is

Rate of Accumulation of Internal and Kinetic Energy = Rate of Internal and Kinetic Energy In by Convection

− Rate of Internal and Kinetic Energy Out by Convection

+ Net Rate of Heat Addition by Conduction − Net Rate of Work Done by the System on the Surroundings (6.11.1)

This statement is not complete for it neglects nuclear terms, radiation terms, electromagnetic terms, and reaction terms. This later reaction term must often be included in chemical engineering problems as a generation term.

The accumulation of total energy in the differential volume $\Delta x \Delta y \Delta z$ is

$$\frac{\partial}{\partial t}\left(\rho \hat{U} + \frac{1}{2}\rho v^2\right) \Delta x \Delta y \Delta z$$

where $\hat{U}$ is the internal energy per unit mass and v is the magnitude of the local fluid velocity. The convective terms are

$$v_x(\rho\hat{U} + 1/2\rho v^2)\Delta z \Delta y \mid_x - v_x(\rho\hat{U} + 1/2\rho v^2)\Delta z \Delta y \mid_{x+\Delta x}$$

$$
\begin{aligned}
&+\; v_y(\rho \hat{U} + 1/2\rho v^2)\Delta x \Delta z \,|_y - v_y(\rho \hat{U} + 1/2\rho v^2)\Delta x \Delta z \,|_{y+\Delta y}\\
&+\; v_z(\rho \hat{U} + 1/2\rho v^2)\Delta x \Delta y \,|_z - v_z(\rho \hat{U} + 1/2\rho v^2)\Delta x \Delta y \,|_{z+\Delta z}
\end{aligned}
$$

The conduction terms are

$$
\begin{aligned}
& q_x \Delta y \Delta z \,|_x - q_x \Delta y \Delta z \,|_{x+\Delta x}\\
&+\; q_y \Delta x \Delta z \,|_y - q_y \Delta x \Delta z \,|_{y+\Delta y}\\
&+\; q_z \Delta x \Delta y \,|_z - q_z \Delta x \Delta y \,|_{z+\Delta z}
\end{aligned}
$$

where q_x, q_y, and q_z are the components of the conductive heat flux vector $\mathbf{q}$.

The work done by the fluid element against its surroundings consists of work against volume forces (gravity) and work against surface forces (pressure, viscous). The gravity term is

$$\rho \Delta x \Delta y \Delta z (v_x g_x + v_y g_y + v_z g_z)$$

the pressure term is

$$
\begin{aligned}
& p v_x \Delta y \Delta z \,|_x - p v_x \Delta y \Delta z \,|_{x+\Delta x}\\
&+\; p v_y \Delta x \Delta z \,|_y - p v_y \Delta x \Delta z \,|_{y+\Delta y}\\
&+\; p v_z \Delta x \Delta z \,|_z - p v_z \Delta x \Delta y \,|_{z+\Delta z}
\end{aligned}
$$

and the viscous terms are

$$
\begin{aligned}
& \Delta y \Delta z(\tau_{xx}v_x + \tau_{xy}v_y + \tau_{xz}v_z)\,|_x - \Delta y \Delta z(\tau_{xx}v_x + \tau_{xy}v_y + \tau_{xz}v_z)\,|_{x+\Delta x}\\
&+\; \Delta x \Delta z(\tau_{yx}v_x + \tau_{yy}v_y + \tau_{yx}v_z)\,|_y - \Delta x \Delta z(\tau_{yx}v_x + \tau_{yy}v_y + \tau_{yx}v_z)\,|_{y+\Delta y}\\
&+\; \Delta x \Delta y(\tau_{zx}v_x + \tau_{zy}v_y + \tau_{zz}v_z)\,|_z - \Delta x \Delta y(\tau_{zx}v_x + \tau_{zy}v_y + \tau_{zz}v_z)\,|_{z+\Delta z}
\end{aligned}
$$

Dividing by the volume element and taking the limit as $\Delta x \Delta y \Delta z$ goes to zero gives

$$
\begin{aligned}
\frac{\partial}{\partial t}\left(\rho \hat{U} + \frac{1}{2}\rho v^2\right) =& -\left[\frac{\partial}{\partial x}v_x\left(\rho\hat{U} + \frac{1}{2}\rho v^2\right) + \frac{\partial}{\partial y}v_y\left(\rho\hat{U} + \frac{1}{2}\rho v^2\right)\right.\\
&\left. +\; \frac{\partial}{\partial z}v_z\left(\rho\hat{U} + \frac{1}{2}\rho v^2\right)\right] - \left(\frac{\partial q_x}{\partial x} + \frac{\partial q_y}{\partial y} + \frac{\partial q_z}{\partial z}\right)\\
&+\; (\rho(v_x g_x + v_y g_y + v_z g_z)) - \left(\frac{\partial}{\partial x}pv_x + \frac{\partial}{\partial y}pv_y + \frac{\partial}{\partial z}pv_z\right)\\
&-\; \left[\frac{\partial}{\partial x}(\tau_{xx}v_x + \tau_{xy}v_y + \tau_{xz}v_z) + \frac{\partial}{\partial y}(\tau_{yx}v_x + \tau_{yy}v_y + \tau_{yz}v_z)\right.\\
&\left. +\; \frac{\partial}{\partial z}(\tau_{zx}v_x + \tau_{zy}v_y + \tau_{zz}v_z)\right] \quad (6.11.2)
\end{aligned}
$$

Microscopic Balances

This is the equation for the conservation of total energy. Subtracting the mechanical energy equation (6.9.2) from equation (6.11.2) gives the *thermal* energy equation

$$\rho \frac{D\hat{U}}{DT} = -(\nabla \cdot \mathbf{q}) - p(\nabla \cdot \mathbf{v}) - \boldsymbol{\tau} : \nabla \mathbf{v}) \qquad (6.11.3)$$

Putting in the thermodynamics relation for the internal energy

$$d\hat{U} = \left(\frac{\partial \hat{U}}{\partial \hat{V}}\right)_T d\hat{V} + \left(\frac{\partial \hat{U}}{\partial T}\right)_{\hat{V}} dT \qquad (6.11.4)$$

gives

$$\rho \frac{D\hat{U}}{Dt} = \left[-p + T\left(\frac{\partial p}{\partial T}\right)_\rho\right](\nabla \cdot \mathbf{v}) + \rho \hat{C}_v \frac{DT}{Dt} \qquad (6.11.5)$$

When included in the thermal energy equation, we get the thermal energy equations given in Tables 6.7, 6.8, and 6.9 for various geometries.

Table 6.7: Components of the Energy Flux q
(Source: Bird et al., 1960)

Rectangular	Cylindrical	Spherical
$q_x = -k\dfrac{\partial T}{\partial x}$	$q_r = -k\dfrac{\partial T}{\partial r}$	$q_r = -k\dfrac{\partial T}{\partial r}$
$q_y = -k\dfrac{\partial T}{\partial y}$	$q_\theta = -k\dfrac{1}{r}\dfrac{\partial T}{\partial \theta}$	$q_\theta = -k\dfrac{1}{r}\dfrac{\partial T}{\partial \theta}$
$q_z = -k\dfrac{\partial T}{\partial z}$	$q_z = -k\dfrac{\partial T}{\partial z}$	$q_\phi = -k\dfrac{1}{r\sin\theta}\dfrac{\partial T}{\partial \phi}$

6.12 HEAT CONDUCTION THROUGH COMPOSITE CYLINDRICAL WALLS

We are interested in computing the steady–state heat flux for conduction through composite material cylindrical walls. A diagram of the system is given in Figure 6.5.

Table 6.8: Equation of Thermal Energy in Terms of Energy and Momentum Fluxes (Source: Bird et al., 1960)

Rectangular coordinates:
$$\rho \hat{C}_v \left(\frac{\partial T}{\partial t} + v_x \frac{\partial T}{\partial x} + v_y \frac{\partial T}{\partial y} + v_z \frac{\partial T}{\partial z} \right) =$$

$$-\left(\frac{\partial q_x}{\partial x} + \frac{\partial q_y}{\partial y} + \frac{\partial q_z}{\partial z} \right) - T \left(\frac{\partial p}{\partial T} \right)_\rho \left(\frac{\partial v_x}{\partial x} + \frac{\partial v_y}{\partial y} + \frac{\partial v_z}{\partial z} \right)$$

$$-\left\{ \tau_{xx} \frac{\partial v_x}{\partial x} + \tau_{yy} \frac{\partial v_y}{\partial y} + \tau_{zz} \frac{\partial v_z}{\partial z} \right\} - \left\{ \tau_{xy} \left(\frac{\partial v_x}{\partial y} + \frac{\partial v_y}{\partial x} \right) \right.$$

$$\left. + \tau_{xz} \left(\frac{\partial v_x}{\partial z} + \frac{\partial v_z}{\partial x} \right) + \tau_{yz} \left(\frac{\partial v_y}{\partial z} + \frac{\partial v_z}{\partial y} \right) \right\}$$

Cylindrical coordinates:
$$\rho \hat{C}_v \left(\frac{\partial T}{\partial t} + v_r \frac{\partial T}{\partial r} + \frac{v_\theta}{r} \frac{\partial T}{\partial \theta} + v_z \frac{\partial T}{\partial z} \right) =$$

$$-\left(\frac{1}{r} \frac{\partial}{\partial r} r q_r + \frac{1}{r} \frac{\partial q_\theta}{\partial \theta} + \frac{\partial q_z}{\partial z} \right) - T \left(\frac{\partial p}{\partial T} \right)_\rho \left(\frac{1}{r} \frac{\partial}{\partial r} r v_r + \frac{1}{r} \frac{\partial v_\theta}{\partial \theta} + \frac{\partial v_z}{\partial z} \right)$$

$$-\left\{ \tau_{rr} \frac{\partial v_r}{\partial r} - \tau_{\theta\theta} \frac{1}{r} \left(\frac{\partial v_\theta}{\partial \theta} + v_r \right) + \tau_{zz} \frac{\partial v_z}{\partial z} \right\} - \left\{ \tau_{\theta r} \left[r \frac{\partial}{\partial r} \left(\frac{v_\theta}{r} \right) + \frac{1}{r} \frac{\partial v_r}{\partial \theta} \right] \right.$$

$$\left. + \tau_{rz} \left(\frac{\partial v_z}{\partial r} + \frac{\partial v_r}{\partial z} \right) + \tau_{\theta z} \left(\frac{1}{r} \frac{\partial v_z}{\partial \theta} + \frac{\partial v_\theta}{\partial z} \right) \right\}$$

Spherical coordinates:
$$\rho \hat{C}_v \left(\frac{\partial T}{\partial t} + v_r \frac{\partial T}{\partial r} + \frac{v_\theta}{r} \frac{\partial T}{\partial \theta} + \frac{v_\phi}{r \sin \theta} \frac{\partial T}{\partial \phi} \right) =$$

$$-\left[\frac{1}{r^2} \frac{\partial}{\partial r} r^2 q_r + \frac{1}{r \sin \theta} \frac{\partial}{\partial \theta} (q_\theta \sin \theta) + \frac{1}{r \sin \theta} \frac{\partial q_\phi}{\partial \phi} \right]$$

$$-T \left(\frac{\partial p}{\partial T} \right)_\rho \left[\frac{1}{r^2} \frac{\partial}{\partial r} r^2 v_r + \frac{1}{r \sin \theta} \frac{\partial}{\partial \theta} (v_\theta \sin \theta) + \frac{1}{r \sin \theta} \frac{\partial v_\phi}{\partial \phi} \right]$$

$$-\left\{ \tau_{rr} \frac{\partial v_r}{\partial r} + \tau_{\theta\theta} \left(\frac{1}{r} \frac{\partial v_\theta}{\partial \theta} + \frac{v_r}{r} \right) + \tau_{\phi\phi} \left(\frac{1}{r \sin \theta} \frac{\partial v_\phi}{\partial \phi} + \frac{v_r}{r} + \frac{v_\theta \cot \theta}{r} \right) \right\}$$

$$-\left\{ \tau_{r\theta} \left(\frac{\partial v_\theta}{\partial r} + \frac{1}{r} \frac{\partial v_r}{\partial \theta} - \frac{v_\theta}{r} \right) + \tau_{r\phi} \left(\frac{\partial v_\phi}{\partial r} + \frac{1}{r \sin \theta} \frac{\partial v_r}{\partial \phi} - \frac{v_\phi}{r} \right) \right.$$

$$\left. + \tau_{\theta\phi} \left(\frac{1}{r} \frac{\partial v_\phi}{\partial \theta} + \frac{1}{r \sin \theta} \frac{\partial v_\phi}{\partial \phi} - \frac{\cot \theta}{r} v_\phi \right) \right\}$$

Note: The terms contained in braces {} are associated with viscous dissipation and may usually be neglected, except for systems with large velocity gradients.

Table 6.9: Equation of Thermal Energy (for Newtonian Fluids with Constant ρ, μ, and k) (Source: Bird et al., 1960)

Rectangular coordinates:

$$\rho \hat{C}_v \left(\frac{\partial T}{\partial t} + v_x \frac{\partial T}{\partial x} + v_y \frac{\partial T}{\partial y} + v_z \frac{\partial T}{\partial z} \right) = k \left(\frac{\partial^2 T}{\partial x^2} + \frac{\partial^2 T}{\partial y^2} + \frac{\partial^2 T}{\partial z^2} \right)$$

$$+ 2\mu \left\{ \left(\frac{\partial v_x}{\partial x} \right)^2 + \left(\frac{\partial v_y}{\partial y} \right)^2 + \left(\frac{\partial v_z}{\partial z} \right)^2 \right\}$$

$$+ \mu \left\{ \left(\frac{\partial v_x}{\partial y} + \frac{\partial v_y}{\partial x} \right)^2 + \left(\frac{\partial v_x}{\partial z} + \frac{\partial v_z}{\partial x} \right)^2 + \left(\frac{\partial v_y}{\partial z} + \frac{\partial v_z}{\partial y} \right)^2 \right\}$$

Cylindrical coordinates:

$$\rho \hat{C}_v \left(\frac{\partial T}{\partial t} + v_r \frac{\partial T}{\partial r} + \frac{v_\theta}{r} \frac{\partial T}{\partial \theta} + v_z \frac{\partial T}{\partial z} \right) = k \left[\frac{1}{r} \frac{\partial}{\partial r} \left(r \frac{\partial T}{\partial r} \right) + \frac{1}{r^2} \frac{\partial^2 T}{\partial \theta^2} + \frac{\partial^2 T}{\partial z^2} \right]$$

$$+ 2\mu \left\{ \left(\frac{\partial v_r}{\partial r} \right)^2 + \left[\frac{1}{r} \left(\frac{\partial v_\theta}{\partial \theta} + v_r \right) \right]^2 + \left(\frac{\partial v_z}{\partial z} \right)^2 \right\} + \mu \left\{ \left(\frac{\partial v_\theta}{\partial z} + \frac{1}{r} \frac{\partial v_z}{\partial \theta} \right)^2 \right.$$

$$\left. + \left(\frac{\partial v_z}{\partial r} + \frac{\partial v_r}{\partial z} \right)^2 + \left[\frac{1}{r} \frac{\partial v_r}{\partial \theta} + r \frac{\partial}{\partial r} \left(\frac{v_\theta}{r} \right) \right]^2 \right\}$$

Spherical coordinates:

$$\rho \hat{C}_v \left(\frac{\partial T}{\partial t} + v_r \frac{\partial T}{\partial r} + \frac{v_\theta}{r} \frac{\partial T}{\partial \theta} + \frac{v_\phi}{r \sin \theta} \frac{\partial T}{\partial \phi} \right) =$$

$$k \left[\frac{1}{r^2} \frac{\partial}{\partial r} \left(r^2 \frac{\partial T}{\partial r} \right) + \frac{1}{r^2 \sin \theta} \frac{\partial}{\partial \theta} \left(\sin \theta \frac{\partial T}{\partial \theta} \right) + \frac{1}{r^2 \sin^2 \theta} \frac{\partial^2 T}{\partial \phi^2} \right]$$

$$+ 2\mu \left\{ \left(\frac{\partial v_r}{\partial r} \right)^2 + \left(\frac{1}{r} \frac{\partial v_\theta}{\partial \theta} + \frac{v_r}{r} \right)^2 + \left(\frac{1}{r \sin \theta} \frac{\partial v_\phi}{\partial \phi} + \frac{v_r}{r} + \frac{v_\theta \cot \theta}{r} \right)^2 \right\}$$

$$+ \mu \left\{ \left[r \frac{\partial}{\partial r} \left(\frac{v_\theta}{r} \right) + \frac{1}{r} \frac{\partial v_r}{\partial \theta} \right]^2 + \left[\frac{1}{r \sin \theta} \frac{\partial v_r}{\partial \phi} + r \frac{\partial}{\partial r} \left(\frac{v_\phi}{r} \right) \right]^2 \right.$$

$$\left. + \left[\frac{\sin \theta}{r} \frac{\partial}{\partial \theta} \left(\frac{v_\phi}{\sin \theta} \right) + \frac{1}{r \sin \theta} \frac{\partial v_\theta}{\partial \phi} \right]^2 \right\}$$

Note: The terms contained in braces {} are associated with viscous dissipation and may usually be neglected.

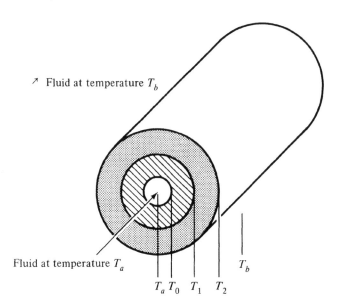

Figure 6.5: Heat Conduction Through Composite Cylindrical Walls.

By writing an energy balance for the solid over a differential volume element $2\pi r \Delta r L$, we get

$$\begin{array}{c}\text{Rate of}\\ \text{Accumulation}\end{array} = \text{Rate In} - \text{Rate Out}$$

$$0 = q_r 2\pi r L \mid_r - q_r 2\pi r L \mid_{r+\Delta r} \quad (6.12.1)$$

Dividing by Δr and taking the limit as $\Delta r \to 0$ gives

$$\frac{dr q_r}{dr} = 0 \quad (6.12.2)$$

which, upon integration, becomes

$$r q_r = \text{constant} = c_1 \quad (6.12.3)$$

Using Fourier's law of heat conduction we have

$$-r k \frac{dT}{dr} = c_1 \quad (6.12.4)$$

which, upon integration, gives

$$T = \frac{c_1}{k} \ln r + c_2 \quad (6.12.5)$$

where c_2 is a constant of integration. If we apply this equation for the two separate solid materials, we have

$$T_0 - T_1 = \frac{c_1}{k_{01}} \ln \frac{r_1}{r_0} \tag{6.12.6}$$

$$T_1 - T_2 = \frac{c_1}{k_{12}} \ln \frac{r_2}{r_1} \tag{6.12.7}$$

Adding equations (6.12.6) and (6.12.7) gives

$$T_0 - T_2 = c_1 \left(\frac{\ln(r_1/r_0)}{k_{01}} + \frac{\ln(r_2/r_1)}{k_{12}} \right) \tag{6.12.8}$$

where $c_1 = rq_r = r_0 q_0$. From Newton's law of cooling, we also know that heat transfer between the solid and gas phases can be given by

$$T_a - T_0 = q_0/h_0 \tag{6.12.9}$$

and

$$T_2 - T_b = q_2/h_2 = q_0 r_0 / h_2 r_2 \tag{6.12.10}$$

where h is the convection heat transfer coefficient. The overall temperature difference can therefore be given by

$$T_a - T_b = q_0 r_0 \left(\frac{1}{h_0 r_0} + \frac{\ln(r_1/r_0)}{k_{01}} + \frac{\ln(r_2/r_1)}{k_{12}} + \frac{1}{h_2 r_2} \right) \tag{6.12.11}$$

The heat flux at the inner wall can be expressed in terms of an overall heat transfer coefficient as

$$q_0 = U_0 (T_a - T_b)$$

where

$$U_0 = 1 \div r_0 \left(\frac{1}{h_0 r_0} + \frac{\ln(r_1/r_0)}{h_{01}} + \frac{\ln(r_2/r_1)}{k_{12}} + \frac{1}{h_2 r_2} \right) \tag{6.12.12}$$

U_0 is defined as the overall heat-transfer coefficient based on the inside area.

The flux at the outer wall is

$$q_2 = U_0 \frac{r_0}{r_2} (T_a - T_b) \tag{6.12.13}$$

6.13 HEAT CONDUCTION WITH CHEMICAL HEAT SOURCE

We consider a tubular packed bed reactor shown in Figure 6.6.

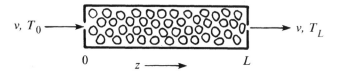

Figure 6.6: Chemical Reactor.

In this system a reactant enters a chemical reactor at $z = 0$ with a superficial velocity of v (cm/sec). The reactor is packed with catalyst particles. The heat of reaction is known as ΔH (cal/g mol). The reaction rate is zero–order in the reactant concentrtion and therefore is a constant down the reactor R (g mol/sec cm^3). Because of the effect of the catalyst packing, both convective and dispersion thermal effects are present in this reactor. We want to compute the temperature profile down the reactor. We write a steady–state energy balance for the differential reactor volume, $A\Delta z$, as

$$0 = \begin{array}{c} \text{Rate of} \\ \text{Energy In} \end{array} - \begin{array}{c} \text{Rate of} \\ \text{Energy Out} \end{array} + \begin{array}{c} \text{Rate of} \\ \text{Generation} \\ \text{of Energy} \end{array}$$

$$0 = \rho v c_p A T \mid_z - \rho v c_p A T \mid_{z+\Delta z} + q A \mid_z - q A \mid_{z+\Delta z} + (\Delta H) R A \, \Delta z \quad (6.13.1)$$

Dividing by the reactor volume, $A\Delta z$, and taking the limit as $\Delta z \to 0$ gives

$$0 = -\rho v c_p \frac{dT}{dz} - \frac{dq}{dz} + (\Delta H)R \quad (6.13.2)$$

We assume that the dispersion mechanism can be given as

$$q = -\tilde{k}\frac{dT}{dz} \quad (6.13.3)$$

The final energy balance is

$$\frac{d^2 T}{dz^2} - \frac{\rho v c_p}{\tilde{k}} \frac{dT}{dz} = -\frac{(\Delta H)R}{\tilde{k}} \quad (6.13.4)$$

or

$$\frac{d^2 T}{dz^2} - A\frac{dT}{dz} = -B \quad (6.13.5)$$

with
$$A = \frac{\rho v c_p}{\tilde{k}} \quad \text{and} \quad B = \frac{(\Delta H) R}{\tilde{k}} \quad (6.13.6)$$

As boundary conditions for the problem we assume that we know the inlet reactor temperature

$$T(z = 0) = T_0 \quad (6.13.7)$$

and we know that once we leave the reactor, the temperature will remain constant, so

$$\frac{dT}{dz}(z = L) = 0 \quad (6.13.8)$$

The differential equation of (6.13.5) is a linear second–order, nonhomogeneous equation. From Appendix A we find that the homogeneous solution is

$$T_h = C_1 e^{Az} + C_2 \quad (6.13.9)$$

where C_1 and C_2 are constants of integration.

The particular solution is

$$T_p = \frac{B}{A} z \quad (6.13.10)$$

The general solution is therefore

$$T = C_1 e^{Az} + \frac{B}{A} z + C_2 \quad (6.13.11)$$

Using the boundary conditions of (6.13.7) and (6.13.8) gives

$$T_0 = C_1 + C_2 \quad (6.13.12)$$

$$C_1 = -\frac{B}{A^2} e^{-AL} \quad (6.13.13)$$

The final temperature profile is calculated as

$$T = T_0 + \frac{B}{A} z + \frac{B}{A^2} \left(e^{-AL} - e^{-A(L-z)} \right) \quad (6.13.14)$$

6.14 MATHEMATICAL MODELING FOR A STYRENE MONOMER TUBULAR REACTOR

We desire to develop a mathematical model for a styrene monomer pilot plant. The process flow sheet is shown in Figure 6.7. Ethylbenzene is passed through a tubular reactor packed with an iron oxide catalyst, and dehydrogenation to styrene takes place. The ethylbenzene feed stream is preheated and mixed with superheated steam to a reactor inlet temperature of over 875 K. Since the dehydrogenation reaction is endothermic, there is a temperature drop of as much as 50 K down the reactor. The residence time of the gas stream in the reactor is typically one second. Superheated steam serves as a diluent, decokes the catalyst extending its life, and supplies sensible heat to keep up the reaction temperature. There are by-product reactions which produce benzene and toluene along with lighter components, methane and ethylene. The latter two can react in water-gas shift reactions to produce more hydrogen, carbon monoxide, then carbon dioxide. A list of the significant chemical reactions appears in Table 6.10. There are ten distinct chemical species, and the rate of reaction, that is, rate of disappearance, of each species is described in Table 6.11. Kinetic expressions have been developed for each reaction rate (Sheel and Crowe, 1969; Wenner and Dybdal, 1948; Clough and Ramirez, 1976) and are given in Table 6.12.

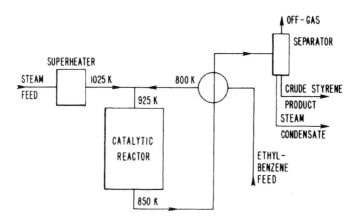

Figure 6.7: Styrene Pilot Plant.

Table 6.10: Significant Chemical Reactions

$C_6H_5CH_2CH_3$ (ethylbenzene)	$\rightleftarrows$	$C_6H_5CHCH_2$ (styrene)	$+$	H_2	
$C_6H_5CH_2CH_3$ (ethylbenzene)	$\rightarrow$	C_6H_6 (benzene)	$+$	C_2H_4 (ethylene)	
$C_6H_5CH_2CH_3$ (ethylbenzene)	$+$ H_2	$\rightarrow$	$C_6H_5CH_3$ (toluene)	$+$	CH_4 (methane)
H_2O	$+$ $1/2\ C_2H_4$ (ethylene)	$\rightarrow$	CO	$+$	$2\ H_2$
H_2O	$+$ CH_4 (methane)	$\rightarrow$	CO	$+$	$3\ H_2$
H_2O	$+$ CO	$\rightarrow$	CO_2	$+$	H_2

Table 6.11: Rates of Disappearance

	Species	Reaction rate (R_j)
1	Ethylbenzene	$R_1 + R_2 + R_3$
2	Styrene	$-R_1$
3	Hydrogen	$-R_1 + R_3 - 2R_4 - 3R_5 - R_6$
4	Benzene	$-R_2$
5	Toluene	$-R_3$
6	Methane	$-R_3 + R_5$
7	Ethylene	$-R_2 + 1/2\ R_4$
8	Carbon monoxide	$-R_4 - R_5 + R_6$
9	Carbon dioxide	$-R_6$
10	Water	$R_4 + R_5 + R_6$

We will develop an appropriate mathematical model for the pilot scale reactor studied by Clough and Ramirez (1976). The mathematical model is derived by simplification of the general conservation equations. The reduction of the general equations is made possible by order–of–magnitude scaling agreements based upon reasonable values for system dependent and independent variables. An important approximation which is implied in the model development is that the flow through the packed bed reactor can be described by both axial and radial dispersion mechanisms.

Table 6.12: Reaction Kinetics Expressions

Ethylbenzene dehydrogenation $\quad k_1=\frac{R_g T w_e}{P_a} \exp(F_1 - E_1/R_g T)$

$R_1 = k_1(C_1 - C_2 C_3/K_e) \qquad K_e = \frac{P_a}{R_g T} \exp(16.12 - 15350/T)$

Dealkylation to benzene

$R_2 = k_2 C_1 \qquad\qquad\qquad k_2 = \frac{R_g T w_e}{P_a} \exp(13.24 - 49675/R_g T)$

Dealkylation to toluene

$R_3 = k_3 C_1 C_3 \qquad\qquad k_3 = \left(\frac{R_g T}{P_a}\right)^2 w_e \exp(0.2961 - 21957/R_g T)$

Water–ethylene shift

$R_4 = k_4 C_{10} \sqrt{C_1} \qquad\quad k_4 = \left(\frac{R_g T}{P_a}\right)^{3/4} w_e \exp(-0.0724 - 24838/R_g T)$

Water–methane shift

$R_5 = k_5 C_6 C_{10} \qquad\qquad k_5 = \left(\frac{R_g T}{P_a}\right)^2 w_e \exp(-2.934 - 15697/R_g T)$

Water–carbon monoxide shift

$R_6 = k_6 C_8 C_{10} \frac{P}{R_g T} \qquad k_6 = \left(\frac{R_g}{P_a}\right)^3 w_e \exp(21.24 - 17585/R_g T)$

w_e = catalyst weight per reactor void volume

6.14.1 Gas Phase Energy Balance

We first start with the general thermal energy balance for the fluid phase within the reactor,

$$\rho C_v \frac{DT}{Dt} = -(\boldsymbol{\nabla} \cdot \boldsymbol{q}) - T\left(\frac{\partial p}{\partial T}\right)_{\hat{V}} (\boldsymbol{\nabla} \cdot \boldsymbol{v}) - (\boldsymbol{\tau} : \boldsymbol{\nabla} \boldsymbol{v}) \qquad (6.14.1)$$

For an ideal gas

$$T\left(\frac{\partial p}{\partial T}\right)_{\hat{V}} = p \qquad (6.14.2)$$

and assuming viscous dissipation effects are negligible gives

$$\rho C_v \frac{DT}{Dt} = -(\boldsymbol{\nabla} \cdot \boldsymbol{q}) - p(\boldsymbol{\nabla} \cdot \boldsymbol{v}) \qquad (6.14.3)$$

We need to add the local sources and sinks of thermal energy to this equation:

Q_c = heat transfer between the catalyst and fluid phases
Q_r = heat of reaction

This gives for the general thermal energy balance

$$\rho C_v \frac{DT}{Dt} = -(\boldsymbol{\nabla} \cdot \boldsymbol{q}) - p(\boldsymbol{\nabla} \cdot \boldsymbol{v}) + Q_c + Q_r \qquad (6.14.4)$$

We assume the thermal dispersion model

$$\mathbf{q} = -\tilde{\mathbf{k}}\,\nabla T \tag{6.14.5}$$

Equation (6.14.4) becomes upon expanding the substantial derivative,

$$\rho C_v \frac{\partial T}{\partial t} = -\rho C_v (\mathbf{v}\cdot\nabla T) + \tilde{\mathbf{k}}\nabla^2 T + \nabla^2 T + \nabla\tilde{\mathbf{k}}\cdot\nabla T$$
$$- p(\nabla\cdot\mathbf{v}) + Q_c + Q_r \tag{6.14.6}$$

In cylindrical coordinates we have, assuming $v_r = 0$, $v_\theta = 0$, no θ dependency upon variables, and the heat–transfer model between the fluid phase and the catalyst phase of

$$Q_c = \frac{h\hat{A}}{\epsilon}(T_c - T) \tag{6.14.7}$$

where ϵ is the porosity of the packed bed reactor, the following general thermal energy balance for all tubular reactors

$$\rho C_v \frac{\partial T}{\partial t} = -\rho C_v v + \frac{\tilde{k}_r}{r}\frac{\partial}{\partial r}\left(r\frac{\partial T}{\partial r}\right) + \frac{\partial \tilde{k}_r}{\partial r}\frac{\partial T}{\partial r} + \tilde{k}_z \frac{\partial T}{\partial z^2}$$
$$+ \frac{\partial \tilde{k}_z}{\partial z}\frac{\partial T}{\partial z} - p\frac{\partial v}{\partial z} + \frac{h\hat{A}}{\epsilon}(T_c - T) + Q_r \tag{6.14.8}$$

We are now ready to perform an order–of–magnitude analysis on the fluid phase energy equation of (6.14.8). We introduce order–of–one independent and dependent variables defined below:

$$\begin{array}{ccc}
T^* = \dfrac{T}{T_0} & T_c^* = \dfrac{T_c}{T_0} & \tilde{k}_z^* = \dfrac{\tilde{k}_z}{\tilde{k}_{z0}} \\[6pt]
v^* = \dfrac{v}{v_0} & r^* = \dfrac{r}{r_0} & C_v^* = \dfrac{C_v}{C_{v0}} \\[6pt]
z^* = \dfrac{z}{z_0} & p^* = \dfrac{p}{p_0} & t^* = t/t_0 \\[6pt]
\rho^* = \dfrac{\rho}{\rho_0} & \tilde{k}_r^* = \dfrac{\tilde{k}_r}{\tilde{k}_{r0}} &
\end{array} \tag{6.14.9}$$

Here the subscript zero means the reference or characteristic value for each variable.

The dimensionless energy balance in terms of the scaled variables becomes

$$\left(\frac{T_0\rho_0 C_{v_0}}{t_0}\right)\rho^* C_v^* \frac{\partial T^*}{\partial t^*} = -\left(\frac{\rho_0 C_{v_0} v_0 T_0}{z_0}\right)\rho^* C_v^* v^* \frac{\partial T^*}{\partial z^*}$$

$$+\left(\frac{\tilde{k}_{r_0} T_0}{r_0^2}\right)\frac{\tilde{k}_r^*}{r^*}\frac{\partial}{\partial r^*}\left(r^*\frac{\partial T^*}{\partial r^*}\right) + \left(\frac{\tilde{k}_{r_0} T_0}{r_0^2}\right)\frac{\partial \tilde{k}_r^*}{\partial r^*}\frac{\partial T^*}{\partial r^*}$$

$$+\left(\frac{\tilde{k}_{z_0} T_0}{z_0^2}\right)\left(\tilde{k}_z^*\frac{\partial^2 T^*}{\partial z^{*2}} + \frac{\partial \tilde{k}_z^*}{\partial z^*}\frac{\partial T^*}{\partial z^*}\right) - \left(\frac{p_0 v_0}{z_0}\right)p^*\frac{\partial v^*}{\partial z^*}$$

$$+\frac{T_0 h \hat{A}}{\epsilon}(T_c^* - T^*) + Q_r \qquad (6.14.10)$$

To prepare an order–of–magnitude analysis on various terms within the fluid thermal energy balance (6.14.10), all terms will be compared to the convective term. We therefore divide all terms by $\left(\frac{\rho_0 C_{v_0} v_0 T_0}{z_0}\right)$,

$$\left(\frac{z_0}{t_0 v_0}\right)\rho^* C_v^* \frac{\partial T^*}{\partial t^*} = -\rho^* C_v^* v^* \frac{\partial T^*}{\partial z^*} + \left(\frac{\tilde{k}_{z_0}}{z_0 \rho_0 C_{v_0} v_0}\right)\tilde{k}_z^* \frac{\partial^2 T^*}{\partial z^{*2}}$$

$$+\left(\frac{\tilde{k}_{z_0}}{z_0 \rho_0 C_{v_0} v_0}\right)\frac{\partial \tilde{k}_z^*}{\partial z^*}\frac{\partial T^*}{\partial z^*} + \left(\frac{\tilde{k}_{r_0} z_0}{r_0^2 \rho_0 C_{v_0} v_0}\right)\left[\frac{\tilde{k}_r^*}{r^*}\frac{\partial}{\partial r^*}\left(r^*\frac{\partial T^*}{\partial r^*}\right) + \frac{\partial k_r^*}{\partial r^*}\frac{\partial T^*}{\partial r^*}\right]$$

$$-\left(\frac{p_0}{\rho_0 C_{v_0} T_0}\right)p^*\frac{\partial v^*}{\partial z^*} + \left(\frac{z_0 h \hat{A}}{\epsilon \rho_0 C_{v_0} v_0}\right)(T_c^* - T^*)$$

$$+\left\{\left(\frac{z_0}{\rho_0 C_{v_0} v_0 T_0}\right)Q_r\right\} \qquad (6.14.11)$$

All leading coefficients are dimensionless except the last. This is because the heat of reaction has not been made dimensionless since it is a complex function of the other system variables. The entire last term (in braces) is dimensionless and needs to be retained in the process model since the endothermic heat of reaction is an important process mechanism.

Typical pilot plant conditions for the specific reactors studied by Clough and Ramirez (1976) are given in Table 6.13.

Using these characteristic values, the dimenionelss coefficients can be calculated and are used to help eliminate the less important terms of equation (6.14.11). The coefficients are

Microscopic Balances

Table 6.13: Typical Pilot Plant Conditions

T_0	reactor inlet temperature	900 K
p_0	back pressure regulator value	20 psia
C_0	molar density of ethylbenzene at inlet conditions	1.8×10^{-4} g mol/cm^3
v_0	superficial velocity at inlet	17 cm/sec
ρ_0	fluid density at inlet conditions	5×10^{-4} g/cm^3
z_0	reactor length	20 cm
t_0	average reactor residence time, z_0/v_0	1.2 sec
r_0	reactor inside radius	1.3 cm
C_{v_0}	heat capacity at inlet conditions	0.59 cal/g °C
$\tilde{k}_{z0}$	axial thermal dispersion coefficient at inlet conditions	0.0012 cal/cm ° sec
$\tilde{k}_{r0}$	radial thermal dispersion coefficient at inlet conditions	0.00023 cal/cm °C sec
$\hat{A}$	area of catalyst surface per unit volume	15 cm^{-1}
h	heat transfer coefficient	1×10^{-3} cal/cm^2 sec °C
ϵ	porosity of the packed bed reactor	0.36
K	permeability	1.6×10^{-4} cm^2
ρ_c	catalyst density	2.4 g/cm^3
C_{p_c}	catalyst heat capacity	0.2 cal/g °C
$\tilde{k}_c$	catalyst thermal dispersion coefficient	0.0034 cal/cm °C sec
A	reactor cross–sectional area	5.47 cm^2
μ	viscosity	0.01 cp

$$\frac{z_0}{t_0 v_0} = 1.0 \qquad \text{we include this term}$$

$$\frac{\tilde{k}_{z_0}}{z_0 \rho_0 C_{v_0} v_0} = 0.012 \quad \text{we can neglect this term}$$

$$\frac{\tilde{k}_{r_o} z_0}{r_0^2 \rho_0 C_{v_0} v_0} = 0.54 \quad \text{we will neglect the radial dispersion effects}$$

$$\frac{p_0}{\rho_0 C_{v_0} T_0} = 0.124 \quad \text{we will neglect the compression–expansion effects}$$

$$\frac{z_0 h \hat{A}}{\epsilon \rho_0 C_{v_0} v_0} = 166 \qquad \text{we include this term}$$

The scaled equation for this reactor is therefore

$$\rho^* C_v^* \frac{\partial T^*}{\partial t^*} = -\rho^* C_v^* v^* \frac{\partial T^*}{\partial z^*} + 166(T_c^* - T^*) + \{4.43 Q_r\} \qquad (6.14.12)$$

6.14.2 Catalyst Bed Energy Balance

We now develop the solid phase catalyst bed thermal energy equation. Since the catalyst bed is stationary, the general energy balance becomes

$$\rho_c C_{p_c} \frac{\partial T_c}{\partial t} = -(\nabla \cdot \mathbf{q}) - \left(\frac{\epsilon}{1-\epsilon}\right) Q_c \qquad (6.14.13)$$

which in cylindrical coordinates is expressed as

$$\rho_c C_{p_c} \frac{\partial T_c}{\partial t} = \tilde{k}_c \frac{\partial^2 T}{\partial z^2} + \frac{\tilde{k}_c}{r} \frac{\partial}{\partial r}\left(r \frac{\partial T_c}{\partial r}\right) - \left(\frac{\epsilon}{1-\epsilon}\right) Q_c \qquad (6.14.14)$$

The dimensionless scaled equation becomes

$$\left(\frac{\rho_c C_{p_c} T_0}{t_0}\right) \frac{\partial T_c^*}{\partial t^*} = \left(\frac{\tilde{k}_c T_0}{z_0^2}\right) \frac{\partial^2 T_c^*}{\partial z^{*2}} + \left(\frac{\tilde{k}_c T_0}{r_0^2}\right) \frac{1}{r^*} \frac{\partial}{\partial r^*}\left(r^* \frac{\partial T_c^*}{\partial r^*}\right)$$
$$- \left(\frac{h \hat{A} T_0}{1-\epsilon}\right)(T_c^* - T^*) \qquad (6.14.15)$$

We normalize the equation with respect to the last term. This gives the dimensionless coefficients,

$$\frac{\rho_c C_{p_c}(1-\epsilon)}{t_0 h \hat{A}} = 17 \quad \text{we include this term}$$

$$\frac{\tilde{k}_c(1-\epsilon)}{z_0^2 h \hat{A}} = 0.0004 \quad \text{we neglect the axial dispersion term}$$

$$\frac{\tilde{k}_c(1-\epsilon)}{r_0^2 h \hat{A}} = 0.09 \quad \text{we will include the radial dispersion term}$$

The final scaled catalyst thermal energy balance is

$$17 \frac{\partial T_c^*}{\partial t^*} = 0.09 \frac{1}{r^*} \frac{\partial}{\partial r^*}\left(r^* \frac{\partial T_c^*}{\partial r^*}\right) - (T_c^* - T^*) \qquad (6.14.16)$$

We can now compare the relative time scales for the gas phase and catalyst phase balances. To do this, we normalize the gas phase balance by the heat transfer term so that both catalyst and gas equations are normalized by the same term. This gives

$$0.00602 \rho^* C_v^* \frac{\partial T^*}{\partial t^*} = -0.00602 \rho^* C_v^* v^* \frac{\partial T^*}{\partial z^*} + (T_c^* - T^*) + \{0.0267 Q_r\}$$
$$(6.14.17)$$

Microscopic Balances

Comparing the leading coefficient of the two accumulation terms shows that the ratio of the fluid dynamics to the catalyst dynamics is

$$\frac{\tau_f}{\tau_c} = \frac{0.00602}{17} = 3.54 \times 10^{-4} \qquad (6.14.18)$$

This means that the dynamic response of the fluid phase is much faster than that of the catalyst temperature response.

6.14.3 Equation of Motion

The general equation of motion or momentum balance is

$$\frac{\partial}{\partial t}(\rho \boldsymbol{v}) = -\boldsymbol{\nabla} \cdot \rho \boldsymbol{vv} - \boldsymbol{\nabla} p - \boldsymbol{\nabla} \cdot \boldsymbol{\tau} + \rho \boldsymbol{g} \qquad (6.14.19)$$

which in cylindrical geometry is

$$\frac{\partial}{\partial t}(\rho v_z) = -\frac{\partial}{\partial z}(\rho v_z^2) - \frac{\partial p}{\partial z} - \left(\frac{1}{r}\frac{\partial}{\partial r}(r\,\tau_{rz}) + \frac{1}{r}\frac{\partial \tau_{\theta z}}{\partial \theta} \frac{\partial \tau_{zz}}{\partial z}\right) \qquad (6.14.20)$$

For flow through a packed bed, the viscous terms are given by Darcy's law

$$\left(\frac{1}{r}\frac{\partial}{\partial}(r\,\tau_{rz}) + \frac{1}{r}\frac{\partial \tau_{\theta z}}{\partial \theta} + \frac{\partial \tau_{zz}}{\partial z}\right) = \frac{\mu}{K}v_z \qquad (6.14.21)$$

where K is the permeability of the packed bed.

Therefore the equation of motion is

$$\frac{\partial}{\partial t}(\rho v_z) = -\frac{\partial}{\partial z}(\rho v_z^2) - \frac{\partial p}{\partial z} - \frac{\mu}{K}v_z \qquad (6.14.22)$$

Using the Equation of Continuity

$$\frac{\partial \rho}{\partial t} = -\frac{\partial}{\partial z}(\rho v_z) \qquad (6.14.23)$$

gives

$$\frac{\partial v_z}{\partial t} = -v_z \frac{\partial v_z}{\partial z} - \frac{1}{\rho}\frac{\partial p}{\partial z} - \frac{\mu}{\rho K}v_z \qquad (6.14.24)$$

Scaling and normalizing with respect to the convective term yields

$$\left(\frac{z_0}{t_0 v_0}\right)\frac{\partial v^*}{\partial t^*} = -v^*\frac{\partial v^*}{\partial z^*} - \left(\frac{p_0}{v_0^2 \rho_0}\right)\frac{1}{\rho^*}\frac{\partial p^*}{\partial z^*} - \left(\frac{\mu z_0}{K r_0 \rho_0}\right)\frac{v^*}{\rho^*} \qquad (6.14.25)$$

Using typical values gives for the coefficients

$$\frac{z_0}{t_0 v_0} = 0.98 \qquad \frac{\mu z_0}{K v_0 \rho_0} = 1.36 \times 10^3 \qquad \frac{p_0}{v_0^2 \rho_0} = 9.5 \times 10^6 \qquad (6.14.26)$$

The values imply that the pressure coefficient dominates the momentum balance and therefore
$$\frac{\partial p^*}{\partial z^*} = 0 \tag{6.14.27}$$
or
$$p^* = \text{constant} \tag{6.14.28}$$

6.14.4 Material Balances

The component material balances in cylindrical geometry are

$$\frac{\partial C_i}{\partial t} + \frac{\partial (v_z C_i)}{\partial z} = \frac{\tilde{D}_r}{r}\frac{\partial}{\partial r}\left(r\frac{\partial C_i}{\partial r}\right) + \tilde{D}_z\frac{\partial^2 C_i}{\partial z^2} - R_i \tag{6.14.29}$$

$i=1, N$ where N = number of chemical species

Using characteristic values, the scaled dimensionless equation is

$$\left(\frac{z_0}{v_0 t_0}\right)\frac{\partial C_i^*}{\partial t^*} = -\frac{\partial (v^* C_i^*)}{\partial z^*} - \left\{\frac{z_0 R_i}{v_0 C_0}\right\} \tag{6.14.30}$$

or

$$0.98\frac{\partial C_i^*}{\partial t^*} = -\frac{\partial (v^* C_i^*)}{\partial z^*} - \left\{\frac{z_0 R_i}{v_0 C_0}\right\} \tag{6.14.31}$$

6.14.5 Steady–State Model Solution

We want to compute the steady–state temperature and composition profiles for the styrene tubular reactor. The final model equations are

$$0 = -\rho C_p v \frac{\partial T}{\partial z} + Q_c + Q_r \tag{6.14.32}$$

$$0 = \frac{\tilde{k}_c}{r}\frac{\partial}{\partial r}\left(r\frac{\partial T_c}{\partial r}\right) - Q_c\left(\frac{\epsilon}{1-\epsilon}\right) \tag{6.14.33}$$

$$0 = -\frac{\partial (v C_i)}{\partial z} - R_i \tag{6.14.34}$$

At any axial position z, the value of the fluid temperature T is a constant and not a function of radial position. This is because equations (6.14.32) and (6.14.34) only have axial derivatives and no radial derivatives appear. This means that we can solve for the catalyst temperature from equation (6.14.33) since it is only a function of the radial position.

Equation (6.14.33) can be solved analytically for the catalyst temperature at any axial position as a function of radial position. The differential equation is

$$\frac{\tilde{k}_c}{r}\frac{d}{dr}\left(r\frac{dT_c}{dr}\right) - \frac{h\hat{A}}{(1-\epsilon)}(T_c - T) = 0 \tag{6.14.35}$$

Microscopic Balances

with the boundary conditions

$$\frac{dT_c}{dr} = 0 \quad \text{at } r = 0 \tag{6.14.36}$$

$$T_c = T_w \quad \text{at } r = R \tag{6.14.37}$$

If we make the following substitutions

$$s = \frac{r}{R} \quad Y = \frac{T_c - T}{T_w - T} \quad \beta^2 = \frac{R^2 h \hat{A}}{\tilde{k}_c (1 - \epsilon)} \tag{6.14.38}$$

then

$$\frac{d^2 Y}{ds^2} + \frac{1}{s}\frac{dY}{ds} - \beta^2 Y = 0 \tag{6.14.39}$$

with

$$\frac{dY}{ds} = 0 \quad \text{at } s = 0 \tag{6.14.40}$$

$$Y = 1 \quad \text{at } s = 1 \tag{6.14.41}$$

This is a Modified Bessel Equation (Wylie, 1960) which is a special linear second–order differential equation with nonconstant coefficients. The solution is given as

$$Y(s) = C_1' \, I_0(\beta s) + C_2' \, K_0(\beta s) \tag{6.14.42}$$

where

$$I_p(x) = \sum_{k=0}^{\infty} \left(\frac{x}{2}\right)^{2k+p} / k!(k+p)! \tag{6.14.43}$$

$$K_n(x) = I_n(x) \int \frac{dx}{x \, I_n^2(x)} \tag{6.14.44}$$

Using the $s = 0$ boundary condition

$$0 = \frac{dY}{ds} = C_1' \frac{dI_0}{ds} + C_2' \frac{dK_0}{ds} \tag{6.14.45}$$

It has been shown (Wylie, 1960) that

$$\frac{d}{dx} I_0(x) = I_{-1}(x) \tag{6.14.46}$$

and

$$\frac{d}{dx} K_0(x) = -K_{-1}(x) \tag{6.14.47}$$

Also $I_{-1}(0) = 0$ and $K_{-1}(0) = \infty$

Therefore the $s = 0$ boundary condition implies that $C_2' = 0$

At $s = 1$
$$Y = 1 = C'_1 I_0(\beta) \quad (6.14.48)$$

Therefore, the general solution is
$$Y(s) = \frac{I_0(\beta s)}{I_0(\beta)} \quad (6.14.49)$$

or
$$(T_c - T) = \frac{I_0\left[\sqrt{\frac{h\hat{A}}{\bar{k}_c(1-\epsilon)}}\, r\right]}{I_0\left[\sqrt{\frac{h\hat{A}}{\bar{k}_0(1-\epsilon)}}\, R\right]} (T_w - T) \quad (6.14.50)$$

We can now integrate the gas phase energy balance (6.14.32) over the radial cross–sectional area for flow,
$$A\epsilon = \pi R^2 \epsilon \quad (6.14.51)$$

or
$$da = 2\pi \epsilon r \, dr \quad (6.14.52)$$

This gives
$$\int_{A\epsilon} -(\rho C_p v)\frac{dT}{dz} da + \int_{A\epsilon} \frac{h\hat{A}}{\epsilon}(T_c - T) da + \int_{A\epsilon} Q_r \, da = 0 \quad (6.14.53)$$

or
$$-\epsilon \pi R^2 (\rho C_p v)\frac{dT}{dz} + 2\pi \int_0^R h\hat{A}(T_c - T) r \, dr + \epsilon \pi R^2 Q_r = 0 \quad (6.14.54)$$

Actually at $r = R$ an additional heat source must be considered. This is the heat transfer between the fluid and the exposed reactor wall. This can be modeled as
$$Q_w = h_w A_{we}(T_w - T) \quad (6.14.55)$$

where h_w is the heat transfer coefficient and A_{we} is the exposed wall area.

We can now use our analytical solution for T_c as a function of radial position to evaluate
$$\int_0^R (T_c - T) r \, dr = \frac{(T_w - T)}{I_0(\psi R)} \int_0^R I_0(\psi r) r \, dr \quad (6.14.56)$$

with
$$\psi = \sqrt{\frac{h\hat{A}}{\bar{k}_c(1-\epsilon)}} \quad (6.14.57)$$

Again using the properties of Bessel Functions (Wylie, 1960), we get

$$\int_0^R (T_c - T) r\, dr = \frac{R}{\sqrt{\psi}} \frac{I_1(R\psi)}{I_0(R\psi)} (T_w - T) \qquad (6.14.58)$$

The fluid energy balance integrated over the radial cross-section is therefore

$$-\rho C_p v \frac{dT}{dz} + \left[\frac{2}{\epsilon R \sqrt{\psi}} \frac{I_1(R\psi)}{I_0(R\psi)}\right] h\hat{A}(T_w - T) + Q_w + Q_r = 0 \qquad (6.14.59)$$

or

$$-\rho C_p v \frac{dT}{dz} + (h\hat{A}\beta_r + h_w A_{we})(T_w - T) + Q_r = 0 \qquad (6.14.60)$$

where β_r is an effectiveness factor defined as

$$\beta_r = \frac{2}{\epsilon R\psi} \frac{I_1(R\psi)}{I_0(R\psi)} \qquad (6.14.61)$$

Equation (6.14.60) can now be integrated with the material balance equation

$$\frac{d(vC_i)}{dz} = -R_i \qquad (6.14.62)$$

With the boundary conditions that the inlet temperature and inlet compositions are known, these equations can be integrated numerically using the IMSL numerical integration routines introduced in Chapter 4. Typical numerical results obtained by Clough and Ramirez (1976) are shown in Figure 6.8. Here we observe the temperature drop down the reactor due to the endothermic dehydrogenation reactions, the steady monotonic rise in styrene conversion down the reactor (styrene concentration per initial ethylbenzene concentration), but an internal maximum in the styrene selectivity (styrene concentration/styrene concentration + benzene concentration + toluene concentration). With this model it is possible to perform an optimization study for the reaction system to maximize profitability. Profitability is defined as the value of the product styrene minus the loss in profit caused by producing by-products, minus the utility cost of generating steam. Such an optimization study has been performed by Clough and Ramirez (1976), who showed that the major control for the reactor is the steam-to-ethylbenzene ratio.

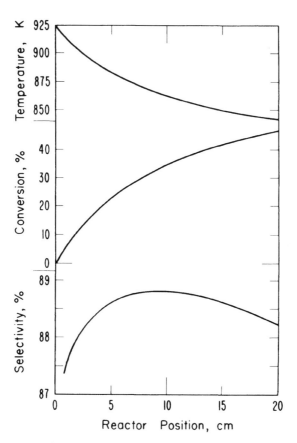

Figure 6.8: Reactor Profiles.

PROBLEMS

6.1. An artificial kidney is used to remove toxic materials from the blood stream by the process of dialysis (mass transfer) across porous cellophane membranes (shown in Figure 6.9).

    a. Develop the steady-velocity profile $v_z = v_z(y)$ in terms of $v_{z,max}$. Use a shell balance approach. The artificial kidney has width W.

    b. Write a mass balance for the toxic component A on an element $\Delta y \Delta z W$ in volume. Assume diffusion is important only in the y direction and that mass is only transported by bulk flow in the z direction.

    c. What are the boundary conditions for this mass balance problem?

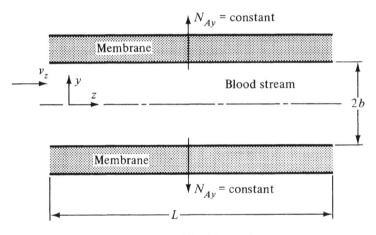

Figure 6.9: Problem 6.1.

6.2. A student has left a volumetric flask (shown in Figure 6.10) partially filled with mercury. Compute an expression for the rate of diffusion of mercury vapor into the room. The sides of the flask are tapered so that the stagnant diffusion section is a frustum of a cone (r is a linear function of z). A shell balance approach should be used.

6.3. A representative slurry grinder is shown in Figure 6.11. Assume that the flow is laminar and that the radial velocity is given by the volumetric flow rate charged divided by the cross–sectional area, that is, $v_r = Q/2\pi r h$, the gradients of v_θ with respect to r are small, and that v_θ is not a function of r. Note however that v_θ is a function of z. Compute by use of the general equations the v_θ profile. Be sure to indicate why

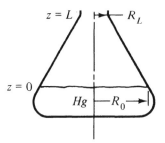

Figure 6.10: Problem 6.2.

you can eliminate the appropriate terms. You should be able to show that

$$\frac{d^2 v_\theta}{dz^2} = E v_\theta \quad \text{where} \quad E = \left(\frac{1}{r^2} + \frac{Q\rho}{2\pi r^2 h \mu}\right)$$

Solve this differential equation with the appropriate boundary conditions.

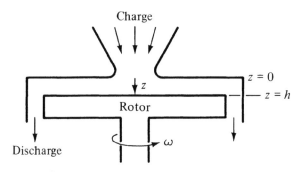

Figure 6.11: Problem 6.3.

6.4. Prove that the forms of Fick's law given in equations (6.2.6) and (6.2.9) are equivalent.

6.5. Temperature Profile in a Porous Reactor Tube

Figure 6.12 shows a schematic representation of a porous tubular reactor. The reactant gas and diluent enter the pores of the tube and flow radially towards the center of the tube. Heat is generated in the solid by means of a constant electric current flowing through the solid, and heat is removed by the endothermic reaction.

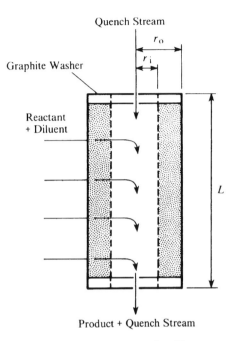

Figure 6.12: Porous Tube Reactor.

Write differential thermal energy balances on both the solid phase and the gas phase in the porous reactor.

You can assume

1. $J = $ a constant volumetric rate of heat generation in the solid due to electric current, cal/hr m^3.

2. The heat–transfer rate per unit volume between the solid and gas phase can be modeled by use of a heat–transfer coefficient, h

$$Q = ha(T_2 - T_f)$$

where $a = $ surface area per volume of porous reactor.

3. A constant porosity of the porous reactor

$$\epsilon = \frac{m^3 (\text{voids})}{m^3 (\text{solids } + \text{ voids})}$$

4. The mass flow rate in the reactor does not depend upon reaction conditions.

5. $R = $ a constant endothermic volumetric reaction heat effect, cal/hr m^3.

a. What do you think are appropriate boundary conditions for this problem? Justify your choice.

Develop an appropriate analytical solution for the case when:

b. The solid temperature is a constant.

c. The cylindrical geometry can be approximated by Cartesian coordinates because curvature effects are small for thin tubes.

6.6. Styrene Tubular Reactor

Develop composition and temperature profiles for the Clough–Ramirez tubular reactor to make styrene.

The water species can be computed assuming the gas phase is an ideal gas. Also the superficial velocity can be computed by knowing the mass flow rate of the ethylbenzene feed stream. These expressions also allow for the computation of the spatial derivative of the velocity.

The fluid heat capacity is the average value based upon mass fraction

$$\bar{C}_p = \frac{\sum_{j=1}^{n} C_j M_j C_{p_j}}{\sum_{j=1}^{n} C_j M_j}$$

The steam to ethylbenzene to ethylbenzene feed ratio is 1.6 g/g and the ethylbenzene feed rate is 480 kg/day. The inlet reactor temperature of the stream and ethylbenzene is 925 K.

6.7. Model of Permeability Reduction Due to Surfactant Adsorption

One model of flow through porous media (Ramirez and Riley, 1984) is that of flow in a cylindrical pore. A surfactant is adsorbed onto the surface of the pore and the film thickness of the adsorbed layer is δ. Assuming that the film thickness, δ, is small compared to the pore radius, R, the velocity profile across δ can be approximated by a linear profile. Show that the velocity profile across the cylindrical pore is

$$v_z(r) = -\frac{(P_0 - P_L)}{4\mu_b L} r^2 + \frac{(P_0 - P_L)}{2L} \left[\frac{R-\delta}{\beta} + \frac{(R-\delta)^2}{2\mu_b} \right]$$

$$R - \delta \geq r \geq 0$$

and

$$v_z(r) = \frac{(R-r)(P_0 - P_L)}{\delta} \frac{}{2\beta L}(R-\delta) \qquad R \geq r \geq R-\delta$$

where β = the slip coefficient (poise/cm) and

$$\delta\beta = \mu_w \qquad \mu_w = \text{viscosity at the wall}$$

Compute the average velocity for this model.

If $v_{z,max} = 1.89 \times 10^{-2}$ cm/sec when $\mu_b = 0.01$ poise, $R = 52.3 \times 10^{-5}$ cm, $\beta = \infty$, and $\delta = 0$, compute the maximum velocity when $\delta = 1 \times 10^{-5}$ and $\mu_w = 2.77$ poise. Also compute the velocity profile for each case. Compare these results. What is the effect of surfactant adsorption based upon this model?

6.8. A Second Model for Permeability Reduction Due to Surfactant Adsorption

Ramirez and Riley (1984) have proposed a second model to explain permeability reduction due to the adsorption of surfactants. They assume that in the presence of surfactants, the fluid viscosity varies with concentration across the film thickness δ reaching a maximum value at the wall of a cylindrical pore and falling off to the viscosity of the bulk at $r = R - \delta$. What is the general expression for v_z as a function of r in terms of $(P_0 - P_L)$, L, and $\mu(r)$.

To determine the exact profile, you need to have an expression for the viscosity across the film thickness. For surfactants

$$\mu(C_A) = 0.63 e^{16.5 C_A} + 0.37$$

with

$$\mu \text{ in } cp \quad \text{and} \quad C_A \text{ in g mol/cc}$$

This means that we need to develop an expression for the concentration profile across the film thickness. At steady state surfactant is transported due to two mechanisms. One is diffusion, and the other is a flux due to the force of attraction at the wall, which tends to draw the surfactant towards the surface. This latter flux can be modeled as

$$J_{wall} = \frac{C_1}{2\pi r L} \quad \text{where } C_1 = \text{constant}$$

Solve for the steady–state concentration profile under these conditions and discuss in detail how you would compute the average velocity for a cylindrical pore.

6.9 Develop the material balance equations that describe the steady–state axial concentration profiles in a flat plate membrane dialyzer used as an artificial kidney. A schematic is shown below:

The inlet blood side with concentration, C_{Bo} is known as well as the flow rate on both the blood and dialysate sides. The inlet dialysate concentration is zero. The geometry of the dialyzer is also known. The flux of urea across the membrane can be described by the membrane permeability which is given by the following equation valid at all axial positions,

$$N_m = P(C_B - C_D)$$

where P is the membrane permeability (cm/min).

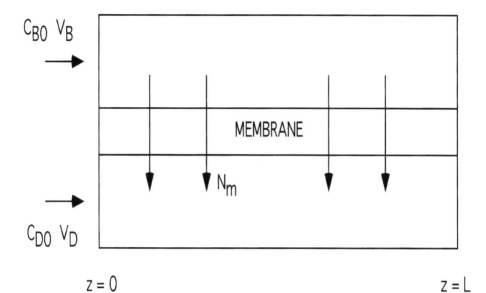

Derive expressions that describe both the blood and dialysate concentrations as a function axial position. You can assume that both the blood and dialysate sides are uniform in both the x and y directions. There are only gradients in the z direction. The mass average velocity on both the blood and dialysate sides are known and are constant.

Develop models for cases when axial diffusion is important and can be neglected. Be sure to write down the boundary conditions needed for both models.

Solve the diaylsis model when diffusion is not important.

System parameters are

$$\begin{aligned}
P &= 0.53 \text{ cm/min} \\
L &= 85 \text{ cm (length)} \\
W &= 60 \text{ cm (width)} \\
h_b &= .008 \text{ cm (blood side height)} \\
h_d &= .135 \text{ cm (dialysate side height)} \\
Q_B &= 206 \text{ cc/min} \\
Q_D &= 240;\ 350;\ 450;\ 500 \text{ cc/min} \\
C_{BO} &= 2.8 \text{ g/l}
\end{aligned}$$

Plot the concentration profiles. How much urea is recovered for each dialysate flow rate?

REFERENCES

Bird, R. B., Stewart, W. E., and Lightfoot, E. N., *Transport Phenomena*, Wiley, New York (1960).

Clough, D. E. and Ramirez, W. F., "Mathematical Modeling and Optimization of the Dehydrogenation of Ethylbenzene to Form Styrene," *AIChE Journal* **22**, No. 4, 1097 (1976).

Friedman, F. and Ramirez, W. F., "A Single–Phase Model of Mechanisms Effecting Miscible Surfactant Oil Recovery," *Chem. Eng. Sci.* **32**, 687 (1977).

Ramirez, W. F. and Riley, K. F., "Effect of Surfactants on Mass Dispersion and Permeability in Porous Media," *Chem. Eng. Commun.* **25**, 363–378 (1984).

Sheel, J. C. P. and Crowe, C. M., "Simulation and Optimization of an Existing Ethylbenzene Dehydrogenation Reactor," *Can. J. Chem. Eng.* **47**, 183 (1969).

Slattery, J., *Momentum, Energy, and Mass Transfer in Continua*, McGraw-Hill, New York (1972).

Wenner, R. R. and Dybdal, E. C., "Catalytic Dehydrogenation of Ethylbenzene," *Chem. Eng. Progr.* **44**, No. 4, 275 (April, 1948).

Whitaker, S., *Ind. & Eng. Chem. Fund.*, **12**, 14 (1962).

Wylie, C. R., *Advanced Engineering Mathematics*, McGraw–Hill, New York (1960).

Chapter 7

SOLUTION OF SPLIT BOUNDARY–VALUE PROBLEMS

This chapter discusses numerical techniques for solving split boundary–value problems. Split boundary–value problems arise from the description of distributed systems in which part of the boundary information needed to solve a set of differential equations is at one boundary of the system and part at another boundary. An example would be a counter–current heat exchanger where the inlet temperatures are known at either end of the exchanger.

Three basic numerical techniques are introduced. These are shooting techniques, quasilinearization with the use of the principle of superposition, and the method of adjoints.

7.1 DIGITAL IMPLEMENTATION OF SHOOTING TECHNIQUES: TUBULAR REACTOR WITH DISPERSION

Consider a tubular chemical reactor in which the first–order reaction $A \to B$ is carried out as illustrated in Figure 7.1.

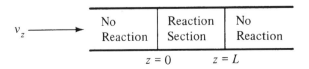

Figure 7.1: Tubular Reactor.

The velocity profile is assumed to be plug flow, that is, v_z is a constant and not a function of the radius of the tube. We will allow for the existence of axial dispersion due to turbulent mixing effects in the reactor. Writing a material balance around a differential element of the reactor gives

$$v_z \frac{dC_A}{dz} = \tilde{D}_{Az} \frac{d^2C_A}{dz^2} - kC_A \qquad (7.1.1)$$

where C_A = molar concentration of component A
k = first–order reaction rate constant
$\tilde{D}_{Az}$ = axial dispersion coefficient

The boundary conditions for this problem are found from flux balances at the inlet, $z = 0$, and the exit, $z = L$, of the reactor.

At $z = 0$, there is no diffusion in the fore section of the reactor but there is dispersion in the reaction zone. A flux balance takes the form

$$N_A(z = 0^-) = N_A(z = 0^+) \qquad (7.1.2)$$

or

$$v_z\, C_{A_0} = v_z\, C_A(0^+) - \tilde{D}_{Az} \frac{dC_A(0^+)}{dz} \qquad (7.1.3)$$

The concentration in the fore section C_{A_0} is known. Note that the flux in the fore section is only due to convection while that in the reactor itself has both a convective and dispersion contribution.

At $z = L$, when there is dispersion in the aft section, we have the same mechanism at $z = L^-$ and $z = L^+$. Since the reaction is terminated, the concentration gradient must be zero:

$$dC_A(L)/dz = 0 \qquad (7.1.4)$$

These two boundary conditions, equations (7.1.3) and (7.1.4), are known as the Danckwerts boundary conditions (Danckwerts, 1953). We will rewrite the problem in terms of dimensionless variables

$$C_A^* = \frac{C_A}{C_{A_0}} \qquad z^* = \frac{z}{L} \qquad Pe = \frac{v_z L}{\tilde{D}_{Az}} \qquad R = \frac{kL^2}{\tilde{D}_{Az}}$$

where Pe is the Peclet and R a dimensionless reaction rate group. The describing differential equation is

$$\frac{d^2 C_A^*}{dz^{*2}} - Pe \frac{dC_A^*}{dz^*} - R\, C_A^* = 0 \qquad (7.1.5)$$

with boundary conditions

$$1 = C_A^*(0^+) - \frac{1}{Pe} \frac{dC_A^*(0^+)}{dz^*} \qquad \text{at } z^* = 0 \qquad (7.1.6)$$

$$\frac{dC_A^*}{dz^*} = 0 \qquad \text{at } z^* = 1 \qquad (7.1.7)$$

Solution of Split Boundary-Value Problems

An information–flow diagram for the solution to the basic differential equation (7.1.1) is shown in Figure 7.2. This figure shows the two equivalent first–order differential equations that make up the state variable description of the original second–order equation. The solution would be straightforward if the initial values $C_A^*(0^+)$ and $x_1^*(0^+)$ were the known boundary conditions for the problem.

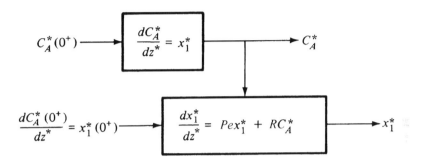

Figure 7.2: Information–Flow Diagram for Tubular Reactor with Dispersion.

To solve this set of two first–order equations with the actual Danckwerts boundary conditions given in equations (7.1.6) and (7.1.7), we will use a shooting technique. We can cast the actual problem into an initial value problem as shown in Figure 7.2 by assuming a value for $dC_A^*(0^+)/dz^*$ and therefore computing a value for $C_A^*(0^+)$ from B.C.1, equation (7.1.6). The proper guess for the initial slope $dC_A^*(0^+)/dz^*$ is the one that results in B.C.2, equation (7.1.7), being satisfied, namely, $dC_A^*(L)/dz^* = 0$ at $z^* = 1$.

Another way to solve the problem would be to guess a value for $C_A^*(z^* = 1)$ and use B.C.2., equation (7.1.7), directly as the other initial value. The correct guess for $C_A^*(z^* = 1)$ is the one that upon integrating backwards through the reactor satisfies B.C.1, namely, equation (7.1.6). To perform a backward integration, a new spatial variable is defined

$$y^* = 1 - z^* \tag{7.1.8}$$

The variable y^* now goes from 0 to 1 as the variable z^* goes from 1 to 0.
The mathematical model in terms of the spatial variable, y^*, is

$$\frac{dC_A^*}{dy^*} = -x_1^* \tag{7.1.9}$$

$$\frac{dx_1^*}{dy^*} = -Pe\, x_1^* - R\, C_A^* \tag{7.1.10}$$

with the boundary conditions

$$x_1^*(y^* = 0) = 0 \tag{7.1.11}$$

$$1 = C_A^*(y^*=1) - \frac{1}{Pe} x_1^*(y^*=1) \qquad (7.1.12)$$

The backward integration shooting technique is to assume the value of $C_A^*(y^*=0)$ and use equation (7.1.11). The problem is now integrated from $y^* = 0$ to $y^* = 1$. Various values of $C_A^*(y^*=0)$ are tried until the computed values of $C_A^*(y=1)$ and $x_1^*(y=1)$ satisfy the actual boundary condition, equation (7.1.12).

Some physical problems give stable convergence properties via forward integration and others via backward integration.

Why stable solution strategies can be obtained by converting an equation into an initial–value problem at $z^*=0$ and not at $z^*=1$ or vice versa can be shown by considering the following differential equation

$$d^2y/dt^2 - 100y = 0 \qquad (7.1.13)$$

which has the general solution

$$y = \alpha e^{10t} + \beta e^{-10t} \qquad (7.1.14)$$

with the initial condition $y(0) = A$ and $\dfrac{dy}{dt}(0) = m$.

We can solve for the constants of integration, α and β, and determine the sensitivity of the solution y to changes in m with A held constant.

$$\frac{d\alpha}{dm} = \frac{1}{20} \quad \text{and} \quad \frac{d\beta}{dm} = -\frac{1}{20} \qquad (7.1.15)$$

Similarly, with the boundary conditions $y(0) = A$ and $y(1) = B$; if we hold A constant and vary B, we have

$$\frac{d\alpha}{dB} = \frac{1}{e^{10} - e^{-10}} \quad \text{and} \quad \frac{d\beta}{dB} = -\frac{1}{e^{10} - e^{-10}} \qquad (7.1.16)$$

The coefficients α and β are much more sensitive to errors in m than in B. It can therefore be concluded that convergence properties depend upon the sensitivity of the constants of integration to various sets of boundary conditions.

For tubular reactors with the Danckwerts boundary conditions, the backward integration method gives stable convergence properties, while the forward integration procedure is unstable.

File **ex71.m** has been created to solve for the backwards integration of dispersion in a tubular reactor. The differential equations are defined in file **model71.m** (Figure 7.3b). This problem assumes that $Pe = 100$ and $R = 200$. File **ex71.m** solves the problem for three different guesses for $C_A^*(y^* = 0)$ or $C_A^*(z^* = 1)$. A plot of the right hand side of the missing boundary condition (equation (7.16)) versus the initial guess for $C_A^*(z^* = 1)$ is given in Figure 7.4a and shows that the convergence is linear which means that actually only two guesses for C_{AO} are actually needed in ex71.m. The reason that the convergence is linear is because the original differential equation and boundary conditions are linear. The correct exit dimensionless concentration $C_A^*(z^* = 1)$ that satisfies the inlet Danckwert's boundary condition is $C_A^*(z^* = 1) = .1406$.

Figure 7.4b gives both the concentration (C_A^*) profile and the concentration gradient (x_1^*) profile for this reactor. It should be noticed that the gradient is very steep near the exit. It is this behavior that requires backward integration for this problem.

Figure 7.3: MATLAB Programs to Solve Dispersions in a Tubular Reactor.

7.3a file ex71.m

```
% This program uses the shooting technique to solve the tubular
% reactorproblem in section 7.1. The equation set and boundary
% conditions in Eqs. (7.1.9) - (7.1.12).
%
I = 1;

for CAO = 0.00:0.1:1,
    x10 = 0;
    c0 = [CAO x10]';
    clear t c; % this is used to make sure that there is no
               % problem with vector lengths
    [t, c] = ode45n('model71',[ 0 1], c0);
    clear CA x1;
    CA = c(:, 1);
    x1 = c(:, 2);
    L = length(CA);
    index(I) = CA(L) - 0.01*x1(L);
    CAOplot(I) = CAO;
    boundary(I) = 1.0;
    I = I+1;

end

plot(CAOplot, index, '-', CAOplot, boundary, '--');
title('Response of Tubular Reactor Dispersion to Boundary Cond');
xlabel('Exit Dimensionless Concentration (CA(L)/CAO)');
ylabel('CA(0+)/CAO - Daz*x1(0+)/(vz*CAO)');
%print fig74.eps
```

7.3b file model71.m

```
function cdot = ex71(t, c)
%
% this file defines the equation set (7.1.9) and (7.1.10),
% x1=c(2), CA=c(1);
%

cdot(1) = -c(2);
cdot(2) = -100*c(2) - 200.0*c(1);
cdot = cdot(:);
```

310 Computational Methods for Process Simulation

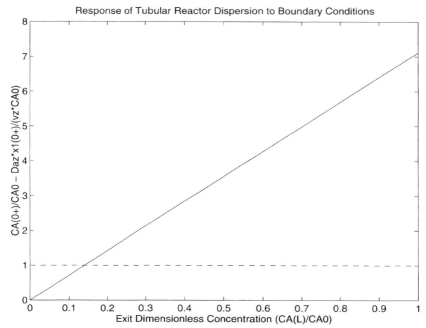

Figure 7.4a: Response of Tubular Reactor Dispersion to Boundary Conditions.

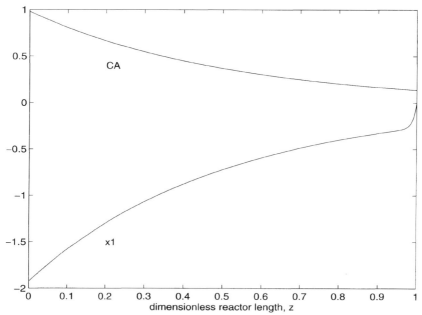

Figure 7.4b: Tubular Reactor with Dispersion Profiles.

7.2 A GENERALIZED SHOOTING TECHNIQUE

Consider a set of state differential equations

$$\dot{x}_1 = f_1(x_1, x_2, x_3, x_4) \tag{7.2.1}$$

$$\dot{x}_2 = f_2(x_1, x_2, x_3, x_4) \tag{7.2.2}$$

$$\dot{x}_3 = f_3(x_1, x_2, x_3, x_4) \tag{7.2.3}$$

$$\dot{x}_4 = f_4(x_1, x_2, x_3, x_4) \tag{7.2.4}$$

with the split boundary conditions

$$\begin{aligned} x_1(0) &= 0 & x_3(1) &= 1 \\ x_2(0) &= 0 & x_4(1) &= 1 \end{aligned} \tag{7.2.5}$$

Assume starting initial guesses for the missing initial conditions as

$$\begin{aligned} x_3(0) &= a \\ x_4(0) &= b \end{aligned} \tag{7.2.6}$$

Now we expand the known final conditions in a linearized Taylor series about the assumed missing initial conditions,

$$x_3(1)\Big|_{n+1} = 1 = x_3(1)\Big|_n + \frac{\partial x_3(1)}{\partial a}\Big|_n (a_{n+1} - a_n) + \frac{\partial x_3(1)}{\partial b}\Big|_n (b_{n+1} - b_n) \tag{7.2.7}$$

where $x_3(1)|_n$ = the calculated value using the assumed a_n and b_n values.

Also,

$$x_4(1)\Big|_{n+1} = x_4(1)\Big|_n + \frac{\partial x_4(1)}{\partial a}\Big|_n (a_{n+1} - a_n) + \frac{\partial x_4(1)}{\partial b}\Big|_n (b_{n+1} - b_n) \tag{7.2.8}$$

To calculate the partials we define "shooting system dynamics" for each unknown initial condition. These equations come from taking partials with respect to the missing initial conditions (a and b), of the describing state differential equations (7.2.1) to (7.2.4).

System 1 (for missing initial condition a)

$$\frac{d}{dt}\left(\frac{\partial x_1}{\partial a}\right) = \frac{\partial f_1}{\partial x_1}\frac{\partial x_1}{\partial a} + \frac{\partial f_1}{\partial x_2}\frac{\partial x_2}{\partial a} + \frac{\partial f_1}{\partial x_3}\frac{\partial x_3}{\partial a} + \frac{\partial f_1}{\partial x_4}\frac{\partial x_4}{\partial a} \tag{7.2.9}$$

$$\frac{d}{dt}\left(\frac{\partial x_2}{\partial a}\right) = \frac{\partial f_2}{\partial x_1}\frac{\partial x_1}{\partial a} + \frac{\partial f_2}{\partial x_2}\frac{\partial x_2}{\partial a} + \frac{\partial f_2}{\partial x_3}\frac{\partial x_3}{\partial a} + \frac{\partial f_2}{\partial x_4}\frac{\partial x_4}{\partial a} \quad (7.2.10)$$

$$\frac{d}{dt}\left(\frac{\partial x_3}{\partial a}\right) = \frac{\partial f_3}{\partial x_1}\frac{\partial x_1}{\partial a} + \frac{\partial f_3}{\partial x_2}\frac{\partial x_2}{\partial a} + \frac{\partial f_3}{\partial x_3}\frac{\partial x_3}{\partial a} + \frac{\partial f_3}{\partial x_4}\frac{\partial x_4}{\partial a} \quad (7.2.11)$$

$$\frac{d}{dt}\left(\frac{\partial x_4}{\partial a}\right) = \frac{\partial f_4}{\partial x_1}\frac{\partial x_1}{\partial a} + \frac{\partial f_4}{\partial x_2}\frac{\partial x_2}{\partial a} + \frac{\partial f_4}{\partial x_3}\frac{\partial x_3}{\partial a} + \frac{\partial f_4}{\partial x_4}\frac{\partial x_4}{\partial a} \quad (7.2.12)$$

and System 2 (for missing initial condition b)

$$\frac{d}{dt}\left(\frac{\partial x_1}{\partial b}\right) = \frac{\partial f_1}{\partial x_1}\frac{\partial x_1}{\partial b} + \frac{\partial f_1}{\partial x_2}\frac{\partial x_2}{\partial b} + \frac{\partial f_1}{\partial x_3}\frac{\partial x_3}{\partial b} + \frac{\partial f_1}{\partial x_4}\frac{\partial x_4}{\partial b} \quad (7.2.13)$$

$$\frac{d}{dt}\left(\frac{\partial x_2}{\partial b}\right) = \frac{\partial f_2}{\partial x_1}\frac{\partial x_1}{\partial b} + \frac{\partial f_2}{\partial x_2}\frac{\partial x_2}{\partial b} + \frac{\partial f_2}{\partial x_3}\frac{\partial x_3}{\partial b} + \frac{\partial f_2}{\partial x_4}\frac{\partial x_4}{\partial b} \quad (7.2.14)$$

$$\frac{d}{dt}\left(\frac{\partial x_3}{\partial b}\right) = \frac{\partial f_3}{\partial x_1}\frac{\partial x_1}{\partial b} + \frac{\partial f_3}{\partial x_2}\frac{\partial x_2}{\partial b} + \frac{\partial f_3}{\partial x_3}\frac{\partial x_3}{\partial b} + \frac{\partial f_3}{\partial x_4}\frac{\partial x_4}{\partial b} \quad (7.2.15)$$

$$\frac{d}{dt}\left(\frac{\partial x_4}{\partial b}\right) = \frac{\partial f_4}{\partial x_1}\frac{\partial x_1}{\partial b} + \frac{\partial f_4}{\partial x_2}\frac{\partial x_2}{\partial b} + \frac{\partial f_4}{\partial x_3}\frac{\partial x_3}{\partial b} + \frac{\partial f_4}{\partial x_4}\frac{\partial x_4}{\partial b} \quad (7.2.16)$$

To integrate these equations we have the following initial conditions:

State System	System 1	System 2
$x_1(0) = 0$	$\frac{\partial x_1}{\partial a}(0) = 0$	$\frac{\partial x_1}{\partial b}(0) = 0$
$x_2(0) = 0$	$\frac{\partial x_2}{\partial a}(0) = 0$	$\frac{\partial x_2}{\partial b}(0) = 0$
$x_3(0) = a$	$\frac{\partial x_3}{\partial a}(0) = 1$	$\frac{\partial x_3}{\partial b}(0) = 0$
$x_4(0) = b$	$\frac{\partial x_4}{\partial a}(0) = 0$	$\frac{\partial x_4}{\partial b}(0) = 1$

Note that the initial conditions for the shooting systems follow directly from the initial conditions on the state system. Solving the state and shooting systems simultaneously gives the information needed to compute new initial conditions a_{n+1} and b_{n+1} from equation (7.2.7) and

(7.2.8). These conditions can be expressed as the following set of linear algebraic equations,

$$\begin{bmatrix} \dfrac{\partial x_3(1)}{\partial a} & \dfrac{\partial x_3(1)}{\partial b} \\ \dfrac{\partial x_4(1)}{\partial a} & \dfrac{\partial x_4(1)}{\partial b} \end{bmatrix} \begin{pmatrix} a_{n+1} - a_n \\ b_{n+1} - b_n \end{pmatrix} = \begin{pmatrix} 1 - x_3(1)|_n \\ 1 - x_4(1)|_n \end{pmatrix}$$
(7.2.17)

Therefore

$$\begin{pmatrix} a_{n+1} \\ b_{n+1} \end{pmatrix} = \begin{bmatrix} \dfrac{\partial x_3(1)}{\partial a} & \dfrac{\partial x_3(1)}{\partial b} \\ \dfrac{\partial x_4(1)}{\partial a} & \dfrac{\partial x_4(1)}{\partial b} \end{bmatrix}^{-1} \begin{pmatrix} 1 - x_3(1)|_n \\ 1 - x_4(1)|_n \end{pmatrix} + \begin{pmatrix} a_n \\ b_n \end{pmatrix}$$
(7.2.18)

This is an iterative scheme. The method converges quadratically like a Newton–Raphson method.

EXAMPLE 7.1 Tubular Reactor Composition Profile:

Let us use the generalized shooting technique to solve for the backward integration of the tubular reactor with dispersion. The describing differential equations and boundary conditions are given by equations (7.1.9) to (7.1.12).

The state functions are

$$f_1 = -x_1^*$$ (7.2.19)

$$f_2 = -R\, C_A^* - Pe\, x_1^*$$ (7.2.20)

We know the initial boundary condition

$$x_1^*(0) = 0$$ (7.2.21)

and we assume the missing initial boundary condition to be

$$C_A^*(0) = a$$ (7.2.22)

From Taylor series expansions of the final boundary values in terms assumed missing initial boundary condition, a, we have

$$C_A^*(1)_{n+1} = C_A^*(1)_n + \left.\dfrac{\partial C_A(1)}{\partial a}\right|_n (a_{n+1} - a_n)$$ (7.2.23)

$$x_1^*(1)_{n+1} = x_1^*(1)_n + \left.\dfrac{\partial x_1^*(1)}{\partial a}\right|_n (a_{n+1} - a_n)$$ (7.2.24)

The shooting system dynamics are

$$\dfrac{d}{dy}\left(\dfrac{\partial C_A^*}{\partial a}\right) = \dfrac{\partial f_1}{\partial C_A^*}\dfrac{\partial C_A^*}{\partial a} + \dfrac{\partial f_1}{\partial x_1^*}\dfrac{\partial x_1^*}{\partial a}$$ (7.2.25)

$$\frac{d}{dy}\left(\frac{\partial x_1^*}{\partial a}\right) = \frac{\partial f_2}{\partial C_A^*}\frac{\partial C_A^*}{\partial a} + \frac{\partial f_2}{\partial x_1^*}\frac{\partial x_1^*}{\partial a} \qquad (7.2.26)$$

Using the state functions of (7.2.19) and (7.2.20), we have $\frac{\partial f_1}{\partial C_A^*} = 0$ and $\frac{\partial f_1}{\partial x_1^*} = -1$; therefore

$$\frac{d}{dy}\left(\frac{\partial C_A^*}{\partial a}\right) = -\frac{\partial x_1^*}{\partial a} \qquad (7.2.27)$$

Also since $\frac{\partial f_2}{\partial C_A^*} = -R$ and $\frac{\partial f_2}{\partial x_1^*} = -Pe$, we have

$$\frac{d}{dy}\left(\frac{\partial x_1^*}{\partial a}\right) = -R\left(\frac{\partial C_A^*}{\partial a}\right) - Pe\frac{\partial x_1^*}{\partial a} \qquad (7.2.28)$$

The appropriate initial conditions are

$$\frac{\partial C_A^*}{\partial a}(0) = 1 \quad \text{and} \quad \frac{\partial x_1^*}{\partial a}(0) = 0 \qquad (7.2.29)$$

Boundary condition (7.1.12) gives

$$1 + \frac{1}{Pe} x_1^*(1)_{n+1} = C_A^*(1)_{n+1} \qquad (7.2.30)$$

Using equations (7.2.23) and (7.2.24), we have

$$1 + \frac{1}{Pe}\left[x_1^*(1)_n + \left.\frac{\partial x_1^*(1)}{\partial a}\right|_n (a_{n+1} - a_n)\right]$$

$$= C_A^*(1)_n + \left.\frac{\partial C_A^*(1)}{\partial a}\right|_n (a_{n+1} - a_n) \qquad (7.2.31)$$

Solving for a_{n+1} gives

$$a_{n+1} = a_n - \frac{\left[1 - C_A^*(1) + \frac{1}{Pe} x_1^*(1)_n\right]}{\left[-\left.\frac{\partial C_A^*(1)}{\partial a}\right|_n + \frac{1}{Pe}\left.\frac{\partial x_1^*(1)}{\partial a}\right|_n\right]} \qquad (7.2.32)$$

File ex72.m (Figure 7.5a) gives the MATLAB program that implements the generalized shooting method for the solution of the composition profile of a tabular reactor with dispersion. Figure 7.5b gives the file model72.m which defines both the model and shooting system dynamic equations. The final concentration profile is plotted in Figure 7.6. Since the problem is linear, the system converges in one iteration.

Figure 7.5a: Main Program, ex72.m

```
% This program uses a generalized shooting technique in section
% 7.2 to solve Eqs. (7.1.9), (7.1.10), (7.2.27), and (7.2.28)
% with boundary conditions (7.2.29). The guess of the missing
% initial boundary condition is updated by Eq. (7.2.32).
%
```

```
max = 20;
CA0 = 0.9;
tol = 0.000001;

for I = 1:max,
    % initial conditions:
    x10 = 0;
    dCA0 = 1;
    dx10 = 0;

    % solve the state and shooting systems simultaneously:
    c0 = [CA0 x10 dCA0 dx10]';
    clear t c;
    [t, c] = ode45n('model72',[ 0  1], c0);
    clear CA x1 dCA dx1;
    CA = c(:, 1);
    x1 = c(:, 2);
    dCA = c(:, 3);
    dx1 = c(:, 4);
    L = length(CA);

    % update CA0:
    delta = (1-CA(L)+.01*x1(L))/(-dCA(L)+.01*dx1(L));
    if abs(delta) < tol
        break;
    end
    CA0 = CA0 - delta;
end

iterations = I
t = 1 - t;
plot(t, CA);
title('Concentration Profile in the Turbular Reactor');
xlabel('Dimensionless Distance (z/L)');
ylabel('Demensionless Concentration (CA/CA0)');
%print fig76.eps;
```

Figure 7.5b: Dynamic model and Shooting System Equations,
model72.m

```
function cdot = ex72(t, c)
%
% this file defines the equation set (7.2.19), (7.2.20), (7.2.27)
% and (7.2.28) with x1=c(2), CA=c(1)
%

cdot(1) = -c(2);
cdot(2) = -100.*c(2) -200.0*c(1);
cdot(3) = -c(4);
cdot(4) = -200.*c(3) - 100.*c(4);
cdot = cdot(:);
```

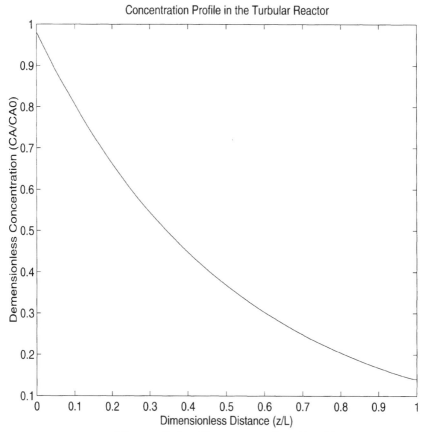

Figure 7.6: Concentration Profile.

7.3 SUPERPOSITION PRINCIPLE AND LINEAR BOUNDARY–VALUE PROBLEMS

The superposition principle states that for *linear* differential equations, a split boundary–value problem can be solved as a linear combination of initial value solutions.

Consider the nth order linear equation

$$y^{(n)} + f_{n-1}\, y^{(n-1)} + f_{n-2}\, y^{(n-2)} + \cdots + f_1\, y^{(1)} + f_0\, y = g \qquad (7.3.1)$$

where $y^{(i)}$ = the ith derivative with respect to the independent variable t
  f_i = functions of t
  g = function of t

For this linear problem we can use the principle of superposition, which allows us to state that the general solution is a particular solution $y_p(t)$ plus a linear combination of n homogeneous solutions, y_{Hi} for $i = 1, \cdots n$.

$$y(t) = y_p(t) + \sum_{i=1}^{n} A_i \, y_{H_i} \tag{7.3.2}$$

where $y(t)$ = the general solution
$y_p(t)$ = a particular solution
$\sum_{i=1}^{n} A_i \, y_{H_i}$ = the linear combination of n homogeneous solutions

The homogeneous solutions have the only requirement that the boundary conditions used be independent. Thus the set of boundary conditions

$$y(0) = 0 \quad y'(0) = 0$$

$$y(0) = 0 \quad y'(0) = 1$$

are *not* an independent set of boundary conditions since the rank of the matrix $\begin{bmatrix} 0 & 0 \\ 0 & 1 \end{bmatrix}$ is one and not two. But the set

$$y(0) = 0 \quad y'(0) = 1$$

$$y(0) = 1 \quad y'(0) = 0$$

are independent since the rank of $\begin{bmatrix} 0 & 1 \\ 1 & 0 \end{bmatrix}$ is two.

Given a boundary–value problem, say,

$$y'' = g(t) \tag{7.3.3}$$

with the boundary conditions

$$y(0) = a \quad y(1) = b \tag{7.3.4}$$

the general solution may be obtained by

1. Generating two solutions to the homogeneous equation ($y'' = 0$) with the requirement that the boundary conditions be independent. This is the only requirement we must meet to solve initial–value problems since the boundary conditions used in applying the

principle of superposition may or may not be related to the actual boundary conditions of the original problem. We choose

$$y(0) = 1 \quad y'(0) = 0$$
$$y(0) = 0 \quad y'(0) = 1$$
(7.3.5)

The solutions to these two initial–value problems can be obtained by either analytic techniques or computer methods.

2. Generating one particular solution—a solution to the nonhomogeneous equation $y'' = g(t)$—using *any* boundary condition: We can therefore also choose an initial–value problem with boundary conditions such as

$$y(0) = 0 \quad y'(0) = 0 \tag{7.3.6}$$

3. At this point, we have three solutions, as shown in Figure 7.7. We therefore know the corresponding final conditions for each solution. The general solution from the principle of superposition, equation (7.3.2), for this problem is given by equation (7.3.7).

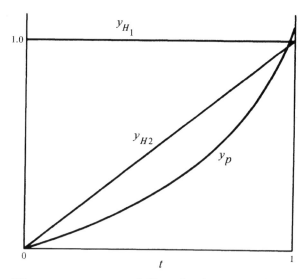

Figure 7.7: Homogeneous and Particular Solution to $y'' = g(t)$.

$$y(t) = y_p + \sum_{i=1}^{2} A_i\, y_{H_i} \tag{7.3.7}$$

We have two coefficients, A_1 and A_2 which are as yet undetermined. These coefficients are chosen so as to satisfy the boundary conditions of the original problem.

$$y(0) = a = y_p(0) + A_1\, y_{H_1}(0) + A_2\, y_{H_2}(0)$$
$$y(1) = b = y_p(1) + A_1\, y_{H_1}(1) + A_2\, y_{H_2}(1) \tag{7.3.8}$$

This represents two linear equations in two unknowns, which can be expressed in matrix form as

$$\begin{bmatrix} y_{H_1}(0) & y_{H_2}(0) \\ y_{H_1}(1) & y_{H_2}(1) \end{bmatrix} \begin{pmatrix} A_1 \\ A_2 \end{pmatrix} = \begin{pmatrix} a - y_p(0) \\ b - y_p(1) \end{pmatrix} \tag{7.3.9}$$

This linear set of equations can be solved for the unknown constants A_1 and A_2. Once these constants are known, the solution is completely specified by equation (7.3.7).

The net result of using the principle of superposition is that a linear boundary–value problem of order n can be solved by obtaining the solution to $n+1$ initial–value problems and solving n linear algebraic equations.

Often, an nth–order differential equation is placed in state–variable form. This is a set of n first–order equations. When deriving equations via material and energy balances, this state–variable form arises naturally. By developing the equations in state–variable form, we must solve the simultaneous sets of linear first–order differential equations. The state variable form is

$$\dot{x}(t) = A(t)x(t) + r(t) \tag{7.3.10}$$

with some of the boundary conditions of x given at t_0, $x_1(t_0)$, and others at t_f, $x_2(t_f)$. To use the principle of superposition for the equation of (7.3.10), we use the following solution strategy.

1. Solve the particular equation set

$$\dot{x}_p(t) = A(t)x_p(t) + r(t) \tag{7.3.11}$$

   with arbitrary initial conditions which can be chosen as

$$x_p(t_0) = 0 \tag{7.3.12}$$

2. Solve n homogeneous equations sets

$$\dot{X}_h(t) = A(t)X_h(t) \tag{7.3.13}$$

where $X_h = [x_{1h} \; x_{2h} \; \cdots \; x_{nh}]$, that is, an augmented matrix formed by the n homogeneous solution vectors,

with independent initial conditions which can be

$$X_h(t_0) = I \tag{7.3.14}$$

where I is the identity matrix.

3. Match the original boundary conditions using the principle of superposition

$$x(t) = x_p(t) + X_h(t) \, c \tag{7.3.15}$$

Applying the superposition principle to the actual boundary conditions, we have

$$\text{B.C.1} \quad x_1(0) = x_{1p}(0) + X_{1h}(0) \, c \tag{7.3.16}$$

where $x_1(0)$ are the boundary conditions specified at $t = t_0$. Note also that $x_{1p}(0) = 0$ when we use equation (7.3.12) and $X_{1h}(0)$ is known from specifying the n independent initial conditions for the homogeneous solutions, equation (7.3.14). Also,

$$\text{B.C.2} \quad x_2(t_f) = x_{2p}(t_f) + X_{2h}(t_f) \, c \tag{7.3.17}$$

where $x_2(t_f)$ are the boundary conditions specified at $t = t_f$. Also, $x_{2p}(t_f)$ is known from solving for the particular solution and $X_{2h}(t_f)$ is known from the solution of the n homogeneous equations.

Combining equations of (7.3.16) and (7.3.17) gives n independent linear algebraic equations which can be solved via matrix techniques for the n unknown coefficients c,

$$c = \begin{bmatrix} X_{1h}(0) \\ X_{2h}(t_f) \end{bmatrix}^{-1} \begin{pmatrix} x_1(0) - x_{1p}(0) \\ x_2(t_f) - x_{2p}(t_f) \end{pmatrix} \tag{7.3.18}$$

Therefore, from the principle of superposition, equation (7.3.15), the solution to the linear split boundary–value problem is known.

7.4 SUPERPOSITION PRINCIPLE: RADIAL TEMPERATURE GRADIENTS IN AN ANNULAR CHEMICAL REACTOR

A catalytic reaction is being carried out at constant pressure in a packed bed annular space between two coaxial cylinders. Compute the radial temperature profile. The entire outer wall ($R_1 = 0.04$ ft) is at a uniform

Solution of Split Boundary–Value Problems

temperature of 800°F. At the inner wall ($R_0 = 0.03$ ft) there is essentially no heat transferred through the surface. The reaction consumes heat (endothermic reaction) at a uniform rate per unit volume of 53,900 Btu/hr ft^3. The thermal conductivity of the gas is 0.3 Btu/hr ft°F. The thermal energy balance can be simplified to

$$0 = k \left[\frac{1}{r} \frac{\partial}{\partial r} \left(r \frac{\partial T}{\partial r} \right) \right] - \Delta H \qquad (7.4.1)$$

When expanded, this equation becomes

$$\frac{\partial^2 T}{\partial r^2} = -\frac{1}{r} \frac{\partial T}{\partial r} + \frac{\Delta H}{k} \qquad (7.4.2)$$

This is the describing equation that must be solved.

Two boundary conditions are known. The temperature at the outer wall is 800°F and the temperature gradient dT/dr at the inner wall is zero.

The following state–variable formulation of equation (7.4.2) is used: Defining

$$DT = dT/dr \qquad (7.4.3)$$

we have

$$\begin{bmatrix} dT/dr \\ dDT/dr \end{bmatrix} = \begin{bmatrix} 0 & 1 \\ 0 & -1/r \end{bmatrix} \begin{pmatrix} T \\ DT \end{pmatrix} + \begin{pmatrix} 0 \\ \Delta H/k \end{pmatrix} \qquad (7.4.4)$$

To solve this problem using the principle of superposition, it will be necessary to generate one particular solution vector and two homogeneous solution vectors, that is

$$\begin{pmatrix} T_p \\ DT_p \end{pmatrix} \text{ and } \begin{pmatrix} T_{H_1} \\ DT_{H_1} \end{pmatrix}, \begin{pmatrix} T_{H_2} \\ DT_{H_2} \end{pmatrix}$$

In matching the boundary conditions, we obtain the following set of linear algebraic equations.

$$\begin{pmatrix} T(r = 0.04 \text{ ft}) \\ DT(r = 0.03 \text{ ft}) \end{pmatrix} = \begin{bmatrix} T_{H_1}(0.04 \text{ ft}) & T_{H_2}(0.04 \text{ ft}) \\ DT_{H_1}(0.03 \text{ ft}) & DT_{H_2}(0.03 \text{ ft}) \end{bmatrix} \begin{pmatrix} C_1 \\ C_2 \end{pmatrix}$$

$$+ \begin{pmatrix} T_p(0.04 \text{ ft}) \\ DT_p(0.03 \text{ ft}) \end{pmatrix} \qquad (7.4.5)$$

MATLAB program ex74.m has been created to solve this problem. File model74.m defines the particular solution differential equations and the homogeneous solution differential equations. These programs are available on the world wide web. Figure 7.8 shows the particular solutions and Figure 7.9 shows two homogeneous solutions with the boundary condition of equations (7.3.12) and (7.3.14). Equation (7.4.5) therefore becomes

$$\begin{pmatrix} 800°F \\ 0 \end{pmatrix} = \begin{bmatrix} 1 & 0.0086 \\ 0 & 1 \end{bmatrix} \begin{pmatrix} C_1 \\ C_2 \end{pmatrix} + \begin{pmatrix} 8.183 \\ 0 \end{pmatrix} \qquad (7.4.6)$$

Solving for C_1 and C_2 gives

$$C_1 = 791.8 \qquad C_2 = 0 \qquad (7.4.7)$$

The general or actual solution to this problem is shown in Figure 7.10.

7.5 QUASILINEARIZATION

Quasilinearization is a technique where nonlinear differential equations are solved by obtaining a sequence of solutions to related linear equations. The method somewhat resembles a generalized Newton–Raphson method. The most important development in the quasilinearization technique was the use of the "maximum operation" to prove that the representation of the original nonlinear equations by a sequence of linear equations converges to the nonlinear equation. This result is due to Bellman, who also used the concept in his development of dynamic programming (Bellman, 1957).

By looking at the easier–to–understand Newton–Raphson technique we introduce two important properties, namely, monotonic convergence and quadratic convergence. In the Newton–Raphson method, we are trying to locate roots to the equation

$$f(x) = 0 \qquad (7.5.1)$$

The function f is expanded in a first–order Taylor series about an approximate root $x = x_0$.

$$f(x) = f(x_0) + f'(x_0)(x - x_0) \qquad (7.5.2)$$

The root $x = r$ corresponds to $f(r) = 0$. As a new approximation to r we solve equation (7.5.2) for x using $f(x) = 0$. This gives

$$x = x_0 - \frac{f(x_0)}{f'(x_0)} \qquad (7.5.3)$$

Solution of Split Boundary–Value Problems 323

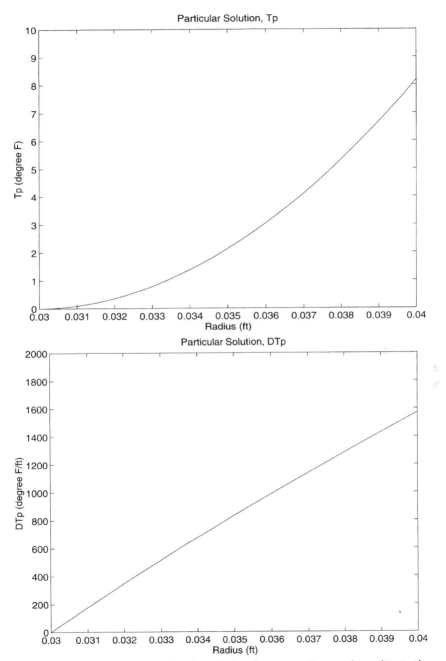

Figure 7.8: Particular Solutions to Equation (7.4.4).

324 Computational Methods for Process Simulation

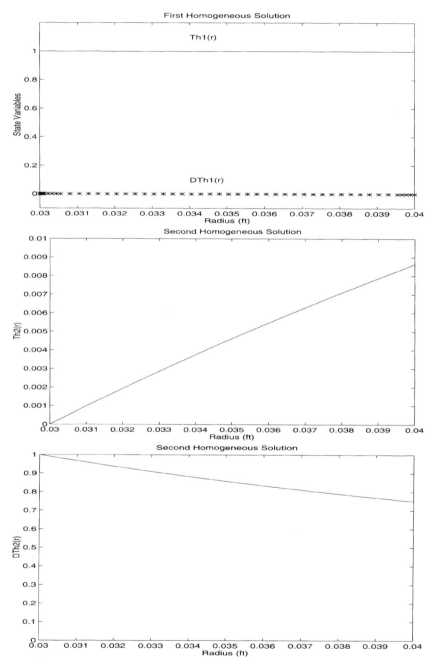

Figure 7.9: Homogeneous Solutions to Equation (7.4.4).

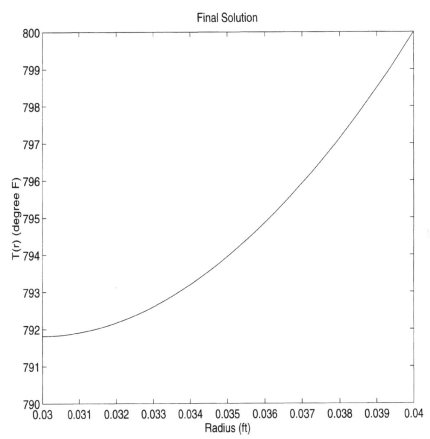

Figure 7.10: Final Solution to Radial Temperature Profile.

Under the assumption that $f(x)$ is a monotonic decreasing function for all x and is strictly convex, that is, $f''(x) > 0$, we may repeat the procedure with x to obtain a new approximation to the root $x = r$. The recurrence relation is therefore

$$x_{n+1} = x_n - \frac{f(x_n)}{f'(x_n)} \tag{7.5.4}$$

Under the above assumptions the sequence x_1, $i = 0, 1, \cdots$, is monotone convergent to the root of the nonlinear expression.

Geometrically the recurrence relation represents the tangent line to the function $f(x)$ at x_n, and x_{n+1} represents the intersection of the tangent line with the x axis, as shown in Figure 7.11.

Again, under the same assumptions for $f(x)$, the rate at which the sequence x_n converges to root r is quadratic. This means that the error

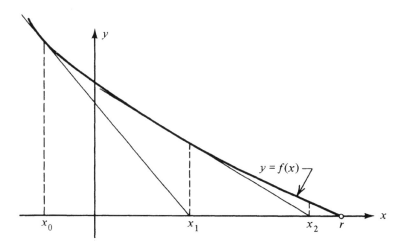

Figure 7.11: Newton–Raphson Convergence.

decreases quadratically

$$(x_{n+1} - r) = (x_n - r)^2 \tag{7.5.5}$$

The quasilinearization technique proceeds in the same manner except not with the function $f(x) = 0$, but with $f(x)$ equal to some nonzero function. For example, let's consider the differential equation

$$d^2x/dt^2 = f(x) \tag{7.5.6}$$

The recurrence relation from the linearization procedure is

$$\frac{d^2 x_{n+1}}{dt^2} = f(x_n) + \frac{\partial f}{\partial x}(x_n)(x_{n+1} - x_n) \tag{7.5.7}$$

Equation (7.5.7) is now a linear differential equation which can be solved using the principle of superposition. Again, the sequence of solutions x_i will approach the root of the nonlinear equation (7.5.6).

Consider the following nonlinear differential equation

$$y\frac{d^2y}{dt^2} + \left(\frac{dy}{dt}\right)^2 + 1 = 0 \tag{7.5.8}$$

with boundary conditions

$$y(t=1) = 1 \qquad y(t=2) = 2 \tag{7.5.9}$$

Equation (7.5.8) can be rearranged to give

$$\frac{d^2y}{dt^2} = -\frac{\left[1 + \left(\frac{dy}{dt}\right)^2\right]}{y} \qquad (7.5.10)$$

Therefore

$$\frac{d^2y}{dt^2} = f\left(y, \frac{dy}{dt}\right) \qquad (7.5.11)$$

where

$$f\left(y, \frac{dy}{dt}\right) = \frac{-\left[1 + \left(\frac{dy}{dt}\right)^2\right]}{y} \qquad (7.5.12)$$

To use quasilinearization we must linearize the nonlinear function given by equation (7.5.12).

$$f(y, y') = f_n + \left.\frac{\partial f}{\partial y}\right|_n (y_{n+1} - y_n) + \left.\frac{\partial f}{\partial y'}\right|_n (y'_{n+1} - y'_n) \qquad (7.5.13)$$

or

$$f(y, y') = \frac{-[1 + (y'_n)^2]}{y_n} + \frac{[1 + (y'_n)^2]}{y_n^2}(y_{n+1} - y_n) - \frac{2y'_n}{y_n}(y'_{n+1} - y'_n) \qquad (7.5.14)$$

Therefore the linear differential equation to be solved is

$$y''_{n+1} = \left(\frac{1 + (y'_n)^2}{y_n^2}\right) y_{n+1} - \left(\frac{2y'_n}{y_n}\right) y'_{n+1} - \frac{2}{y_n} \qquad (7.5.15)$$

The sequence of linear solutions will converge quadratically to the nonlinear solution if convergence is obtained. Experience has shown that almost all problems will converge if reasonable initial guesses of y_0 and y'_0 are used.

7.6 NONLINEAR TUBULAR REACTOR WITH DISPERSION: QUASILINEARIZATION SOLUTION

Consider a second-order chemical reaction $2A \rightarrow B$ taking place in a tubular reactor in which there is axial mixing. A material balance on the reactor gives

$$\frac{1}{Pe}\frac{d^2C}{dX^2} - \frac{dC}{dX} - RC^2 = 0 \qquad (7.6.1)$$

where C = the concentration of reactant A
Pe = the Peclet number = Lv_x/D
R = the reaction rate group = kL^2/D
X = the dimensionless reactor length, x/L
L = the reactor length
v_x = the fluid velocity
D = the dispersion coefficient
k = the rate constant, vol/time mol

The Danckwerts boundary conditions are

$$C_0 = C(0^+) - \frac{1}{Pe} \frac{dC(0^+)}{dX} \quad \text{at } X = 0$$

$$\frac{dC}{dX} = 0 \quad \text{at } X = 1 \qquad (7.6.2)$$

Due to the Danckwerts boundary conditions, we anticipate stability via backward integration. We therefore introduce a new length variable $z = 1 - X$. The differential equation and boundary conditions now become

$$\frac{d^2C}{dz^2} = -Pe \frac{dC}{dz} + PeRC^2 \qquad (7.6.3)$$

B.C.1 $$C_0 = C + \frac{1}{Pe} \frac{dC}{dz} \quad \text{at } z = 1$$

(7.6.4)

B.C.2 $$\frac{dC}{dz} = 0 \quad \text{at } z = 0$$

The only nonlinear term in the differential equation (7.6.3) is due to C^2.

Linearizing the differential equation about an assumed solution C_n gives

$$\frac{d^2C_{n+1}}{dz^2} = -Pe \frac{dC_{n+1}}{dz} + PeRC_n^2 + PeR \left(\frac{\partial C^2}{\partial C}\right)_n (C_{n+1} - C_n) \qquad (7.6.5)$$

or

$$\frac{d^2C_{n+1}}{dz^2} = -Pe \frac{dC_{n+1}}{dz} + PeRC_n^2 + 2PeRC_n (C_{n+1} - C_n) \qquad (7.6.6)$$

which simplifies to

$$\frac{d^2C_{n+1}}{dz^2} = -Pe \frac{dC_{n+1}}{dz} + 2PeRC_nC_{n+1} - PeRC_n^2 \qquad (7.6.7)$$

Solution of Split Boundary–Value Problems

We now place equation (7.6.7) in state–variable form by introducing a new variable

$$DC_{n+1} = dC_{n+1}/dz \qquad (7.6.8)$$

The linear equation set which will be solved via the principle of superposition is

$$\begin{pmatrix} dC_{n+1}/dz \\ dDC_{n+1}/dz \end{pmatrix} = \begin{bmatrix} 0 & 1 \\ 2PeRC_n & -Pe \end{bmatrix} \begin{pmatrix} C_{n+1} \\ DC_{n+1} \end{pmatrix} + \begin{pmatrix} 0 \\ -PeRC_n^2 \end{pmatrix} \qquad (7.6.9)$$

To get the procedure started, we must assume a solution $C_{n=0}(z)$. A good starting function would be just a constant value down the reactor of C_0, the known inlet concentration. Any assumed solution should satisfy the actual boundary conditions (which this particular solution does for the problem under consideration). Therefore, for the first iteration we will solve

$$\begin{pmatrix} dC_{(1)}/dz \\ dDC_{(1)}/dz \end{pmatrix} = \begin{bmatrix} 0 & 1 \\ 2PeRC_{(0)}(z) & -Pe \end{bmatrix} \begin{pmatrix} C_{(1)} \\ DC_{(1)} \end{pmatrix}$$

$$+ \begin{pmatrix} 0 \\ -PeRC_{(0)}^2(z) \end{pmatrix} \qquad (7.6.10)$$

where $C_{(0)}(z) = C_0$

The principle of superposition says that the solution to (7.6.10) is obtained from one particular solution and two homogeneous conditions, that is,

$$\begin{pmatrix} C_{(1)P} \\ DC_{(1)P} \end{pmatrix} \quad \text{and} \quad \begin{pmatrix} C_{(1)H1} \\ DC_{(1)H1} \end{pmatrix}, \begin{pmatrix} C_{(1)H2} \\ DC_{(1)H2} \end{pmatrix}$$

In matching the boundary conditions, we have to solve the following set of linear algebraic equations for the constants C_1 and C_2.

$$\begin{pmatrix} DC_{(1)(z=0)} = 0 \\ C_0 \end{pmatrix} = \begin{bmatrix} DC_{(1)H1}(0) & DC_{(1)H2}(0) \\ C_{(1)H1}(1) + \dfrac{1}{Pe}DC_{(1)}(1) & C_{(1)H2}(1) + \dfrac{1}{Pe}DC_{(1)}(1) \end{bmatrix}$$

$$\begin{pmatrix} C_1 \\ C_2 \end{pmatrix} + \begin{pmatrix} DC_{(2)P}(0) \\ C_{(1)P}(1) + \dfrac{1}{Pe}DC_{(1)P}(1) \end{pmatrix} \qquad (7.6.11)$$

The MATLAB program `ex76.m` has been created to solve this problem. It uses file `model76.m` to define the particular and homogeneous differential equations. These programs are available on the world wide web. The convergence sequence is shown in Figure 7.12. Note that all solutions satisfy the actual boundary conditions and that the convergence to the nonlinear solution is very rapid.

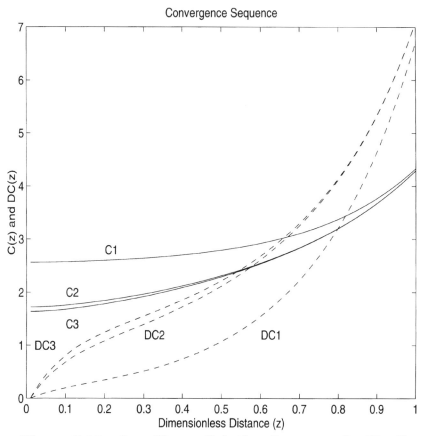

Figure 7.12: **Quasilinear Solution Sequence for Nonlinear Tubular Reactor with Dispersion.**

7.7 THE METHOD OF ADJOINTS

Another efficient method for the solution of linear two–point boundary–value problems is that of the method of adjoints. A good reference for this material is the book by Roberts and Shipman (1972). In the method of adjoints we define the system equations and the adjoint system equations.

Solution of Split Boundary–Value Problems

The basic linear differential equations are the system equations and are given by
$$\dot{x} = A\,x + f(t) \tag{7.7.1}$$
The adjoint system is defined in terms of the homogeneous part of the system equations as
$$\dot{p} = -A^T p \tag{7.7.2}$$
By multiplying the transpose of the adjoint variable vector, p, by the system equation (7.7.1), we have
$$p^T \dot{x} = p^T [A\,x + f] \tag{7.7.3}$$
Multiplying the transpose of the system variable vector, x, by the adjoint system equation (7.7.2) we have
$$x^T \dot{p} = x^T [-A^T p] \tag{7.7.4}$$
Adding equations (7.7.3) and (7.7.4) gives
$$p^T \dot{x} + x^T \dot{p} = p^T f \tag{7.7.5}$$
or
$$\frac{d(p^T x)}{dt} = p^T f \tag{7.7.6}$$
Integrating equation (7.7.6) gives
$$\int_{t_0}^{t_f} \frac{d}{dt}(p^T x)\,dt = \int_{t_0}^{t_f} p^T f\,dt \tag{7.7.7}$$
or
$$p^T(t_f)x(t_f) - p^T(t_0)x(t_0) = \int_{t_0}^{t_f} p^T f\,dt \tag{7.7.8}$$

Equation (7.7.8) is called the Fundamental Property of the Adjoint System. It is a relationship between the initial state and adjoint variables and their values at the final conditions. The right–hand side is only a function of the adjoint variables, f, and the known forcing vector, f, of the system equations. The adjoint variables $p(t)$ can be calculated using equation (7.7.2) and arbitrary initial values.

Let us now assume that the boundary–value problem is of the following form:

x_i are the boundary conditions specified at t_0
x_L are the boundary conditions specified at t_f

We will integrate the adjoint equations given by equation (7.7.2) backwards in time with assumed final conditions. We perform this backward integration m times, where m is the number of states specified at the final time, t_f. The final conditions are arbitrary as long as they are independent, but it is convenient to assume that the set of final conditions are given by

$$P_L(t_f) = \begin{bmatrix} I \\ 0 \end{bmatrix} = [p_1(t_f) \vdots p_2(t_f) \vdots \cdots \vdots p_m(t_f)] \tag{7.7.9}$$

where m is the number of states specified at t_f (i.e., the size of the x_L vector), I is an m by m identity matrix and 0 a $(n-m)$ by m matrix of zeros. Each solution $p_i(t_f)$ is a vector of length n.

Therefore, we solve the adjoint equations (equation (7.7.2)) starting at t_f and going to t_0, m times. Each solution starts with an assumed final costate condition of p_i, which has a zero for all elements except the ith element, which is unity. Each ith unity element corresponds to a known state boundary condition in x_L.

By doing this, it follows that

$$P_L^T(t_f) x(t_f) = x_L(t_f) \tag{7.7.10}$$

Using the Fundamental Property of the Adjoint System (equation (7.7.8)), we have

$$P_L^T(t_f) \, x(t_f) - P_L^T(t_0) \, x(t_0) = \int_{t_0}^{t_f} P_L^T(t) \, f(t) \, dt \tag{7.7.11}$$

Using equation (7.7.10), we have

$$x_L(t_f) - \begin{bmatrix} P_L^i(0) \\ \cdots \\ P_L^j(0) \end{bmatrix}^T \begin{bmatrix} x_i(0) \\ \cdots \\ x_j(0) \end{bmatrix} = \int_{t_0}^{t_f} P_L^T(t) \, f(t) \, dt \tag{7.7.12}$$

where $x_i(0)$ are the known boundary conditions at t_0
and $x_j(0)$ are the unknown states at t_0

Everything in equation (7.7.12) is known except $x_j(0)$, the missing initial state boundary conditions. We therefore solve equation (7.7.12) for $x_j(0)$ as

$$x_j(0) = \left[P_L^j(0)^T\right]^{-1} \left(x_L(t_f) - P_L^i(0)^T x_i(0) - \int_{t_0}^{t_f} P_L^T f(t) \, dt \right) \tag{7.7.13}$$

Let us consider the following example problem

$$\frac{d^2 x}{dt^2} = x + t \tag{7.7.14}$$

with

$$x(0) = 0 \tag{7.7.15}$$

$$x(1) = 1 \tag{7.7.16}$$

In state-variable form, the system equations are

$$\begin{pmatrix} \dot{x}_1 \\ \dot{x}_2 \end{pmatrix} = \begin{bmatrix} 0 & 1 \\ 1 & 0 \end{bmatrix} \begin{pmatrix} x_1 \\ x_2 \end{pmatrix} + \begin{pmatrix} 0 \\ t \end{pmatrix} \tag{7.7.17}$$

with boundary conditions

$$x_1(0) = 0 \tag{7.7.18}$$
$$x_1(1) = 1 \tag{7.7.19}$$

The adjoint system of differential equations is

$$\begin{pmatrix} \dot{p}_1 \\ \dot{p}_2 \end{pmatrix} = - \begin{bmatrix} 0 & 1 \\ 1 & 0 \end{bmatrix} \begin{pmatrix} p_1 \\ p_2 \end{pmatrix} \tag{7.7.20}$$

The partitioned variables for this system are

$$\boldsymbol{x}_i(t_0) = x_1(0) = 0 \tag{7.7.21}$$
$$\boldsymbol{x}_L(t_f) = x_1(1) = 1 \tag{7.7.22}$$
$$\boldsymbol{x}_j(t_0) = x_2(0) \tag{7.7.23}$$
$$\boldsymbol{P}_L(t_f) = \begin{pmatrix} p_1(t_f) \\ p_2(t_f) \end{pmatrix} = \begin{pmatrix} 1 \\ 0 \end{pmatrix} \tag{7.7.24}$$
$$\boldsymbol{P}_L^i(0) = p_1(0) \tag{7.7.25}$$
$$\boldsymbol{P}_L^j(0) = p_2(0) \tag{7.7.26}$$

Equation (7.7.12) for the missing initial condition $\boldsymbol{x}_j(t_0)$ becomes

$$x_2(0) = p_2(0)^{-1} \left(x_1(1) - p_1(0)x_1(0) - \int_{t_0}^{t_f} (p_1 \ p_2) \begin{pmatrix} 0 \\ t \end{pmatrix} dt \right) \tag{7.7.27}$$

or

$$x_2(0) = p_2(0)^{-1} \left(1 - \int_{t_0}^{t_f} p_2 t \, dt \right) \tag{7.7.28}$$

The integral of equation (7.7.28) can be computed a number of different ways. First, once p_2 is known, it can be stored so that the product $p_2 t$ can be numerically integrated by computing the area under a curve. Numerous techniques are available for computing areas under curves. These include quadrature methods, equally spaced base point methods such as the trapezoidal rule, and unequally spaced base point methods using orthogonal polynomials (see Carnahan et al., 1969). MATLAB has several quadrature routines available under the NAG toolbox. Another approach to computing the integral is to convert it to the differential form of

$$\dot{y} = p_2 t \tag{7.7.29}$$

This equation can either be integrated forward in time with the state equations or backward in time with the adjoint equations. The choice of the backward in time integration is more efficient since we do not have to store the values of p_2. Instead y can be computed simultaneously while computing the adjoint variables.

We have created the MATLAB file `ex77.m` for the backward computation of the adjoint variables (equation (7.7.20) with the final conditions of equation (7.7.24)) along with the auxiliary variable y (equation (7.7.29)), the computation of the missing initial state variable (equation (7.7.28)), and the computation of the state–variable profile. File `model77.m` defines the differential equations and file `model77a.m` defines the adjoint differential equations and integral differential equation given by (7.7.29). Figure 7.13 gives the adjoint variable profiles and Figure 7.14 gives the state–variable profiles.

7.8 MODELING OF PACKED BED SUPERHEATERS

Morrow (1988) has presented a model of a packed bed superheater and then verified the model with real–time experiments. Important aspects of this model include the application of scaling arguments to simplify a general model form. In addition, the effect of the superheater endcaps and the conduction of heat along the wall is included in the model. The model developed in this section can be applied to the vaporization and superheating of any fluid.

Industrial–sized superheaters are commonly used to produce steam. These superheaters are usually heated directly from combustion where radiative heat transfer is important. Some mathematical modeling of industrial superheaters has been published, but it usually involves the arrangement of the tube geometries or semi-empirical dynamic models (Bauver et al., 1987 and Kukalov, 1970). The two–phase to single–phase transition point was not investigated in these models. Small–scale processes such as pilot plants often superheat a stream of fluid by wrapping a pipe with electrical wire. There has been a large amount of research on single–phase fluid flow through a packed bed as applied to packed bed reactors. There has also been some work on two–phase fluid flow in heat exchangers (Shoureshi and McLaughlin, 1984).

A packed bed superheater that is heated by electrical wire wrapped around the superheater wall is shown in Figure 7.15.

The model describes the superheater fluid flow in two different regimes. One regime is the single–phase vapor flow through the superheater. These equations are developed by considering linked energy balances (including both radial and axial directions) on the bed packing and fluid. The equations are simplified through rigorous scaling arguments and combined into a single nonlinear partial differential equation. The

Solution of Split Boundary–Value Problems

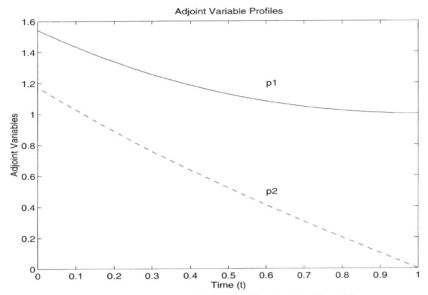

Figure 7.13: Adjoint Variable Profiles.

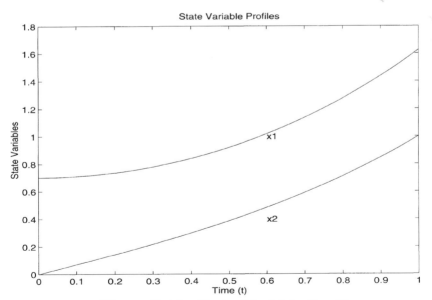

Figure 7.14: State–Variable Profiles.

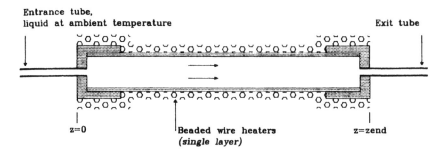

Figure 7.15: Superheater Diagram.

other regime is that of the entering liquid and two–phase fluid flow. This fluid flow can also be described by a single nonlinear partial differential equation.

The equation describing axial conduction of heat along the superheater wall is developed through a shell balance. Finally, a method of including the properties of the large endcaps of the superheater is developed. Including the endcaps in the model introduces discontinuous first spatial derivatives of the superheater wall temperature.

7.8.1 Single–Phase Fluid Flow Energy Balance

The derivation of the single–phase fluid flow equation follows work by Clough and Ramirez (1976) and Clough (1975). The flow of fluid through the packed bed is modeled as if it were passing through a pipe whose volume is the same as the void volume of the superheater. The mechanical dispersion of the fluid in both the radial and axial directions due to the tube packing is superimposed upon this flow. The mechanical dispersion approximation takes the form of a diffusion term in the transport equations. Local heat transfer between the tube packing and fluid is added as an effect internal to the flow. Steady–state equations describing the fluid energy balance, fluid momentum balance, and the energy balance on the superheater packing are presented below.

The fluid thermal energy balance is

$$\rho_v C_{v_f} v \frac{\partial T}{\partial z} + \frac{k_r}{r} \frac{\partial}{\partial r}\left(r \frac{\partial T}{\partial r}\right) + \frac{\partial k_r}{\partial r} \frac{\partial T}{\partial r} + k_{ax} \frac{\partial^2 T}{\partial z^2}$$

$$+ \frac{\partial k_{ax}}{\partial z} \frac{\partial T}{\partial z} - P \frac{\partial v}{\partial z} + \frac{h \hat{A}}{\epsilon}(T_s - T) = 0 \qquad (7.8.1)$$

The packed bed energy balance is

$$k_s \frac{\partial^2 T_s}{\partial z^2} + \frac{k_s}{r}\frac{\partial}{\partial r}\left(r\frac{\partial T_s}{\partial r}\right) - \frac{h\hat{A}}{1-\epsilon}(T_s - T) = 0 \qquad (7.8.2)$$

and the linear momentum balance is

$$v\frac{\partial v}{\partial z} + \frac{1}{\rho_v}\frac{\partial P}{\partial z} + \frac{\mu}{\kappa\rho_v}v = 0 \qquad (7.8.3)$$

where
T	=	fluid temperature	t = time
T_w	=	wall temperature	$\hat{A}$ = surface area of
h	=	heat–transfer coefficient		packing per unit
ρ	=	density		volume of
C_v	=	fluid capacity at		superheater bed
		constant volume	ϵ = void fraction
R	=	superheater inside radius	$k_{ax}k_r$ = thermal diffusion
μ	=	viscosity		coefficients
v	=	fluid velocity	κ = permeability
T_s	=	packing temperature	P = pressure

Details of the derivation are given by Morrow (1988).

The scaling arguments are as described earlier in this text. First, all the variables are made dimensionless by the introduction of scale factors. The differential equations are then transformed by substitution of the dimensionless variables. The dimensionless variables and derivatives are of order one. Typical values for the scale factors and parameters are used, and then the relative importance of the various terms in the differential equation can be determined.

Scale factors are used to define the following dimensionless variables:

$$T^* = T/T_0 \qquad P^* = P/P_0 \qquad T_s^* = T_s/T_0$$
$$z^* = z/z_0 \qquad v^* = v/v_0 \qquad k_{ax}^* = k_{ax}/k_{ax0}$$
$$k_r^* = k_r/k_{r0} \qquad r^* = r/r_0$$

The variables with the zero subscripts are the characteristic values of the variables. The characteristic values and parameter values are given in Table 7.1.

Table 7.1: Summary of One Phase Scaling Parameters

Variable	Value	Description
T_0	700 K	temperature
P_0	1.1×10^5 N/m² (1 atm)	pressure of outlet stream
v_0	0.17 m/sec	average linear velocity
z_0	0.20 m	superheater length
r_0	0.013 m	radius
ϵ	0.6	void fraction
C_{vf}	1.8×10^3 J/kg K	heat capacity of fluid
C_{vs}	1.1×10^3 J/kg K	heat capacity of packing
$\hat{A}$	900 m⁻¹	area of packing/volume of bed
μ	3.7×10^{-5} kg/m sec	viscosity
ρ_s	2.2×10^3 kg/m³	density of packing
ρ_v	0.29 kg/m3	density
κ	7.1×10^{-7} m²	permeability constant
p	0.92	emissivity of packing
h	22 J/m² sec K	local heat–transfer coefficient
k_f	6.7×10^{-2} J/m sec K	thermal conductivity of fluid
k_{ax_0}	0.31 J/m sec K	axial thermal dissipation coefficient
k_{r_0}	0.11 J/m sec K	radial thermal dissipation coefficient
k_s	1.5 J/m sec K	effective thermal conductivity of packed bed

Scaling substitutions are made in equations (7.8.1), (7.8.2), and (7.8.3), resulting in the following equations. The equations are scaled so that the packing/fluid heat transfer term is of order one. The fluid thermal energy balance is

$$\underbrace{\frac{\epsilon \rho_v C_{vf} v_0}{z_0 h \hat{A}}}_{O(0.01)} \cdot v^* \frac{\partial T^*}{\partial z^*} + \underbrace{\frac{\epsilon k_{r_0}}{r_0^2 h \hat{A}}}_{O(0.02)} \cdot \left[\frac{\partial k_r^*}{\partial r^*} \frac{\partial T^*}{\partial r^*} + \frac{k_r^*}{r^*} \frac{\partial}{\partial r^*} \left(r^* \frac{\partial T^*}{\partial r^*} \right) \right] +$$

$$\underbrace{\frac{\epsilon k_{ax_0}}{z_0^2 h \hat{A}}}_{O(0.0002)} \cdot \left[k_{ax}^* \frac{\partial^2 T^*}{\partial z^{*2}} + \frac{\partial k_{ax}^*}{\partial z^*} \frac{\partial T^*}{\partial z^*} \right] - \underbrace{\frac{P_0 \epsilon}{h \hat{A}}}_{O(4)} \cdot P^* \frac{\partial v^*}{\partial z^*} +$$

$$\underbrace{(T_s^* - T^*)}_{O(1)} = 0 \tag{7.8.4}$$

Solution of Split Boundary–Value Problems

The packed bed energy equation is

$$\underbrace{\frac{k_s(1-\epsilon)}{z_0^2 h \hat{A}} \frac{\partial^2 T_s^*}{\partial z^{*2}}}_{O(0.0007)} + \underbrace{\frac{k_s(1-\epsilon)}{r_0^2 h \hat{A}} \frac{1}{r^*} \frac{\partial}{\partial r^*}\left(r^* \frac{\partial T_s^*}{\partial r^*}\right)}_{O(0.2)} - \underbrace{(T_s^* - T^*)}_{O(1)} = 0 \quad (7.8.5)$$

The linear momentum balance is

$$\underbrace{\frac{v_0^2 \rho_v}{P_0} v^* \frac{\partial v^*}{\partial z^*}}_{O(8\times 10^{-8})} + \underbrace{\frac{z_0 \mu v_0}{P_0 \kappa} v^*}_{O(1\times 10^{-5})} + \underbrace{\frac{\partial P^*}{\partial z^*}}_{O(1)} = 0 \quad (7.8.6)$$

The scaling arguments led to several simplifications of the model. These simplifications are

1. Negligible axial dispersion through the superheater bed packing and fluid.

2. Radial heat transfer through the fluid is an order–of–magnitude smaller than the radial heat transfer through the bed packing. A radial effectiveness factor can be defined and the fluid temperature can be averaged at any axial position z. This eliminates the fluid temperature dependence upon radial position. (Clough, 1975; Morrow, 1988).

3. Negligible pressure gradient in the single–phase fluid flow regime. This allows us to modify the derivation of the fluid thermal energy balance by assuming constant pressure.

The equations are combined into a single nonlinear partial differential equation that describes single–phase fluid flow. Details are given by Morrow (1988). The final single–phase equation is

$$-\frac{wC_{pv}}{\epsilon \pi R^2} \frac{\partial T}{\partial z} + \frac{2}{\epsilon R} \cdot h_e(T_w - T) = 0 \quad (7.8.7)$$

where h_e is the effective heat–transfer coefficient. The heat transfer between the wall/fluid, wall/packing and packing/fluid is lumped into this parameter.

7.8.2 Two–Phase Fluid Flow Energy Balance

Liquid enters the superheater and begins boiling within a short distance of the entrance. The boiling (or two–phase) regime is where both the vapor and liquid phases are present. The two phases remain at a constant temperature until all the liquid is vaporized. An assumption must be made about this regime. Boiling is a turbulent process, especially in a packed bed. It will be assumed that the turbulence mixes the fluid well. This implies that radial gradients can be ignored. Also, the liquid and vapor phases must be dispersed; it is assumed that there are not any pockets of superheated vapor or unsaturated liquid. The fluid is modeled as if it were passing through a tube whose volume is the same as the void volume of the superheater.

A shell balance can be written for the two–phase regime in terms of f, the weight fraction of vapor. It is defined as

$$f = \text{fraction of vapor} = \frac{\text{mass of vapor}}{\text{mass of vapor} + \text{mass of liquid}} \quad (7.8.8)$$

The reference state of the fluid is defined to be saturated liquid. The two–phase energy balance is

$$\frac{\Delta H_{vap} w}{\epsilon \pi R^2} \frac{\partial f}{\partial z} + \frac{2 h_e}{\epsilon R}(T_w - T_b) = 0 \quad (7.8.9)$$

where T_b is the fluid boiling temperature. It should be noted that the heat–transfer coefficient h_e in the two–phase section is different than the coefficient in the one–phase section. This coefficient depends on the fraction of vapor, type of packing, fluid velocity and the difference between the wall temperature and fluid temperature. Accurate correlations of the heat–transfer coefficient in a packed bed with boiling are difficult to obtain. Therefore, the heat–transfer coefficient in the two–phase regime will be expressed as a constant times the coefficient in the one–phase regime. As a first approximation, the heat–transfer coefficient in the two–phase regime is assumed to be five times that of the single–phase regime (Tong, 1965).

7.8.3 Superheater Wall Energy Balance

An energy balance on the superheater wall can be developed via a shell balance. It will be assumed that radial temperature gradients in the wall are negligible. The energy flux q_h from the heater (Figure 7.13) is assumed to be a known function of time. The heat transfer between the wall and the fluid phase is modeled by the term $h_e(T_w - T)$. Fourier's law, $q_z = -k_w \frac{\partial T_w}{\partial z}$ is used to describe axial heat conduction in the superheater wall. The derivation leads to the following equation:

$$k_w \frac{\partial^2 T_w}{\partial z^2} - \frac{2R}{a^2 + 2aR} \cdot h_e(T_w - T) - \frac{2(R+a)}{a^2 + 2aR} \cdot q_h = 0 \quad (7.8.10)$$

where a is the thickness of the wall, a function of the axial distance z.

Solution of Split Boundary–Value Problems

Table 7.2: Superheater Wall Area and Thermal Mass

Section	Thermal Mass	Wall/Bed Heat Transfer Area
Pipe	271 J/K	1.15×10^{-2} m^2
Endcaps	246 J/K	5.60×10^{-3} m^2
Total	517 J/K	1.71×10^{-2} m^2

7.8.4 Endcap Model

It is common practice in much theoretical work to ignore end effects. This model will investigate the effect of large endcaps upon the thermal properties of a superheater. Table 7.2 shows the thermal mass of each endcap along with the wall/bed heat transfer area, compared to the total superheater. Thermal mass is defined as the amount of energy necessary to raise the superheater wall section one degree Kelvin. Obviously, the endcaps cannot be ignored.

Figure 7.16 is a diagram of an endcap. There are two distinct zones. A partial differential equation could be written for each zone, but it would depend upon radial position, axial position, and time. A problem arises with this approach. The equations would require information on heat–transfer coefficients in the fluid/wall, solid/wall, solid/solid, and fluid/solid interfaces. This would require knowledge of the fluid velocity field in the endcaps along with correlations for the heat–transfer coefficients. That information is not available. Therefore, writing complete, comprehensive energy balances for the endcaps would not be productive. Simplifications must be made.

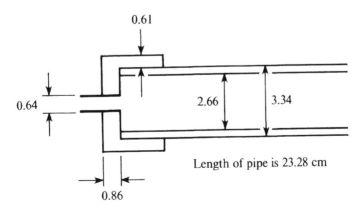

Figure 7.16: Endcap Dimensions.

Table 7.3: Differences Between the Model and Real System

Zone	Heat Transfer Area	Model Area	Model Length	Endcap Bed Vol	Model Volume
2	18.0 cm^2	25.1 cm^2	3.01 cm	11.9 cm^3	16.7 cm^3
3	5.5 cm^2	4.2 cm^2	0.50 cm	0.0 cm^3	2.8 cm^3
Total	23.5 cm^2	29.3 cm^2	3.51 cm	11.9 cm^3	19.5 cm^3

A one–dimensional thermal energy equation that eliminates radial dependence must be written. There are two approaches to modeling the endcaps. The thickness of the wall could be varied or the length of the superheater could be extended to account for the extra mass of the endcaps; i.e., the model will have a thickness or length that does not match the physical system. Either of these methods will affect one of the properties of the superheater, such as

1. Thermal mass
2. Wall thickness
3. Wall/solid/fluid heat–transfer area
4. Volume of the superheater

The thermal mass of the endcaps is an important parameter in the thermal response of the superheater and will be preserved in the model. The wall thickness is directly related to the axial conduction of heat and will also be preserved. These two properties determine the final form of the model. The heat–transfer area could have been preserved at the expense of one of the other properties. This would not have been the best choice. The heat–transfer coefficient in the neighborhood of the endcaps will vary with position in an unknown fashion. This coefficient will have to be estimated experimentally, so using an "artificial" thermal mass or wall thickness for a well–characterized property would not be appropriate. These calculations are presented in Table 7.3.

In summary, we now have a model of the superheater wall which is relatively simple, but only approximates the physical system. The thermal mass and axial conductivity are identical, but the volume and heat–transfer area of the model are slightly different from the physical system. Comparison with experimental data will show if the approximations used are adequate. Figure 7.17 shows the dimensions of the model with endcaps.

The steady–state model is summarized by the following system of ordinary differential equations:

$$\frac{d^2 T_w}{dz^2} = \frac{2R}{k_w(a^2 + 2aR)} h_e(T_w - T) + \frac{2(R+a)}{k_w(a^2 + 2aR)} q_h \quad (7.8.11)$$

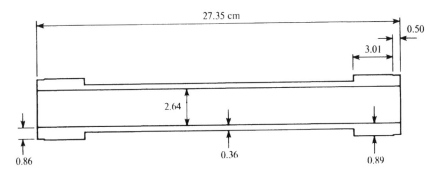

Figure 7.17: Model Dimensions.

$$\frac{dT}{dz} = \frac{2\pi R}{wC_{pv}} h_e(T_w - T) \qquad (7.8.12)$$

$$\frac{df}{dz} = \frac{2\pi R}{w\Delta H_{vap}} h_e(T_w - T_b) \qquad (7.8.13)$$

The wall thickness changes along the length of the reactor, as shown in Figure 7.17. Equation (7.8.13) is applied in the two–phase section of the superheater and equation (7.8.12) is applied in the one–phase section.

This system of equations can be reduced to four first–order linked linear ordinary differential equations with split boundary conditions. The transition point where the two–phase fluid is completely vaporized is unknown. Furthermore, the wall thickness is discontinuous, which means that $\frac{dT_w}{dz}$ is also discontinuous.

7.8.5 Boundary Conditions

Five boundary conditions are required to solve the steady–state equations. At the transition point, the fluid is at its boiling point and the vapor fraction is 1.0 ($f = 1.0$ and $T = T_b$). The other three boundary conditions are the heat flux through the endcaps at either end of the superheater and the temperature (or vapor fraction) of the entering fluid.

The exit of the superheater and the exit tube carrying the hot vapor are well insulated. Therefore, the loss of heat through conduction from the endcap out the exit tube should be negligible; i.e., $\frac{dT_w}{dz}|_{z=z_{end}} = 0$. The same argument is not valid for the heat flux at the entrance because of the temperature gradient from the hot superheater to the cooler entering tube.

The most practical entrance boundary condition assumes that the fluid enters the superheater at its boiling temperature with no heat losses through the ends of the superheater endcaps. The heat necessary to raise the water temperature from ambient to boiling is included in the model. This energy is lumped into the heat of vaporization. The model considers that each element of fluid is raised from ambient temperature to vapor at its boiling point; it does not consider liquid flow.

7.8.6 Solution Method

A shooting method was used to solve the steady-state model. The wall temperature at $z = 0$ is guessed and the boundary conditions of $f = 0$ and $\frac{dT_w}{dz}|_{z=0}$ are applied. The solution is marched forward using the wall equation (7.8.11) and the two-phase equation (7.8.13) until $f = 1.0$. The one-phase equation (7.8.12) is applied until $z = z_{end}$. This procedure is repeated until the boundary conditions at $z = z_{end}$ are met.

For this problem, the shooting method was slightly modified. The known boundary conditions at $z = 0$ are $\frac{dT_w}{dz}|_{z=0} = 0$ (no heat flux through the entrance of the superheater) and $f = 0$. The known boundary condition at $z = z_{end}$ is $\frac{dT_w}{dz}|_{z=z_{end}} = 0$. The wall temperature at $z = 0$ is the "guessed" boundary condition. The differential equations are solved forward from $z = 0$ to $z = z_{end}$. A common method of adjusting the guessed boundary condition assumes a linear relationship between the guessed boundary condition at $z = 0$ and the calculated boundary condition at $z = z_{end}$. However, Figure 7.18 shows that the relationship is highly nonlinear. A different method of adjusting the guessed boundary condition was used by Morrow (1988). The guess of the wall temperature is adjusted according to the sign of $\frac{dT_w}{dz}|_{z=z_{end}}$. Figure 7.16 shows that a high guess of the wall temperature results in $\frac{dT_w}{dz}|_{z=z_{end}}$ value greater than zero. A low guess of the wall temperature results in a negative $\frac{dT_w}{dz}|_{z=z_{end}}$. The desired boundary condition is $\frac{dT_w}{dz}|_{z=z_{end}} = 0$. Therefore, the guess is adjusted by bracketing the desired boundary condition. An initial guess of the wall temperature is made and the sign of $\frac{dT_w}{dz}|_{z=z_{end}}$ is examined. The guess is then moved in the correct direction until the sign of $\frac{dT_w}{dz}|_{z=z_{end}}$ changes. The guess of the wall temperature is then adjusted until $\frac{dT_w}{dz}|_{z=z_{end}} = 0$.

The first derivative of the wall temperature is discontinuous at the points where the pipe wall thickness changes. An energy balance around the point of discontinuity and temperature continuity lead to the following equations:

$$T_1 = T_2 \qquad (7.8.14)$$
$$T_{w1} = T_{w2} \qquad (7.8.15)$$
$$k_w \pi (a_1^2 + 2a_1 R) \frac{dT_w}{dz}\bigg|_{z=z_1} = k_w \pi (a_2^2 + 2a_2 R) \frac{dT_w}{dz}\bigg|_{z=z_2} \qquad (7.8.16)$$

where the subscripts 1 and 2 refer to the regions on either side of the dis-

Solution of Split Boundary–Value Problems

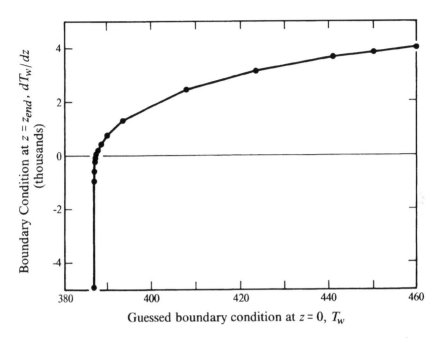

Figure 7.18: Relationship Between the Guessed and Calculated Boundary Conditions.

continuity in wall thickness. Note that the gradient $\frac{dT_w}{dz}$ is discontinuous. The steady–state system of equations (7.8.11), (7.8.12), and (7.8.13) can not be solved directly because of this discontinuity. Instead, the numerical solution is stopped each time a change of thickness is encountered and then re-initialized with new boundary conditions.

An outline of the shooting algorithm follows:

1. Choose a fluid outlet temperature. Use a macroscopic energy balance to determine q_h.

2. Guess $T_w|_{z=0}$.

3. The other boundary conditions are $\frac{dT_w}{dz}|_{z=0} = 0$ and $f|_{z=0} = 0.0$. Apply equation (7.8.11) and the two–phase equation (7.8.13).

4. Step the solution forward in z until the thickness changes.

5. Stop the solution and calculate a new $\frac{dT_w}{dz}$ using equation (7.8.16).

6. Restart the solution. (Repeat steps 4, 5, and 6 when changes in the wall thickness occur.)

7. Continue until $f > 1.0$ at $z = z_t$.

8. Iterate near the point z_t until $f = 1.0$ at the transition point between the single- and two-phase flows.
9. Start the solution with $T = T_b$. Apply equation (7.8.11) and the one-phase equation (7.8.12).
10. Continue until $z = z_{end}$.
11. Compare $\frac{dT_w}{dz}\big|_{z=z_{end}}$ with the known boundary condition.
12. Go to step 2.

7.8.7 Results

The simulator is used to investigate three important aspects of the model: the significance of the endcaps, conductivity in the heat exchanger wall and the adjustable parameter (the heat-transfer coefficient). Model parameters are given in Table 7.4. Simulation shows that the endcaps have little effect on the temperature profiles in the steady-state simulator. The shape of the temperature profiles are very similar and the transition points are within 1 percent of each other. The only significant difference that the addition of the endcaps makes is in the "corner" in the wall temperature profile because of the discontinuity in the wall thickness. This corner can be seen in Figures 7.19, 7.20, and 7.21 at $z = 0.85$.

Table 7.4: Model Parameters

Variable	Value	Description
C_{pv}	2×10^3 J/kg K	heat capacity of fluid
w	1.7×10^{-5} kg/sec	mass flow rate
h_e	35 J/m^2 sec K	effective heat-transfer coefficient
k_w	16.3 J/m sec K	thermal conductivity of wall
q_h	2×10^3 J/m^2	heat from external heaters
ΔH_{vap}	2.5×10^6 J/kg	heat of vaporization

One of the unusual features in this model is that the conductivity of the superheater wall is not ignored. Figures 7.19 and 7.20 show the result of raising the conductivity of the wall. This has a very strong effect on the temperature profiles. A large amount of heat is transferred from the hot end of the superheater to the cold end by conduction. Under typical experimental conditions, there is a wall temperature gradient of up to 15 K/cm, which is a large driving force. A reduction of the wall conductivity of k_w to $0.5 \cdot k_w$ increases the size of the two-phase region by 40 percent. The change in the conductivity of the wall has little effect on the separation between the wall and fluid temperatures.

Only one adjustable parameter is embedded in the model. The heat-transfer coefficient, h_e, must be adjusted experimentally since good correlations are not presently available. Note that it has different values in

Solution of Split Boundary–Value Problems 347

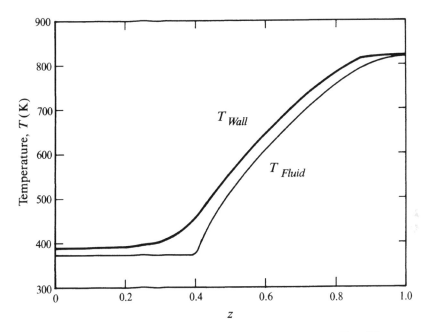

Figure 7.19: Steady–State Temperature Profile with Normal Conductivity k_w.

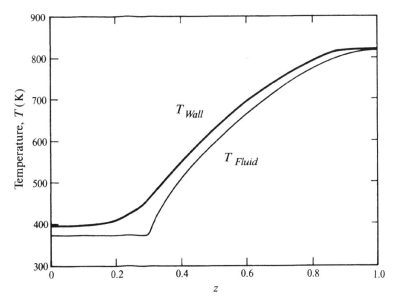

Figure 7.20: Steady–State Temperature Profile with $k_w = 1.5 \; k_w$.

348 Computational Methods for Process Simulation

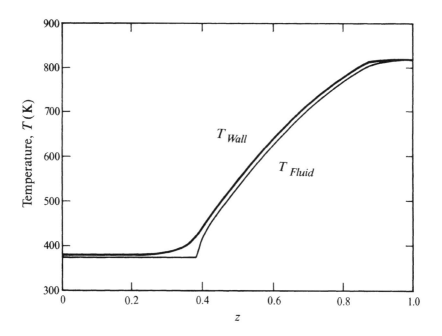

Figure 7.21: Steady–State Temperature Profile with $h_e = 100$ J/m^2 sec K.

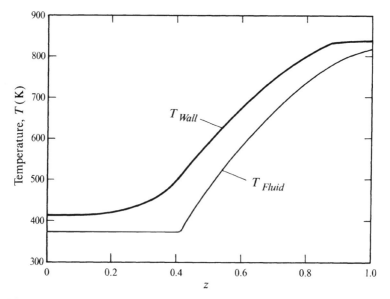

Figure 7.22: Steady–State Temperature Profile with $h_e = 15$ J/m^2 sec K.

the boiling regime and in the one–phase regime. The previous plots used a boiling h_e of five times the vapor phase h_e. Figures 7.19, 7.21, and 7.22 show the temperature profiles for different values of h_e. Note that the value of h_e controls the temperature separation between the wall and the fluid. The size of the two–phase region is only a weak function of h_e. Increasing the heat–transfer coefficient h_e in the two–phase regime decreases the wall/fluid temperature separation in the two–phase region, as expected.

PROBLEMS

7.1. The design of fixed bed catalytic reactors of the tubular type has generally been based upon a one–dimensional model. This model assumes that concentration and temperature gradients occur only in the axial direction, and that the only transport mechanism operating in this direction is the overall flow itself, considered to be of the plug–flow type. The one–dimensional model leads to average values for the temperatures and conversions. The reaction system to be considered is:

$$A \xrightarrow{k_1} B \xrightarrow{k_2} C$$

$$A \xrightarrow{k_3} C$$

This reaction model is fairly representative of the gas–phase air oxidation of o–xylene into phathalic anhydride on V_2O_5 catalysts. A represents o–xylene; B, phathalic anhydride; and C, the final oxidation products to CO and CO_2. Air is used in very large excess. The o–xylene mole fraction is generally kept below 1 percent, to stay under the explosive limit. One way of carrying out this very exothermic process is to use a multitubular reactor, consisting of 2500 tubes of 2.5 cm diameter, 2.5 to 3 m long, packed with catalyst and cooled by a salt bath that transfers the heat of reaction to a steam generator. Owing to the very large excess of oxygen, the rate equations may be considered in a first approximation to be of the pseudo–first–order type.

$$r_A = (k_1 + k_3) N_{A_0} N_0 (1 - y)$$

$$r_B = N_{A_0} N_0 [k_1 (1 - y) - k_1 x]$$

$$r_C = N_{A_0} N_0 [k_2 x + k_3 (1 - y)]$$

with
$$y = x + w$$
$$\ln k_1 = -27000/[1.98(t+T_0)] + 19.837$$
$$\ln k_2 = -31400/[1.98(t+T_0)] + 20.87$$
$$\ln k_3 = -28600/[1.98(t+T_0)] + 18.97$$

where N_{A_0} = inlet mole fraction of o–xylene
N_0 = inlet mole fraction of oxygen
x, w, y = conversions
t = temperature difference between reacting fluid and inlet feed temperature
T_0 = temperature of feed and coolant, °K

The following typical data should be used in the computations. $N_{A_0} = 0.00924$, which corresponds to 44 g mol/m³; $N_0 = 0.208$; $\Delta H_1 = -307$ kcal/g mol; $\Delta H_3 = -1.09$ kcal/g mol for the formation of CO and CO_2; $d_t = 0.025$ m; $d_p = 0.003$ m; $\rho_b = 1300$ kg/m³; $G = 4.684$ kg/m² hr; and $U = 82.7$ kcal/m² hr °C,

where d_t = tube diameter
d_p = particle diameter
ρ_b = catalyst bulk density
G = superficial mass flow velocity
U = heat–transfer coefficient

For inlet temperatures of the range 360–367°C, determine the steady-state profiles for this reactor when the maximum length is 1.2 m. You should note that a "hot spot" can exist and also the reactor can "run away."

7.2. Apply the method of adjoints to solve for the annular reactor profiles of the problem of section 7.5.

7.3 Deckwer et al. (*AIChE Journal* **29**, 915 (1983)) present a convective-dispersion model to describe the gas–liquid mass transfer characteristics of bubble columns. The model in dimensionless form is given by Equation 1 in their paper. Solve the model for countercurrent flow ($a' = 1$) with the Danckwerts' boundary conditions of Equations 6 and 7.

Use the generalized shooting technique and the principle of superposition to obtain solutions. Solve the problem for $Pe_L = 3.105$, $St_L = 2.004$, $C_L^* = .00028$ mol/liter, and $C_{Li} = 0$.

7.4 Solve problem 6.2 numerically using a shooting technique.

7.5 Pinjala et al. (*AIChE Jurnal* **34**, 1663 (1988)) present the following model for a tubular reactor that accounts for the axial dispersion of energy and mass

Mass Balance

$$0 = \frac{1}{Pe_m} \frac{d^2x}{dz^2} - \frac{dx}{dz} - Da\, x \exp\left(-\frac{1}{y}\right)$$

Energy Balance

$$0 = \frac{1}{Pe_h}\frac{d^2y}{dz^2} - \frac{dy}{dz} - \beta\, Da\, x \exp\left(-\frac{1}{y}\right) + U(y_w - y)$$

where x = dimensionless composition
y = dimensionless temperature
y_w = known wall dimensionless temperature
Pe_m, Pe_h, Da, β and U are all known dimensionless constants

The boundary conditions are

$$-\frac{dx}{dz} = Pe_m(1-x) \qquad z = 0$$

$$-\frac{dy}{dz} = Pe_h(y_f - y) \qquad z = 0$$

$$\frac{dy}{dz} = \frac{dx}{dz} = 0 \qquad z = 1$$

where y_f = dimensionless feed temperature.

Develop the generalized shooting technique for solving this problem. Write down the shooting system dynamics *for this problem* and the updating equations for unknown initial conditions. Discuss a computer implementation.

REFERENCES

Bauver, W. P., Friedman, C. H., and McGowan, J. G., *24th National Heat Transfer Conference and Exhibition*, ASME Heat Transfer Division (HTD) **75**, 105–112 (1987).

Bellman, R., *Dynamic Programming*, Princeton University Press, Princeton, N.J. (1957).

Carnahan, B., Luther, H. A., and Wilkes, J. O., *Applied Numerical Methods*, Chapter 2, Wiley, New York (1969).

Clough, D. E., *Optimization and Control of the Dehydrogenation of Ethylbenzene to Form Styrene*, Ph.D. Thesis, University of Colorado, Boulder (1975).

Clough, D. E. and Ramirez, W. F., "Mathematical Modeling and Optimization of the Dehydrogenation of Ethylbenzene to Form Styrene," *AIChE Journal* **22**, No. 4, 1097 (1976).

Danckwerts, P. V., *Chem. Eng. Sci.* **2**, 10 (1953).

Kukalov, G. T., *Thermal Engr.* **17**, 72, Feb. (1970).

Morrow, A. B., *Dynamics of Packed Bed Superheaters*, M.S. Thesis, University of Colorado, Boulder (1988).

Roberts, S. M. and Shipman, J. S., *Two Point Boundary Value Problems: Shooting Methods*, Elsevier, New York (1972).

Shoureshi, R. and McLaughlin, K., *Proceedings of the 1984 American Control Conference*, San Diego, Calif. (June 1984).

Tong, L. S., *Boiling Heat Transfer and Two–Phase Flow*, Wiley, New York (1965).

Chapter 8

SOLUTION OF PARTIAL DIFFERENTIAL EQUATIONS

This chapter discusses finite–difference techniques for the solution of partial differential equations. Techniques are presented for pure convection problems, pure diffusion or dispersion problems, and mixed convection–diffusion problems. Each case is illustrated with common physical examples. Special techniques are introduced for one– and two–dimensional flow through porous media. The method of weighted residuals is also introduced with special emphasis given to orthogonal collocation.

8.1 TECHNIQUES FOR CONVECTION PROBLEMS

For pure convective problems, the continuous time and spatial domains can be represented by a series of discrete finite points in each domain by a centered–difference technique (see Figure 8.1).

For the centered–difference technique, the finite–difference analogs are written about the half point $(i+1/2, n+1/2)$. The reason for this is that this centered differencing develops second–order correct analogs for both the variable and its first derivatives. To get the first–order time partial, we develop Taylor series expansions for the functions $u(t_i, x_{n+1/2})$ and $u(t_{i+1}, x_{n+1/2})$ about the half point $u(t_{i+1/2}, x_{n+1/2})$; then we have

$$u_{i,n+1/2} = u_{i+1/2,n+1/2} + \frac{\partial u}{\partial t}(i+1/2, n+1/2)(t_i - t_{i+1/2})$$
$$+ \frac{\partial^2 u}{\partial t^2}(i+1/2, n+1/2)\frac{(t_i - t_{i+1/2})^2}{2} + \text{HOT} \quad (8.1.1)$$

$$u_{i+1,n+1/2} = u_{i+1/2,n+1/2} + \frac{\partial u}{\partial t}(i+1/2, n+1/2)(t_{i+1} - t_{i+1/2})$$

$$+\frac{\partial^2 u}{\partial t^2}(i+1/2, n+1/2)\frac{(t_{i+1}-t_{i+1/2})^2}{2} + \text{HOT} \quad (8.1.2)$$

Subtracting equation (8.1.1) from (8.1.2) gives

$$u_{i+1,n+1/2} - u_{i,n+1/2} = \frac{\partial u}{\partial t}(i+1/2, n+1/2)(t_{i+1}-t_i) + \text{HOT} \quad (8.1.3)$$

or

$$\frac{\partial u}{\partial t}(i+1/2, n+1/2) = \frac{(u_{i+1,n+1/2} - u_{i,n+1/2})}{\Delta t} \quad (8.1.4)$$

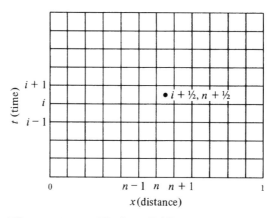

Figure 8.1: Finite–Difference Grid.

This is a second–order correct analog for $\partial u/\partial t$. What is meant by second–order correct is that the linearization procedure truncates after the second–order terms in the Taylor series. By averaging the $n + 1/2$ point, we can obtain an expression only in terms of the grid points.

$$\frac{\partial u}{\partial t}(i+1/2, n+1/2) = \frac{1}{2\Delta t}(u_{i+1,n+1} + u_{i+1,n} - u_{i,n+1} - u_{i,n}) \quad (8.1.5)$$

Similarly, we can obtain a second–order expression for the distance derivative as

$$\frac{\partial u}{\partial x}(i+1/2, n+1/2) = \frac{1}{2\Delta x}(u_{i+1,n+1} + u_{i,n+1} - u_{i+1,n} - u_{i,n}) \quad (8.1.6)$$

One of two methods is usually used to obtain a finite–difference analog for the variable itself. The first just averages the four neighbors of $u_{i+1/2,n+1/2}$ to obtain

$$u_{i+1/2,n+1/2} = 1/4(u_{i+1,n+1} + u_{i+1,n} + u_{i,n+1} + u_{i,n}) \quad (8.1.7)$$

Von Rosenberg (1969) points out that it can be advantageous to use the analog

$$u_{i+1/2, n+1/2} = 1/2(u_{i+1,n+1} + u_{i,n}) \quad (8.1.8)$$

where the grid size ratio $\Delta x/\Delta t$ is taken as the velocity of the perturbed fluid. This analog essentially propagates a convective discontinuity forward one Δt and therefore works better than the complete neighbor average of equation (8.1.7).

8.2 UNSTEADY–STATE STEAM HEAT EXCHANGER: EXPLICIT CENTERED–DIFFERENCE PROBLEM

We want to develop the dynamic response of a steam–heated plug–flow heat exchanger as shown in Figure 8.2.

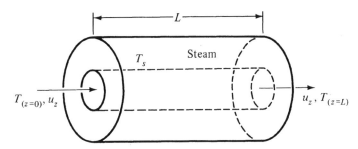

Figure 8.2: Steam Heat Exchanger.

We are interested in computing the dynamic response of the axial temperature profile of the fluid flowing in the inner tube for a step change in the inlet temperature $T(z = 0)$. The steam shall maintain a constant wall temperature of T^s in the exchanger. The describing differential equation is obtained by writing an energy balance around the tube and yields

$$\frac{\partial T}{\partial t} = -v_z \frac{\partial T}{\partial z} + \frac{4U}{\rho c_p D}(T^s - T) \quad (8.2.1)$$

For this problem,
B.C. 1 $T(z)$ at $t = 0$ is a known steady–state profile.
B.C. 2 T at $z = 0$ for all $t > 0$ is the new known inlet temperature, T_0.

To solve this problem, we will form a finite–difference grid for the two continuous independent variables, t and z as shown in Figure 8.3. We will divide the distance grid into 20 elements. The Δt size is chosen so that the ratio $\Delta z/\Delta t$ is the fluid velocity v_z. Known grid points (from the

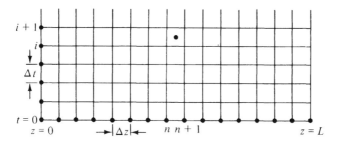

Figure 8.3: Finite–Difference Grid for Steam Heat Exchanger.

boundary conditions) are shown as black dots. We will finite–difference the partial differential equation in order to solve for the unknown grid points. Since this is a pure convective problem, we use the centered–difference second–order correct analogs of Section 8.1. The resulting finite–difference equation is

$$\frac{1}{2\Delta t}(T_{i+1,n+1} + T_{i+1,n} - T_{i,n+1} - T_{i,n}) =$$
$$-\frac{v_z}{2\Delta z}(T_{i+1,n+1} + T_{i,n+1} - T_{i+1,n} - T_{i,n})$$
$$+\frac{4U}{\rho c_p D}[T^s - 1/2(T_{i+1,n+1} + T_{i,n})] \qquad (8.2.2)$$

Putting in $n=0$ and $i=0$, the only unknown temperature is $T_{1,1}$ since $T_{0,0}$, $T_{0,1}$, and $T_{1,0}$ are all known boundary conditions. Therefore equation (8.2.2) can be solved explicitly for $T_{1,1}$. Knowing $T_{1,1}$, we can then get $T_{1,2}$. The process is repeated until the entire z profile is computed for the time level $i + 1 = 1$. We then can move to a new time level $i + 1 = 2$ and compute the distance grid points sequentially. This explicit method is continued for as many time levels as necessary until the new steady–state level is reached. MATLAB program ex82.m has been created to solve this problem. It is given in Figure 8.4. The dynamic response of the system is given in Figure 8.5.

Figure 8.4: MATLAB Program for Dynamic Response of Steam Heat Exchanger.

```
% This program uses the centered-difference technique to solve the
% unsteady-state steam heat exchanger problem (explicit) in section 8.2.
%
global Cp rho vz Tsteam D U;
```

```
% simulation parameters
%
Cp = 1.0;
D = 0.5;
rho = 62.4;
vz = 3;
U = 0.2;
TT0 = 75;
Tsteam = 250;
L = 12;
N = 40;                 % divide the distance grid into N elements.
delta_z = L/N;
delta_t = delta_z/vz;


% the initial steady state temp. can also be solved analytically

for I = 2:(N+1),
    T(1,I) = Tsteam + (TT0-Tsteam)*exp(-4*U*(I-1)*delta_z/(rho*Cp*D*vz));
end

T(1,1) = 75;
z = [0:delta_z:12]';
plot(z, T(1,:), '+');
title('Dynamic Response of Steam Heat Exchanger to Step Disturbance');
xlabel('Reactor Length (ft)');
ylabel('Water Temperature (degree F)');
pause;
hold on;

% define the step change in inlet temp.
%
for K = 2:(N+1),
    T(K,1) = 60;
end

% constants for iteration:
%
a = 1/(2*delta_t);
b = vz/(2*delta_z);
c = 2*U/(rho*Cp*D);

% loop for time iteration
%
for i = 1:N,

    % loop for distance iteration
    %
    for n = 1:N,
        T(i+1, n+1) = (2*c*Tsteam + (b-a)*T(i+1,n) + (-b+a)*T(i,n+1) ...
                    + (b+a-c)*T(i,n))/(a+b+c);
    end
end
```

```
plot(z, T(3,:), ':');
plot(z, T(11,:), '-.');
plot(z, T(21,:), '-');
plot(z, T(31,:), '--');
plot(z, T(41,:), 'o');
text(0.5, 92, '+++    t=0')
text(0.5, 90, '...    t=0.2 sec')
text(0.5, 88, '-.-    t=1.0 sec')
text(0.5, 86, '___    t=2.0 sec')
text(8, 65, '- - -    t=3.0 sec')
text(8, 63, 'ooo    t=4.2 sec')

hold off;
print -deps fig85.eps;
```

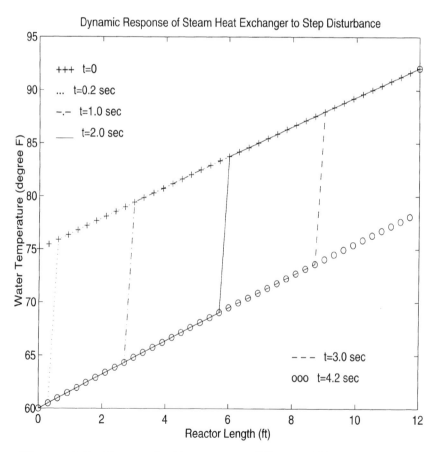

Figure 8.5: Dynamic Response of Steam Heat Exchanger to Step Disturbance.

8.3 UNSTEADY–STATE COUNTERCURRENT HEAT EXCHANGER: IMPLICIT CENTERED–DIFFERENCE PROBLEM

The system to be considered is the dynamic response of a countercurrent plug–flow heat exchanger as shown in Figure 8.6.

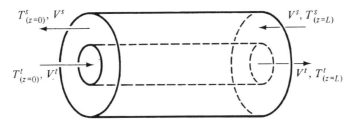

Figure 8.6: Countercurrent Heat Exchanger.

We are interested in computing the dynamic response of the axial temperature profile for both the tube–side (superscript t) and the shell–side (superscript s) fluids to a step change in the inlet tube temperature, $T^t(z=0)$.

The energy balance on the tube side gives

$$\frac{\partial T^t}{\partial t} + V^t \frac{\partial T^t}{\partial x} = P^t(T^s - T^t) \tag{8.3.1}$$

where

$$P^t = US/c_p^t \rho^t A^t \tag{8.3.2}$$

and

T^t = tube–side temperature
T^s = shell–side temperature
V^t = tube–side velocity
U = overall heat–transfer coefficient
S = inside perimeter of inner pipe
A^t = cross–sectional area available for flow on tube side

The energy balance on the shell side gives

$$\frac{\partial T^s}{\partial t} - V^s \frac{\partial T^s}{\partial x} = P^s(T^t - T^s) \tag{8.3.3}$$

where

$$P^s = US/c_p^s \rho^s A^s \tag{8.3.4}$$

and V^s = shell–side velocity.

The boundary conditions for this problems are

B.C. 1 $T^t(x)$ and $T^s(x)$ at $t = 0$ are a known steady–state profile.

B.C. 2 T^t at $x = 0$ for all $t > 0$ is the new, known inlet tube temperature.

B.C. 3 T^s at $x = L$ for all $t > 0$ is the old, known inlet shell temperature.

Notice that the spatial boundary conditions are split between information at $x = 0$ and $x = L$. As usual, split boundary information leads to difficulties.

To solve this problem, we form a finite–difference grid for the two continuous independent variables, t and x, as shown in Figure 8.7. We will divide the distance grid into 20 elements. The Δt size is chosen so that the ratio of $\Delta x / \Delta t$ is the tube (disturbed) fluid velocity V^t. When both the tube and shell temperatures are known from the given boundary conditions, we use solid black dots in the diagram. Where the tube temperature is the only known boundary condition, the left side of the dots is filled. If the right side is filled, the shell temperature is specified. We finite–difference the partial differential equation in order to solve for the dependent variables at the unknown grid points. This is a pure convective problem so we use the centered–difference second–order analogs of Section 8.1. Since the tube–side temperature is perturbed in this problem, we use the Von Rosenberg analog

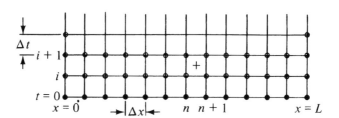

Figure 8.7: Finite–Difference Grid for Countercurrent Heat Exchanger.

$$T^t_{i+1/2, n+1/2} = (T^t_{i+1,n+1} + T^t_{i,n})/2 \tag{8.3.5}$$

The shell–side temperature is not forcefully disturbed so the conventional form for T^s is used,

$$T^s_{i+1/2, n+1/2} = (T^s_{i+1,n+1} + T^s_{i+1,n} + T^s_{i,n+1} + T^s_{i,n})/4 \tag{8.3.6}$$

Solution of Partial Differential Equations

The resulting finite-difference equations are

$$\frac{1}{2\Delta t}(T^t_{i+1,n+1} + T^t_{i+1,n} - T^t_{i,n+1} - T^t_{i,n})$$

$$+V^t\frac{1}{2\Delta x}(T^t_{i+1,n+1} + T^t_{i,n+1} - T^t_{i+1,n} - T^t_{i,n}) =$$

$$P^t[(T^s_{i+1,n+1} + T^s_{i+1,n} + T^s_{i,n+1} + T^s_{i,n})/4$$

$$-(T^t_{i+1,n+1} + T^t_{i,n})/2] \qquad (8.3.7)$$

$$\frac{1}{2\Delta t}(T^s_{i+1,n+1} + T^s_{i,n+1} - T^s_{i+1,n} - T^s_{i,n})$$

$$-V^s\frac{1}{2\Delta x}(T^s_{i+1,n+1} + T^s_{i,n+1} - T^s_{i+1,n} - T^s_{i,n}) =$$

$$P^s[(T^t_{i+1,n+1} + T^t_{i,n})/2$$

$$-(T^s_{i+1,n+1} + T^s_{i+1,n} + T^s_{i,n+1} + T^s_{i,n})/4] \qquad (8.3.8)$$

Putting in $n = 0$ and $i = 0$, we have three unknown temperatures $T^t_{1,1}$, $T^s_{1,1}$, and $T^s_{1,0}$. There are only two algebraic equations so there is one more unknown variable than algebraic equation. Therefore, we cannot explicitly solve for the shell and tube temperatures at the new time level. However, if we formulate the entire set of equations for all distance grid points with the time level $i = 0$, we get 40 equations in 40 unknowns. The equations are linear so we can use matrix techniques to solve for all the shell and tube temperatures at the new time level simultaneously. This simultaneous solution strategy is called an implicit method. The matrix formulation is shown in Figure 8.8, where

$$a^t = \frac{1}{2\Delta t} - V^t\frac{1}{2\Delta x} \qquad a^s = 0$$

$$b^t = -P^t/4 \qquad b^s = \frac{1}{2\Delta t} + V^s\frac{1}{2\Delta x} + P^s/4$$

$$c^t = \frac{1}{2\Delta t} + V^t\frac{1}{2\Delta x} + P^t/2 \qquad c^s = -P^s/2$$

$$d^t = -P^t/4 \qquad d^s = \frac{1}{2\Delta t} - V^s\frac{1}{2\Delta x} + P^s/4$$

and

$$f_1 = \frac{1}{2\Delta t} + \frac{V^t}{2\Delta x} - \frac{P^t}{2} \qquad f_2 = \frac{1}{2\Delta t} - \frac{V^t}{2\Delta x}$$

$$f_3 = \frac{1}{2\Delta t} - \frac{V^s}{2\Delta x} - P^s/4 \qquad f_4 = \frac{1}{2\Delta t} + \frac{V^s}{2\Delta x} - P^s/4$$

$$\begin{bmatrix} b^t c^t d^t & & & & & & & \\ b^s c^s d^s & & & & & & & \\ a^t b^t c^t d^t & & & & & & & \\ a^s b^s c^s d^s & & & & & & & \\ & a^t b^t c^t d^t & & & & & & \\ & a^s b^s c^s d^s & & & & & & \\ & & \cdot & & & & & \\ & & & \cdot & & & & \\ & & & & a^t b^t c^t d^t & & & \\ & & & & a^s b^s c^s d^s & & & \\ & & & & & a^t b^t c^t & & \\ & & & & & a^s b^s c^s & & \end{bmatrix} \begin{bmatrix} T^s_{i+1,0} \\ T^t_{i+1,1} \\ T^s_{i+1,1} \\ T^t_{i+1,2} \\ T^s_{i+1,2} \\ \cdot \\ \cdot \\ \cdot \\ T^t_{i+1,19} \\ T^s_{i+1,19} \\ T^t_{i+1,20} \end{bmatrix} = \begin{bmatrix} f_1 T^t_{i,0} + f_2 T^t_{i,1} - b^t(T^s_{i,0} + T^s_{i,1}) \\ -a^t T^t_{i+1,0} \\ f_3 T^s_{i,0} + f_4 T^s_{i,1} - c^s T^t_{i,0} - a^s T^t_{i+1,0} \\ f_1 T^t_{i,1} + f_2 T^t_{i,2} - b^t(T^s_{i,1} + T^s_{i,2}) \\ f_3 T^s_{i,1} + f_4 T^s_{i,2} - c^s T^t_{i,1} \\ \cdot \\ \cdot \\ f_1 T^t_{i,n} + f_2 T^t_{i,n+1} - b^t(T^s_{i,n} + T^s_{i,n+1}) \\ f_3 T^s_{i,n} + f_4 T^s_{i,n+1} - c^s T^t_{i,n} \\ \cdot \\ \cdot \\ f_1 T^t_{i,19} + f_s T^t_{i,20} - b^t(T^s_{i,20} + T^s_{i+1,20}) \\ -d^t T^s_{i+1,20} \\ f_3 T^s_{i,19} + f_4 T^s_{i,20} - c^s T^t_{i,19} - d^s T^s_{i+1,20} \end{bmatrix}$$

Figure 8.8: Matrix Formulation.

Notice that all the nonzero entries in the matrix are grouped around the diagonal elements. In fact, we have no more than two elements on either side of the diagonal for this problem. In order to solve for the unknown time level temperatures, this matrix must be inverted. Very efficient algorithms have been developed for the solution of linear algebraic equations with band type matrices (Von Rosenberg, 1969).

Consider the set of simultaneous equations of the form

$$A(J, M)X_{J-M} + A(J, M-1)X_{J-(M-1)}$$
$$+ \cdots + A(J, 2)X_{J-2} + A(J, 1)X_{J-1}$$
$$+ B(J)X_J + C(J, 1)X_{J+1} + C(J, 2)X_{J+2}$$
$$+ \cdots + C(J, M)X_{J+M} = D(J) \tag{8.3.9}$$

where each $A(J, K)$ is a vector of coefficients to the left of the diagonal of length equal to the number of unknowns, N; $B(J)$ is a vector of diagonal coefficients of length N; and $C(J, K)$ is a vector of coefficients to the right of the diagonal of length N. The maximum value of K is M, which is the maximum number of nonzero entries either to the left or right of the diagonal element.

For our countercurrent heat–exchanger problem, $M = 2$, and Figure 8.9 gives the entries for each of the A, B, and C vectors.

The band algorithm proceeds in a straightforward fashion according to the following equations for $J = 1, \ldots, N$, where $N =$ the number of

$$A(J,2) = \begin{pmatrix} 0 \\ 0 \\ 0 \\ a^s \\ 0 \\ a^s \\ 0 \\ a^s \\ \vdots \\ 0 \\ a^s \\ 0 \\ a^s \end{pmatrix} \quad A(J,1) = \begin{pmatrix} 0 \\ b^s \\ a^t \\ b^s \\ a^t \\ b^s \\ \vdots \\ a^t \\ b^s \\ a^t \\ b^s \end{pmatrix} \quad B(J) = \begin{pmatrix} b^t \\ c^s \\ b^t \\ c^s \\ \vdots \\ b^t \\ c^s \\ b^t \\ c^s \end{pmatrix}$$

$$C(J,1) = \begin{pmatrix} c^t \\ d^s \\ c^t \\ d^s \\ \vdots \\ d^s \\ c^t \\ 0 \end{pmatrix} \quad C(J,2) = \begin{pmatrix} d^t \\ 0 \\ d^t \\ 0 \\ d^t \\ 0 \\ \vdots \\ d^t \\ 0 \\ 0 \\ 0 \end{pmatrix}$$

Figure 8.9: Band Coefficients of the Countercurrent Heat Exchanger.

unknown variables.

$$AL(J,K) = A(J,K) - \sum_{P=K+1}^{M} AL(J,P) \cdot W(J-P, P-K) \quad (8.3.10)$$

$$BETA(J) = B(J) - \sum_{P=1}^{M} AL(J,P) \cdot W(J-P, P) \quad (8.3.11)$$

$$W(J,K) = \left[C(J,K) - \sum_{P=K+1}^{M} AL(J, P-K) \cdot W(J-[P-K], P) \right]$$
$$\div BETA(J)$$
$$(8.3.12)$$

$$GAM(J) = \left[D(J) - \sum_{P=1}^{M} AL(J,P) \cdot GAM(J-P) \right] \div BETA(J)$$
$$(8.3.13)$$

The solution vector is then obtained via the backward solution $J = N, \ldots, 1$, from

$$X(J) = GAM(J) - \sum_{P=1}^{M} W(J,P) \cdot X(J+P) \quad (8.3.14)$$

Figure 8.10 gives MATLAB m-file Band.m for implementing the band algorithm. An alternative formulation of the problem is possible in MATLAB. Rather than using the band algorithm directly, we can use the sparse matrix capability of MATLAB. The matrix to be inverted can be defined and stored in the sparse or band format (see Chapter 2, section 2.2.3.3). This option is particularly efficient in MATLAB.

Using the sparse matrix formulation, the shell and tube temperatures for all distance grid points are computed for an unknown time level, $t + \Delta t$. This procedure is continued until enough time steps are taken so that a steady–state condition is reached. A MATLAB program ex83.m (Figure 8.11) has been developed to solve this problem. The dynamic response results for this countercurrent heat exchanger are shown in Figure 8.12. It should be noted that some numerical oscillations are present on the shell side. This is due to not matching the convective flow on this side but using the nearest neighbor assumption (Eqn. (8.3.6)).

Figure 8.10: m-file Band.

```
function x = band (a,b,c,d,n,m)
%
% This file implements the band inversion algorithm
%
w = zeros(n,m);
al = zeros(n,m);
beta = zeros(n,1);
gam = zeros(n,1);
for J = 1:n
  for IK = 1:m
    K = (m+1) - IK;
    sum = 0;
    KP1 = K+1;
    if ((K+1)-m) <= 0;
      for IP = KP1:m
        JLP = J-IP;
        MPK = IP -K;
        if (IP - J) < 0;
          sum = sum +al(J,IP) *w(JLP,MPK);
        else
        end
      end
    else
      al(J,K) = a(J,K) -sum;
    end
  end
%
% Computation of Beta
%
sum = 0;
  for IP = 1:m
    JLP = J - IP;
    if JLP > 0
```

```
           sum = sum+al(J,IP)*w(JLP,IP);
      else
      end
   end
 beta(J) = b(J) - sum;
 if beta(J) == 0
    beta(J)
 else
 end
%
% Computation of w
%
for K = 1:m
   sum = 0;
   IK = K+1;
   if (m-IK) >= 0
    for IP = IK:m
       IMK = IP - K;
       JLPMK = J - IMK;
       if (J-IMK) > 0
          sum = sum+al(J,IMK) * w(JLPMK,IP);
       else
       end
      end
   else
     w(J,K) = (c(J,K) - sum)/beta(J);
   end
end
%
% Computation of gam
%
sum = 0;
for IP = 1:m
   JLP = J-IP;
   if (IP-J) < 0
      sum = sum+al(J,IP) * gam(JLP);
   else
   end
end
   gam(J) = (d(J) - sum)/beta(J);
end
%
% Computation of solution vector
%
for IJ = 1:n
   J = n+1-IJ;
   sum = 0;
   for IP = 1:m
      JPP = J +IP;
      if (n-JPP) >= 0
         sum = sum+ w(J,IP) *x(JPP);
      else
      end
```

```
    end
    x(J) = gam(J) - sum;
end
```

Figure 8.11: MATLAB file ex83.m which uses the sparse matrix formulation to solve for heat exchanger dynamics.

```
% This program uses implicit centered-difference technique to solve the
% unsteady-state countercurrent heat exchanger problem in section 8.3.
%

global Pt Ps Vt Vs;

% simulation parameters
%
Pt = 0.2; Ps = 0.6;     % no values specified in the book
Vt = 2.5; Vs = 8;       % these numbers are assumed
L = 10;                 % length of the heat exchanger
N = 20;                 % number of grid points
delta_x = L/N;          % divide the distance grid into N elements.
delta_t = delta_x/Vt;
time = 20;              % simulation time in minutes
iteration = time/delta_t;

% solve for the initial steady state temperature profile. In order to
% get temp. values on the N distance grid points, integration is
% divided into N small steps.
% Note that the initial steady state problem is a split boundary value
% problem and superposition technique is used.
%
% generate one particular solution:
%
x0 = 0.0; xf = L;
Tp0 = [45 50]';
for i = 2:1:N+1,
    xf = x0 + delta_x;
    [xp, Tp] = ode45('Model83', x0, xf, Tp0, 1.e-8, 1);
    Tpt(i) = Tp(length(Tp),1);
    Tps(i) = Tp(length(Tp),2);
    x(i) = xp(length(xp));
    Tp0 = [Tpt(i) Tps(i)]';
    x0 = xf;
end
x(1) = 0;
Tpt(1) = 45;
Tps(1) = 50;
%plot(x, Tpt, '-', x, Tps, '-.');
%title('Particular Solution');
%pause;

% generate two homogeneous solutions. In this special case, the
% differential equations used for the particular solution and
% homogeneous solutions are the same, which is defined in Model83.m.
```

```
%
x0 = 0.0; xf = L;
Th10 = [45 0]';
for i = 2:1:N+1,
    xf = x0 + delta_x;
    [xh1, Th1] = ode45('Model83', x0, xf, Th10, 1.e-8, 1);
    Th1t(i) = Th1(length(Th1),1);
    Th1s(i) = Th1(length(Th1),2);
    x(i) = xh1(length(xh1));
    Th10 = [Th1t(i) Th1s(i)]';
    x0 = xf;
end
x(1) = 0;
Th1t(1) = 45;
Th1s(1) = 0;
%plot(x, Th1t, '-', x, Th1s, '-.');
%title('Homogeneous Solution#1');
%pause;

x0 = 0.0; xf = L;
Th20 = [0 50]';
for i = 2:1:N+1,
    xf = x0 + delta_x;
    [xh2, Th2] = ode45('Model83', x0, xf, Th20, 1.e-8, 1);
    Th2t(i) = Th2(length(Th2),1);
    Th2s(i) = Th2(length(Th2),2);
    x(i) = xh2(length(xh2));
    Th20 = [Th2t(i) Th2s(i)]';
    x0 = xf;
end
x(1) = 0;
Th2t(1) = 0;
Th2s(1) = 50;
%plot(x, Th2t, '-', x, Th2s, '-.');
%title('Homogeneous Solution#2');
%pause;

% solve linear equation set to get the constants that satisfy the real
% boundary conditions.
%
A = [Th1t(1)   Th2t(1);
     Th1s(N+1) Th2s(N+1)];

B = [45-Tpt(1) 75-Tps(N+1)]';

const = A\B;

% compose the final solution.
%
Tt(1,:) = Tpt + const(1)*Th1t + const(2)*Th2t;
Ts(1,:) = Tps + const(1)*Th1s + const(2)*Th2s;

plot(x, Tt(1,:));
```

```
title('Tube side temperature');
xlabel('length, ft');
ylabel('temperature, deg F');
%print;
pause;
plot(x, Ts(1,:));
title('Shell side temperature');
xlabel('length, ft');
ylabel('temperature, deg F');
%print;
pause;

% define the constants used in the matrices:
%
at = 1/(2*delta_t) - Vt/(2*delta_x);
as = 0;
bt = -Pt/4;
bs = 1/(2*delta_t) + Vs/(2*delta_x) + Ps/4;
ct = 1/(2*delta_t) + Vt/(2*delta_x) + Pt/2;
cs = -Ps/2;
dt = -Pt/4;
ds = 1/(2*delta_t) - Vs/(2*delta_x) + Ps/4;
f1 = 1/(2*delta_t) + Vt/(2*delta_x) - Pt/2;
f2 = 1/(2*delta_t) - Vt/(2*delta_x);
f3 = 1/(2*delta_t) - Vs/(2*delta_x) - Ps/4;
f4 = 1/(2*delta_t) + Vs/(2*delta_x) - Ps/4;
%
% define the sparse a matrix
%
r1 = [1 1 1 2 2 2 3 3 3 4 4 4 5 5 5 6 6 6 6];
c1 = [1 2 3 1 2 3 2 3 4 5 2 3 4 5 4 5 6 7 4 5 6 7];
v1 = [bt ct dt bs cs ds at bt ct dt as bs cs ds at bt ct dt as bs cs ds];
r2 = [ 7 7 7 7 8 8 8 8 9 9 9 9 10 10 10 10 11 11 11 11 12 12 12 12];
c2 = [ 6 7 8 9 6 7 8 9 10 11 8 9 10 11 10 11 12 13 10 11 12 13];
v2 = [at bt ct dt as bs cs ds at bt ct dt as bs cs ds at bt ct dt];
v22 = [as bs cs ds];
r3 = [13 13 13 13 14 14 14 14 15 15 15 15 16 16 16 16 ];
c3 = [12 13 14 15 12 13 14 15 14 15 16 17 14 15 16 17];
v3 = [at bt ct dt as bs cs ds at bt ct dt as bs cs ds ];
r4 =[17 17 17 17 18 18 18 18 19 19 19 19 20 20 20 20];
c4 =[16 17 18 19 16 17 18 19 18 19 20 21 18 19 20 21];
v4 =[at bt ct dt as bs cs ds at bt ct dt as bs cs ds];
r5 =[21 21 21 21 22 22 22 22 23 23 23 23 24 24 24 24];
c5 =[20 21 22 23 20 21 22 23 22 23 24 25 22 23 24 25];
v5 =[at bt ct dt as bs cs ds at bt ct dt as bs cs ds];
r6 =[25 25 25 25 26 26 26 26 27 27 27 27 28 28 28 28];
c6 =[24 25 26 27 24 25 26 27 26 27 28 29 26 27 28 29];
v6 = v5;
r7 =[29 29 29 29 30 30 30 30 31 31 31 31 32 32 32 32];
c7 =[28 29 30 31 28 29 30 31 30 31 32 33 30 31 32 33];
v7 = v6;
r8 =[33 33 33 33 34 34 34 34 35 35 35 35 36 36 36 36];
c8 =[32 33 34 35 32 33 34 35 34 35 36 37 34 35 36 37];
```

Solution of Partial Differential Equations

```
    v8 = v7;
    r9 =[37 37 37 37 38 38 38 38 39 39 39 40 40 40];
    c9 =[36 37 38 39 36 37 38 39 38 39 40 38 39 40];
    v9 =[at bt ct dt as bs cs ds at bt ct as bs cs];
    r =[r1 r2 r3 r4 r5 r6 r7 r8 r9];
    c =[c1 c2 c3 c4 c5 c6 c7 c8 c9];
    v =[v1 v2 v22 v3 v4 v5 v6 v7 v8 v9];

    aa = sparse(r,c,v,40,40);

% define B.C.2:
%
    Ttnew = 35;

% define B.C.3:
%
    Tsold = 75;
%
% generate the right hand side vector
%
    for j=1:iteration,
        B(1) = f1*Tt(j,1) + f2*Tt(j,2) - bt*(Ts(j,1)+Ts(j,2)) - at*Ttnew;
        B(2) = f3*Ts(j,1) + f4*Ts(j,2) - cs*Tt(j,1);

%       B(3) = f1*Tt(j,2) + f2*Tt(j,3) - bt*(Ts(j,2)+Ts(j,3));
%       B(4) = f3*Ts(j,2) + f4*Ts(j,3) - cs*Tt(j,2);
        for k=3:2:37,
            B(k) = f1*Tt(j,(k+1)/2) + f2*Tt(j,(k+1)/2+1) ...
                 - bt*(Ts(j,(k+1)/2)+Ts(j,(k+1)/2+1));
            B(k+1) = f3*Ts(j,(k+1)/2) + f4*Ts(j,(k+1)/2+1) - cs*Tt(j,(k+1)/2);
        end

        B(39) = f1*Tt(j,20) + f2*Tt(j,21) - bt*(Ts(j,21)+Tsold) - dt*Tsold;
        B(40) = f3*Ts(j,20) + f4*Ts(j,21) - ds*Tsold - cs*Tt(j,20);
%
% generate the solution vector
%
        TT = aa\B;
%
% update the right hand side
%
        Tt(j+1,1) = 35;
        for m=1:N,
            Tt(j+1,m+1) = TT(2*m);
            Ts(j+1,m) = TT(2*m-1);
        end
        Ts(j+1,m+1) = 75;
    end

    for n=1:5:iteration+1
        plot(x, Tt(n,:));
        hold on;
    end
```

```
title('Dynamic Response of Tube Side to Step Disturbance');
xlabel('Length (ft)');
ylabel('Tube Side Water Temperature (degree F)');
print -deps fig812a.eps
pause;
hold off;

for n=1:5:iteration+1
    plot(x, Ts(n,:));
    hold on;
end
title('Dynamic Response of Shell Side to Step Disturbance');
xlabel('Length (ft)');
ylabel('Shell Side Water Temperature (degree F)');
print -deps fig812b.eps
hold off;

clear Tt, Tp;
```

8.4 TECHNIQUES FOR DIFFUSIVE PROBLEMS

Diffusive problems are characterized by second–order spatial derivatives with equations of the form

$$\frac{\partial u}{\partial t} = \frac{\partial^2 u}{\partial x^2} \tag{8.4.1}$$

In order to develop second–order correct finite–difference analogs, a Crank-Nicolson analog is used. This analog is centered in time spacings and is on the grid point for distance as shown in Figure 8.13.

To get the first–order time partial, we develop Taylor series expansions for the functions $u(t_i, x_n)$ and $u(t_{i+1}, x_n)$ about the Crank–Nicolson finite–difference point $u(t_{i+1/2}, x_n)$.

Following a similar derivation as presented in Section 8.1, we have

$$\frac{\partial u}{\partial t}(i+1/2,\ n) = \frac{u_{i+1,n} - u_{i,n}}{\Delta t} \tag{8.4.2}$$

The second–order correct second–spatial derivative analog is obtained by developing Taylor series expansions for $u(t_{i+1/2}, x_{n+1})$ and $u(t_{i+1/2}, x_{n-1})$ about the point $u(t_{i+1/2}, x_n)$, thus:

$$\begin{aligned}
u_{i+1/2,n+1} = &\ u_{i+1/2,n} + \frac{\partial u}{\partial x}(i+1/2,n)(x_{n+1} - x_n) \\
&+ \frac{\partial^2 u}{\partial x^2}(i+1/2,n)\frac{(x_{n+1} - x_n)^2}{2} \\
&+ \frac{\partial^3 u}{\partial x^3}(i+1/2,n)\frac{(x_{n+1} - x_n)^3}{3 \cdot 2} + \text{HOT} \tag{8.4.3}
\end{aligned}$$

Solution of Partial Differential Equations 371

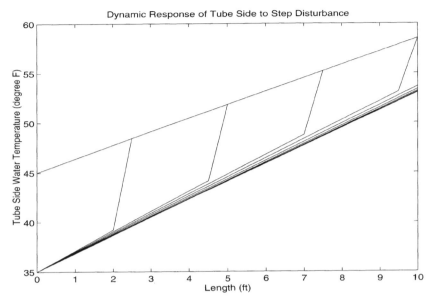

Figure 8.12a: Dynamic Response of Countercurrent Heat Exchange: Tube Side.

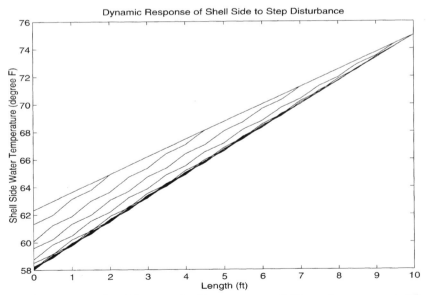

Figure 8.12b: Dynamic Response of Countercurrent Heat Exchange: Shell Side.

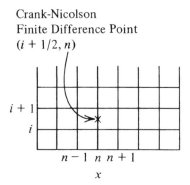

Figure 8.13: Crank–Nicolson Grid.

$$u_{i+1/2,n-1} = u_{i+1/2,n} + \frac{\partial u}{\partial x}(i+1/2,n)(x_n - x_{n-1})$$
$$+ \frac{\partial^2 u}{\partial x^2}(i+1/2,n)\frac{(x_n - x_{n-1})^2}{2}$$
$$+ \frac{\partial^3 u}{\partial x^3}(i+1/2,n)\frac{(x_n - x_{n-1})^3}{3 \cdot 2} + \text{HOT} \quad (8.4.4)$$

Adding equations (8.4.3) and (8.4.4) and solving for $\partial^2 u/\partial x^2$ gives the second–order correct analog:

$$\frac{\partial^2 u}{\partial x^2}(i+1/2,n) = \frac{u_{i+1/2,n+1} - 2u_{i+1/2,n} + u_{i+1/2,n-1}}{(x_{n+1} - x_n)^2} \quad (8.4.5)$$

In terms of the grid points i and $i+1$, the analog is

$$\frac{\partial^2 u}{\partial x^2}(i+1/2,n) = \frac{u_{i+1,n+1} + u_{i,n+1} - 2u_{i+1,n} - 2u_{i,n} + u_{i+1,n-1} + u_{i,n-1}}{2(x_{n+1} - x_n)^2}$$
$$(8.4.6)$$

The Crank-Nicolson analog for the variable $u_{i+1/2,n}$ is

$$u_{i+1/2,n} = 1/2(u_{i+1,n} + u_{i,n}) \quad (8.4.7)$$

For the Crank-Nicolson analog, the ratio of $\Delta x^2/\Delta t$ is usually taken equal to the coefficient of the diffusive term. Therefore, for equation (8.4.1), $\Delta t/\Delta x^2 = 1$ is used for efficiency and accuracy although the analog is stable for all ratios of $\Delta t/\Delta x^2$.

8.5 UNSTEADY–STATE HEAT CONDUCTION IN A ROD

Let us consider the problem of simulating the unsteady–state heat conduction in a long rod as shown in Figure 8.14.

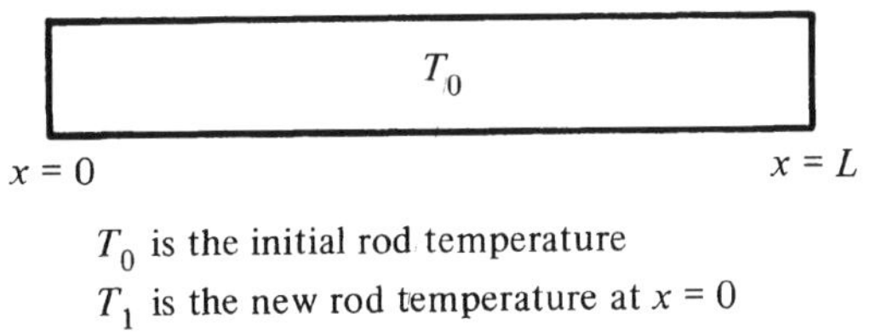

Solution of Partial Differential Equations

```
            ┌─────────────── T₀ ───────────────┐
            │                                   │
            └───────────────────────────────────┘
         x = 0                               x = L
```

T_0 is the initial rod temperature
T_1 is the new rod temperature at $x = 0$

Figure 8.14: Heat Conduction in Rod.

Applying the microscopic energy balance to this system gives

$$\frac{\partial T}{\partial t} = \alpha \frac{\partial^2 T}{\partial x^2} \tag{8.5.1}$$

with boundary conditions

$$T(x, 0) = T_0 \tag{8.5.2}$$

$$T(0, t) = T_1 \tag{8.5.3}$$

$$T(L, t) = T_0 \tag{8.5.4}$$

This being a diffusive system or a parabolic partial differential equation, we apply the Crank–Nicolson finite–difference analogs to equation (8.5.1) and obtain

$$\frac{T_{i+1,n} - T_{i,n}}{\Delta t} = \alpha \left(\frac{T_{i+1,n+1} + T_{i,n+1} - 2T_{i+1,n} - 2T_{i,n} + T_{i+1,n-1} + T_{i,n-1}}{2(x_{n+1} - x_n)^2} \right) \tag{8.5.5}$$

Figure 8.13 shows the finite–difference grid for this heat–transfer problem. We divide the distance grid into 20 elements. The Δt size is chosen so that $\Delta x^2 / \Delta t = \alpha$. When the rod temperature is specified via the boundary condition, solid black dots are indicated in Figure 8.15.

Putting $n = 1$ and $i = 0$, we have two unknown temperatures $T_{1,1}$ and $T_{1,2}$ but just one algebraic relation. However, if we formulate the entire set of relations for this new time level, we get 19 equations in 19 unknowns. Since the equations are linear, we can express this set of equations in the matrix notation shown in Figure 8.16, where

$$a = -\frac{1}{2} \frac{\alpha \Delta t}{\Delta x^2} \tag{8.5.6}$$

$$b = 1 + \frac{\alpha \Delta t}{\Delta x^2} \tag{8.5.7}$$

$$c = -\frac{1}{2} \frac{\alpha \Delta t}{\Delta x^2} \tag{8.5.8}$$

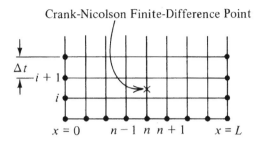

Figure 8.15: Finite–Difference Grid.

$$\begin{bmatrix} bc & & & & & & & \\ abc & & & & & & & \\ & abc & & & & & & \\ & & abc & & & & & \\ & & & \ddots & & & & \\ & & & & abc & & & \\ & & & & & abc & & \\ & & & & & & ab \end{bmatrix} \begin{bmatrix} T_1 \\ T_2 \\ T_3 \\ \cdot \\ \cdot \\ \cdot \\ \cdot \\ T_{19} \end{bmatrix} = \begin{bmatrix} -aT_0 + d_1 \\ d_2 \\ d_3 \\ \cdot \\ \cdot \\ \cdot \\ d_{18} \\ -cT_{20} + d_{19} \end{bmatrix}$$

Figure 8.16: Matrix for Unsteady–State Heat Conduction Simulation.

$$d_n = \left(1 - \frac{\alpha \Delta t}{\Delta x^2}\right) T_{i,n} - a(T_{i,n-1} + T_{i,n+1}) \qquad (8.5.9)$$

The matrix of Figure 8.16 is a band matrix and the band algorithm provides an efficient calculational sequence for solution. File `ex85.m`, available on the world wide web, presents a MATLAB solution to the problem. Figure 8.17 shows the dynamic response curves for this heat conduction problem.

8.6 TECHNIQUES FOR PROBLEMS WITH BOTH CONVECTIVE AND DIFFUSION EFFECTS: THE STATE–VARIABLE FORMULATION

Partial differential equations which contain both diffusive and convective terms pose special problems for numerical solution via finite differencing. The basic partial differential equation of interest is given by

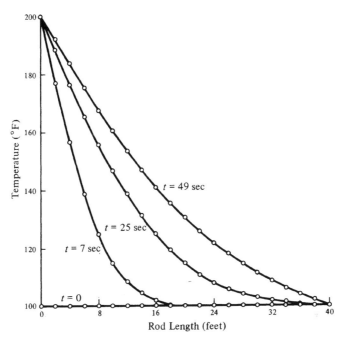

Figure 8.17: Dynamic Response Curves for Unsteady–State Heat Conduction.

$$U_2 \frac{\partial^2 c}{\partial z^2} - U_1 \frac{\partial c}{\partial z} = \frac{\partial c}{\partial t} \qquad (8.6.1)$$

where the coefficient U_2 specifies the amount of diffusive characteristic and U_1 the amount of convective characteristic. McCracken, Leefe, and Weaver (1970) have studied and compared three alternatives for finite differencing this type of mixed diffusive-convective problem. A summary of the finite–difference analogs considered is given in Table 8.1. The Crank–Nicolson and centered analogs have been presented previously. The state–variable formulation introduces a new variable, x, defined by

$$x = \partial c / \partial z \qquad (8.6.2)$$

Therefore, the original second–order differential equation is transformed into a coupled set of two first–order state–variable equations,

$$\frac{\partial c}{\partial z} = x \quad \text{and} \quad U_2 \frac{\partial x}{\partial z} - U_1 x = \frac{\partial c}{\partial t} \qquad (8.6.3)$$

Table 8.1: Summary of Finite–Difference Analogs

Analog	Formulation	Derivative estimates
Crank-Nicolson	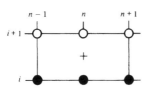	$\left.\dfrac{dc}{dz}\right]_n \cong \dfrac{\bar{c}_{n+1} - \bar{c}_{n-1}}{2\Delta z}$ $\left.\dfrac{d^2c}{dz^2}\right]_n \cong \dfrac{\bar{c}_{n+1} - 2\bar{c}_n + \bar{c}_{n-1}}{(\Delta z)^2}$
Centered difference	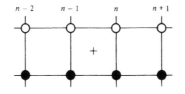	$\left.\dfrac{dc}{dz}\right]_{n-\frac{1}{2}} \cong \dfrac{\bar{c}_n - \bar{c}_{n-1}}{\Delta z}$ $\left.\dfrac{d^2c}{dz^2}\right]_{n-\frac{1}{2}} \cong \dfrac{\bar{c}_{n+1} - \bar{c}_n - \bar{c}_{n-1} + \bar{c}_{n-2}}{2(\Delta z)^2}$
State variable	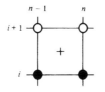	$\left.\dfrac{dc}{dz}\right]_{n-\frac{1}{2}} \cong \dfrac{\bar{c}_n - \bar{c}_{n-1}}{\Delta z}$ $\left.\dfrac{d^2c}{dz^2}\right]_{n-\frac{1}{2}} \cong \left.\dfrac{dx}{dz}\right]_{n-\frac{1}{2}} \cong \dfrac{\bar{x}_n - \bar{x}_{n-1}}{\Delta z}$

Key: ● Known values
     ○ Solution values
     + Base point for finite-difference equation

$\bar{c}_n$ Designates $\dfrac{(c_{n,i+1} + c_{n,i})}{2}$

i = time index
n = space index

Since this new formulation has only first–order derivatives, the centered–difference analog can be applied for a second–order correct estimate as given in Table 8.1.

In order to compare the merits of the different finite–difference schemes, McCracken, Leefe, and Weaver studied solutions with different U_2/U_1 ratios for a modified step forcing (Table 8.2) with 50 space increments and Δz and Δt both set to 0.01.

For the pure convective case ($U_2 = 0$), the Crank–Nicolson formulation suffered rather severe oscillations, while the centered–difference (which is equivalent to the state–variable formulation for this case) performed well.

In the pure diffusion case the centered–difference scheme results were encumbered by severe oscillation, while the Crank–Nicolson performed well. The state–variable scheme followed the Crank–Nicolson results well to U_2 values of 10^{-6} when oscillations became severe. Table 8.3 presents a summary of the pure diffusion results.

Table 8.2: **Smoothed Step Input for Test Solutions**

Time Increment	Dependent Variable C
1	0.450
2	0.900
3	1.350
4	1.800
5	2.688
6	3.575
7	4.463
8	5.350
9, et seq.	5.500

Table 8.3: **Performance for Pure Diffusion ($U_1 = 0$) (Source: McCracken et al., 1970)**

Derivative Analog	Crank–Nicolson	Centered Difference	State Variable
$U_2 = 0.1$	Concurs with state variable	Severe oscillation	Concurs with Crank–Nicolson
$U_2 = 0.5 \times 10^{-2}$	Concurs with state variable	Oscillation	Concurs with Crank–Nicolson
$U_2 = 0.1 \times 10^{-5}$	Continued good performance	Oscillation	Incidence of oscillation

Difficulties are encountered by both the centered-difference and Crank-Nicolson schemes when both U_1 and U_2 play a role in the problem formulation. Results are given in Table 8.4. These results also show the flexible and well–behaved characteristics of the state–variable scheme for problems when both convection and diffusion are of consequence.

Table 8.4: **Performance of Alternative Analogs**
(Source: McCracken et al., 1970)

Derivative Analog		Crank–Nicolson	State Variable	Centered Difference
U_1	U_2			
1	0.1	Results for these analogs agree and appear well–behaved.		Consequential oscillation
1	0.5×10^{-2}	Results for these analogs agree and appear well–behaved.		Consequential oscillation
1	0.1×10^{-3}	Severe oscillation.	Results for these analogs agree and appear well–behaved.	
			State–variable is somewhat preferred, being totally free of oscillations.	

8.7 MODELING OF MISCIBLE FLOW OF SURFACTANT IN POROUS MEDIA

Friedman and Ramirez (1977) have presented a model for single–phase miscible surfactant flow in porous media. The mechanisms of dispersion, adsorption, and dead–space volume contributions are included. A mathematical model was constructed in order to investigate the concurring mechanisms in single–phase flow of a miscible surfactant in porous rock. The physical system considered was a linear displacement of oil by an oil–soluble surfactant through a Berea sandstone core. The core is considered to consist of two fluid compartments: a main channel, and a dead–space volume in which there is no bulk flow. The only transport between the dead–space volume and the main channel is through pure diffusion. The final model arrived at after using volume averaging concepts (Slattery, 1972) treats the complex geometry of porous media, and order–of–magnitude scaling (Friedman and Ramirez, 1977) to simplify the model is as follows.

Main channel:

$$U_1 \frac{\partial \langle c^* \rangle^m}{\partial t^*} + \frac{\partial \langle c^* \rangle^m}{\partial z^*} = K \frac{\partial^2 \langle c^* \rangle^m}{\partial z^{*2}} - \frac{3DLV_d}{2R^2 v_0 V_m}(\langle c^* \rangle^m - \langle c^* \rangle^d) \qquad (8.7.1)$$

where U_1 = $1 + (A_{mr}/V_m M)$
$\langle c^* \rangle^m$ = dimensionless main–channel surfactant concentration
$\langle c^* \rangle^d$ = dimensionless dead–space volume surfactant concentration
M = adsorption rate parameter
K = dispersion coefficient
A_{mr} = rock main–channel fluid–surface area in main channel
V_d = dead–space volume in an REV (representative element of volume)
V_m = main–channel fluid volume in an REV
L = length of rock–core system
D = diffusivity
R = radius of dead–space volume pore
v_0 = injection velocity
z^* = dimensionless axial distance
t^* = dimensionless time

The initial condition for the problem is,

$$\langle c^* \rangle^m = 0 \quad \text{at } t^* = 0 \qquad (8.7.2)$$

and the boundary conditions are

$$\langle c^* \rangle^m - \frac{1}{Pe} \frac{\partial \langle c^* \rangle^m}{\partial z^*} = 1 \quad \text{at } z^* = 0$$

$$\frac{\partial \langle c^* \rangle^m}{\partial z^*} = 0 \quad \text{at } z^* = 1 \qquad (8.7.3)$$

For the dead–space volume phase, the material balance is

$$U_2 \frac{\partial \langle c^* \rangle^d}{\partial t^*} = \frac{3}{2} \frac{DL}{R^2 v_0}(\langle c^* \rangle^m - \langle c^* \rangle^d) \qquad (8.7.4)$$

with

$$U_2 = 1 + 3/RM \qquad (8.7.5)$$

and the initial condition

$$\langle c^* \rangle^d = 0 \quad \text{at } t^* = 0 \qquad (8.7.6)$$

Using the state–variable formulation, the model can be stated as

$$\frac{\partial \langle c^* \rangle^m}{\partial z^*} = x \qquad (8.7.7)$$

$$U_1 \frac{\partial \langle c^\star \rangle^m}{\partial t^\star} + x = K \frac{\partial x}{\partial z^\star} - \frac{3DLV_d}{2R^2 v_0 V_m}(\langle c^\star \rangle^m - \langle c^\star \rangle^d) \qquad (8.7.8)$$

$$U_2 \frac{\partial \langle c^\star \rangle^d}{\partial t^\star} = \frac{3}{2}\frac{DL}{R^2 v_0}(\langle c^\star \rangle^m - \langle c^\star \rangle^d) \qquad (8.7.9)$$

I.C. : $\quad \langle c^\star \rangle^m = \langle c^\star \rangle^d = 0 \quad$ at $t^\star = 0 \qquad (8.7.10)$

B.C. 1. $\quad \langle c^\star \rangle^m - \dfrac{1}{Pe} x = 1 \quad$ at $z^\star = 0 \;\; t^\star > 0$

B.C. 2 $\quad\quad x = 0 \quad$ at $z^\star = 1 \;\; t^\star > 0 \qquad (8.7.11)$

Four–point finite–difference analogs are used as shown in Figure 8.18. The analogs are

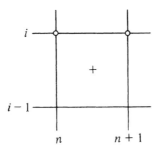

Figure 8.18: A Four–Point Analog.

$$\left(\frac{\partial \langle c^\star \rangle^m}{\partial z^\star}\right)_{i-1/2,n+1/2} = \frac{\langle c^\star \rangle^m_{i,n+1} + \langle c^\star \rangle^m_{i,n} - \langle c^\star \rangle^m_{i-1,n+1} - \langle c^\star \rangle^m_{i-1,n}}{2\Delta z^\star}$$

$$(8.7.12)$$

$$\left(\frac{\partial^2 \langle c^\star \rangle^m}{(\partial z^\star)^2}\right)_{i-1/2,n+1/2} = \left(\frac{\partial x}{\partial z^\star}\right)_{i-1/2,n+1/2}$$

$$= \frac{x_{i,n+1} + x_{i,n} - x_{i-1,n+1} - x_{i-1,n}}{2\Delta z^\star} \qquad (8.7.13)$$

when n is the time index; i is the space index. The other derivatives become

$$\left(\frac{\partial \langle c^\star \rangle^m}{\partial t^\star}\right)_{i-1/2,n+1/2} = \frac{1}{2\Delta t}(\langle c^\star \rangle^m_{i,n+1} + \langle c^\star \rangle^m_{i-1,n+1} - \langle c^\star \rangle^m_{i,n} - \langle c^\star \rangle^m_{i-1,n})$$

$$(8.7.14)$$

$$\left(\frac{\partial \langle c^\star \rangle^d}{\partial t^\star}\right)_{i-1/2,n+1/2} = \frac{1}{2\Delta t}(\langle c^\star \rangle^d_{i,n+1} + \langle c^\star \rangle^d_{i-1,n+1} - \langle c^\star \rangle^d_{i,n} - \langle c^\star \rangle^d_{i-1,n}) \quad (8.7.15)$$

$$\langle c^\star \rangle^m = [\langle c^\star \rangle^m_{i-1,n+1} + \langle c^\star \rangle^m_{i,n+1} + \langle c^\star \rangle^m_{i-1,n} + \langle c^\star \rangle^m_{i,n}]/4 \quad (8.7.16)$$

$$\langle c^\star \rangle^d = [\langle c^\star \rangle^d_{i-1,n+1} + \langle c^\star \rangle^d_{i,n+1} + \langle c^\star \rangle^d_{i,n} + \langle c^\star \rangle^d_{i-1,n}]/4 \quad (8.7.17)$$

Using these analogs, the discrete finite differenced equations are

For the main channel:

$$\left(\frac{U_1}{\Delta t} - \frac{V_1}{2}\right)\langle c^\star \rangle^m_{i,n+1} + \left(\frac{U_1}{\Delta t} - \frac{V_1}{2}\right)\langle c^\star \rangle^m_{i-1,n+1} + \left(\frac{1}{2} - \frac{K}{\Delta z}\right)x_{i,n+1}$$
$$+ \left(\frac{1}{2} + \frac{K}{\Delta z}\right)x_{i-1,n+1} + \frac{1}{2}V_1 \langle c^\star \rangle^d_{i,n+1} + \frac{1}{2}V_1 \langle c^\star \rangle^d_{i-1,n+1}$$
$$= \left(\frac{U_1}{\Delta t} + \frac{1}{2}V_1\right)\langle c^\star \rangle^m_{i,n} + \left(\frac{U_1}{\Delta t} + \frac{1}{2}V_1\right)\langle c^\star \rangle^m_{i-1,n} + \left(\frac{K}{\Delta z} - \frac{1}{2}\right)x_{i,n}$$
$$+ \left(-\frac{K}{\Delta z} - \frac{1}{2}\right)x_{i-1,n} - \frac{1}{2}V_1 \langle c^\star \rangle^d_{i,n} - \frac{1}{2}V_1 \langle c^\star \rangle^d_{i-1,n} \quad (8.7.18)$$

For the dead–end pores:

$$-\frac{V_2}{2}\langle c^\star \rangle^m_{i,n+1} - \frac{V_2}{2}\langle c^\star \rangle^m_{i-1,n+1} + \left(\frac{U_2}{\Delta t} + \frac{1}{2}V_2\right)\langle c^\star \rangle^d_{i,n+1}$$
$$+ \left(\frac{U_2}{\Delta t} + \frac{V_2}{2}\right)\langle c^\star \rangle^d_{i-1,n+1} = \frac{V_2}{2}\langle c^\star \rangle^m_{i,n} + \frac{V_2}{2}\langle c^\star \rangle^m_{i-1,n}$$
$$+ \left(\frac{U_2}{\Delta t} - \frac{V_2}{2}\right)\langle c^\star \rangle^d_{i,n} + \left(\frac{U_2}{\Delta t} - \frac{V_2}{2}\right)\langle c^\star \rangle^d_{i-1,n} \quad (8.7.19)$$

The auxiliary equation, defined by equation (8.7.7), is

$$\frac{1}{\Delta z}\langle c^\star \rangle^m_{i,n+1} - \frac{1}{\Delta z}\langle c^\star \rangle^m_{i-1,n+1} - \frac{1}{2}x_{i,n+1} - \frac{1}{2}x_{i-1,n+1}$$
$$= -\frac{1}{\Delta z}\langle c^\star \rangle^m_{i,n} + \frac{1}{\Delta z}\langle c^\star \rangle^m_{i-1,n}$$
$$+ \frac{1}{2}x_{i,n} + \frac{1}{2}x_{i-1,n} \quad (8.7.20)$$

Here

$$V_1 = -\frac{3DLV_d}{2R^2 v_0 V_m} \quad (8.7.21)$$

$$V_2 = \frac{3}{2}\frac{DL}{R^2 v_0} \quad (8.7.22)$$

The equations above are valid for space points inside the boundaries. The boundary conditions are also discretized, to yield

$$\langle c^\star \rangle^d - \frac{1}{Pe}x_1 = 1 \quad \text{at } z^\star = 0 \quad (8.7.23)$$

and
$$x_M = 0 \qquad \text{at } z^\star = 1 \qquad (8.7.24)$$

In order for the system to be completely specified, we must specify one additional boundary condition:

$$\langle c^\star \rangle^m = \langle c^\star \rangle^d \qquad \text{at } z^\star = 0 \qquad (8.7.25)$$

This set of linear algebraic equations can be solved on a digital computer using the band inversion subroutine. The total width of the band is equal to 7.

Figure 8.19 shows the main–channel concentration profile as a function of dimensionless time (pore volumes injected) for a dispersion coefficient of $K = 0.1$.

Figure 8.20 displays the influence of a constant dispersion coefficient on the main–channel concentration profile for a dead–space pore ratio of V_d/V of 1 percent and a dimensionless time of 0.6. As expected, the profiles become sharper as the dispersion coefficient is lowered.

Figure 8.21 shows the influence of dead–space volume on the main–channel concentration. The influence of dead–space volume becomes significant for V_d/V percentages larger than 10 percent. For fired Berea cores, the maximum realistic value that can be expected is around 20 percent. Therefore, depending upon the actual percentage of dead–space volume, its effect may or may not be significant for particular porous media.

Figure 8.22 shows the effect of adsorption. Based upon these simulations, the adsorption of sulfonates on fired Berea appears to be a highly significant model parameter.

8.8 UNSTEADY–STATE RESPONSE OF A NONLINEAR TUBULAR REACTOR [1]

Simulation of the adiabatic tubular reactor consists of the solution of the partial differential equations which describe the system. The nonlinear nature of these equations makes solution difficult. A powerful technique developed in recent years for the solution of nonlinear partial differential equations is that of quasilinearization.

Although the application of quasilinearization requires a detailed and complicated derivation of equations, the basic technique is straightforward. First, one must consider a finite–differenced solution space and a marching–like solution. Here, the solution domain is of two dimensions: dimensionless distance from 0 to 1 and open–ended time from the initial

---

[1] This section is reprinted from D.E. Clough and W. F. Ramirez, "Stability of Tubular Reactors," *Simulation* **16**, May 1971. Reprinted with permission. ©1971 Simulation Councils, Inc.

Solution of Partial Differential Equations

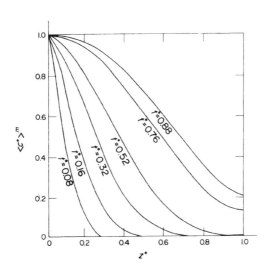

Figure 8.19: Concentration vs. z^* for $K = 0.1$ and $V_d/V = 1$ percent.

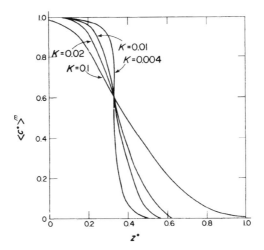

Figure 8.20: Influence of Dispersion.

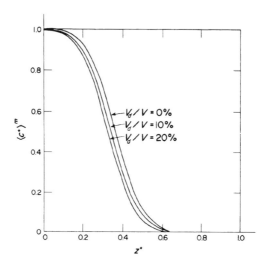

Figure 8.21: Influence of Dead–Space Volume.

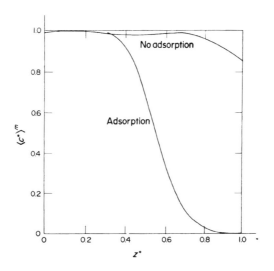

Figure 8.22: Influence of Adsorption.

time. A complete spatial profile at a given time is considered as a solution level and will be further described as either known (that is, given or solved) or unknown. The specific objective of the algorithm is then to solve iteratively for the next unknown level utilizing the previously solved levels.

To initiate the algorithm, a guess for the unknown level solution is needed. The known previous solution level is sufficient. The partial differential equations are linearized about this guess and then finite–differenced and solved. This solution yields an improved estimate to the solution of the nonlinear equations. The linearization is now carried about this new estimate and the process is iterated to desired convergence. The last estimate is taken as the solution for the previously unknown level and used as the initial guess for the next unknown level. The nonlinear solution is marched out in time as far as desired.

In order to solve the linearized finite–differenced equations, it is only necessary to invert a high–order coefficient matrix which has a diagonal band form.

The nonlinear equations for the adiabatic tubular reactor with diffusion and first-order reaction are given by Clough and Ramirez (1971) as

$$\rho c_p \frac{\partial T}{\partial \Theta} = \lambda \frac{\partial^2 T}{\partial z^2} - \rho v c_p \frac{\partial T}{\partial z} + (-\Delta H) k C \exp(-E/RT) \tag{8.8.1}$$

$$\frac{\partial C}{\partial \Theta} = D \frac{\partial^2 C}{\partial z^2} - v \frac{\partial C}{\partial z} - kC \exp(-E/RT)$$

with the boundary conditions being

$$-\lambda \frac{\partial T}{\partial z}(0) = \rho c_p v [T_0 - T(0)]; \qquad \frac{\partial T}{\partial z}(L) = 0$$

$$-D \frac{\partial C}{\partial z}(0) = v[C_0 - C(0)]; \qquad \frac{\partial C}{\partial z}(L) = 0 \tag{8.8.2}$$

The equations are nondimensionalized and simplified with the following definitions:

$$x = z/L \qquad u_1 = T/T_0 \qquad u_2 = C/C_0$$
$$D\Theta/L^2 = t \qquad r_1 = vL/\alpha \qquad r_2 = vL/D$$
$$q = E/RT_0 \qquad B_1 = \frac{(-\Delta H)kC_0 L^2}{\rho c_p T_0 D} \qquad B_2 = kL^2/D$$
$$\alpha = \lambda/\rho c_p$$

The resulting equations are

$$\frac{\partial u_1}{\partial t} = \frac{\alpha}{D} \frac{\partial^2 u_1}{\partial x^2} - r_2 \frac{\partial u_1}{\partial x} + B_1 u_2 \exp(-q/u_1)$$

$$\frac{\partial u_2}{\partial t} = \frac{\partial^2 u_1}{\partial x^2} - r_2 \frac{\partial u_2}{\partial x} - B_2 u_2 \exp(-q/u_1) \qquad (8.8.3)$$

along with

$$-\frac{\partial u_1}{\partial x}(0) = r_1[1 - u_1(0)] \qquad \frac{\partial u_1}{\partial x}(1) = 0$$

$$\qquad\qquad\qquad\qquad\qquad\qquad\qquad\qquad (8.8.4)$$

$$-\frac{\partial u_2}{\partial x}(0) = r_2[1 - u_2(0)] \qquad \frac{\partial u_2}{\partial x}(1) = 0$$

In the linearization of the above equations about u_1^0 and u_2^0, it is convenient to define

$$F = \frac{q u_2^0}{(u_1^0)^2} \exp(-q/u_1^0) \quad G = \exp(-q/u_1^0) \quad \text{and} \quad M = -\frac{q u_2^0}{u_1^0} \exp(-q/u_1^0)$$

Then the linearized equations take the form

$$\frac{\partial u_1}{\partial t} = \frac{\alpha}{D} \frac{\partial^2 u_1}{\partial x^2} - r_2 \frac{\partial u_1}{\partial x} + B_1 F u_1 + B_1 G u_2 + B_1 M$$

$$\qquad\qquad\qquad\qquad\qquad\qquad\qquad\qquad (8.8.5)$$

$$\frac{\partial u_2}{\partial t} = \frac{\partial^2 u_2}{\partial x^2} - r_2 \frac{\partial u_2}{\partial x} - B_2 F u_1 - B_2 G u_2 - B_2 M$$

with the boundary conditions

$$-\frac{\partial u_1}{\partial x}(0) = r_1[1 - u_1(0)] \qquad \frac{\partial u_1}{\partial x}(1) = 0$$

$$\qquad\qquad\qquad\qquad\qquad\qquad\qquad\qquad (8.8.6)$$

$$-\frac{\partial u_2}{\partial x}(0) = r_2[1 - u_2(0)] \qquad \frac{\partial u_2}{\partial x}(1) = 0$$

These linearized equations can now be finite–differenced. Three important indexes are used here:

i: time level, from 0 to end of solution
j: spatial location, from 0 to 1
n: indicates the solution iteration about which the linearization is made.

Variables that do not carry an n superscript are results of solution at the previous time (i) level. The superscript $n+1$ always indicates the next iteration level to be solved.

The Crank–Nicolson finite–difference analogs are used and are written based on point (i, j). When the linearized equations are finite–differenced about the point (i, j), and the unknown level variables are solved for, one obtains for the energy balance

$$\left\{-\frac{\alpha}{D}\frac{1}{2\Delta x^2} - \frac{r_2}{4\Delta x}\right\} u_1^{n+1}(i+1, j-1)$$
$$+ \left\{\frac{\alpha}{D}\frac{1}{\Delta x^2} + \frac{1}{\Delta t} - \frac{B_1 F}{2}\right\} u_1^{n+1}(i+1, j)$$
$$+ \left\{\frac{B_1 G}{2}\right\} u_2^{n+1}(i+1, j) + \left\{-\frac{\alpha}{D}\frac{1}{2\Delta x^2} + \frac{r_2}{4\Delta x}\right\} u_1^{n+1}(i+1, j+1)$$
$$= \left\{\frac{\alpha}{D}\frac{1}{2\Delta x^2} + \frac{r_2}{4\Delta x}\right\} u_1(i, j-1) + \left\{-\frac{\alpha}{D}\frac{1}{\Delta x^2} + \frac{B_1 F}{2} + \frac{1}{\Delta t}\right\} u_1(i, j)$$
$$+ \left\{\frac{B_1 G}{2}\right\} u_2(i, j) + \left\{\frac{\alpha}{D}\frac{1}{2\Delta x^2} - \frac{r_2}{4\Delta x}\right\} u_1(i, j+1) + B_1 M \quad (8.8.7)$$

and, for the material balance,

$$\left\{-\frac{1}{2\Delta x^2} - \frac{r_2}{4\Delta x}\right\} u_2^{n+1}(i+1, j-1) + \left\{\frac{B_2 F}{2}\right\} u_1^{n+1}(i+1, j)$$
$$+ \left\{\frac{1}{\Delta t} + \frac{1}{\Delta x^2} + \frac{B_2 G}{2}\right\} u_2^{n+1}(i+1, j)$$
$$+ \left\{-\frac{1}{2\Delta x^2} + \frac{r_2}{4\Delta x}\right\} u_2^{n+1}(i+1, j+1)$$
$$= \left\{\frac{1}{2\Delta x^2} + \frac{r_2}{4\Delta x}\right\} u_2(i, j-1) + \left\{\frac{B_2 F}{2}\right\} u_1(i, j)$$
$$+ \left\{\frac{1}{\Delta t} - \frac{1}{\Delta x^2} - \frac{B_2 G}{2}\right\} u_2(i, j)$$
$$+ \left\{\frac{1}{2\Delta x^2} - \frac{r_2}{4\Delta x}\right\} u_2(i, j+1) - B_2 M \quad (8.8.8)$$

As the based point is varied across the spatial domain, a set of simultaneous linear equations is generated which may be represented in diagonal band matrix form. However, at the base points $(i, 1)$ and (i, N), it is difficult to write the analogs since they require points outside the solution domain, that is, at $(i,0)$, $(i+1,0)$, $(i,N+1)$, and $(i+1, N+1)$. This problem is handled by writing the analogs using the "fictitious" points, and then the points outside the spatial domain are eliminated via use of the boundary conditions. In this way, the boundary conditions are incorporated in the solution.

When the boundary conditions are finite-differenced, solved for the fictitious points, and substituted into the energy and material balance equations, the following results, written about base point $(i, 1)$:

$$\left\{\frac{\alpha}{D}\frac{1}{\Delta x^2} + \frac{1}{\Delta t} - \frac{B_1 F}{2} - \Delta x r_1 \left[-\frac{\alpha}{D}\frac{1}{2\Delta x^2} - \frac{r_2}{4\Delta x}\right]\right\} u_1^{n+1}(i+1, 1)$$
$$+ \left\{\frac{B_1 G}{2}\right\} u_2^{n+1}(i+1, 1) + \left\{-\frac{\alpha}{D}\frac{1}{\Delta x^2}\right\} u_1^{n+1}(i+1, 2) =$$

$$\left\{-\frac{\alpha}{D}\frac{1}{\Delta x^2}+\frac{B_1 F}{2}+\frac{1}{\Delta t}-3\Delta x r_1\left[\frac{\alpha}{D}\frac{1}{2\Delta x^2}+\frac{r_2}{4\Delta x}\right]\right\}u_1(i,1)$$
$$+\left\{\frac{B_1 G}{2}\right\}u_2(i,1)+\left\{\frac{\alpha}{D}\frac{1}{\Delta x^2}\right\}u_1(i,2)+B_1 M$$
$$+4\Delta x r_1\left[\frac{\alpha}{D}\frac{1}{2\Delta x^2}+\frac{r_2}{4\Delta x}\right] \qquad (8.8.9)$$

and

$$\left\{\frac{B_2 F}{2}\right\}u_1^{n+1}(i+1,1)+\left\{\frac{1}{\Delta t}+\frac{1}{\Delta x^2}+\frac{B_2 G}{2}\right.$$
$$\left.-\Delta x r_2\left[-\frac{1}{2\Delta x^2}-\frac{r_2}{4\Delta x}\right]\right\}u_2^{n+1}(i+1,1)$$
$$+\left\{-\frac{1}{\Delta x^2}\right\}u_2^{n+1}(i+1,2)=$$
$$\left\{-\frac{B_2 F}{2}\right\}u_1(i,1)+\left\{\frac{1}{\Delta t}-\frac{1}{\Delta x^2}-\frac{B_2 G}{2}\right.$$
$$\left.-3\Delta x r_2\left[\frac{1}{2\Delta x^2}+\frac{r_2}{4\Delta x}\right]\right\}u_2(i,1)+\left\{\frac{1}{\Delta x^2}\right\}u_2(i,2)$$
$$-B_2 M+4\Delta x r_2\left[\frac{1}{2\Delta x^2}+\frac{r_2}{4\Delta x}\right] \qquad (8.8.10)$$

When written about base point (i, N),

$$\left\{-\frac{\alpha}{D}\frac{1}{\Delta x^2}\right\}u_1^{n+1}(i+1,N-1)$$
$$+\left\{\frac{\alpha}{D}\frac{1}{\Delta x^2}+\frac{1}{\Delta t}-\frac{B_1 F}{2}\right\}u_1^{n+1}(i+1,N)$$
$$+\left\{-\frac{B_1 G}{2}\right\}u_2^{n+1}(i+1,N)=$$
$$\left\{\frac{\alpha}{D}\frac{1}{\Delta x^2}\right\}u_1(i,N-1)$$
$$+\left\{-\frac{\alpha}{D}\frac{1}{\Delta x^2}+\frac{B_1 F}{2}-\frac{1}{\Delta r}\right\}u_1(i,N)$$
$$+\left\{\frac{B_1 G}{2}\right\}u_2(i,N)+B_1 M \qquad (8.8.11)$$

and

$$\left\{-\frac{1}{\Delta x^2}\right\}u_2^{n+1}(i+1,N-1)+\left\{\frac{B_2 F}{2}\right\}u_1^{n+1}(i+1,N)$$
$$+\left\{\frac{1}{\Delta t}+\frac{1}{\Delta x^2}-\frac{B_2 G}{2}\right\}u_2^{n+1}(i+1,N)=$$
$$\left\{\frac{1}{\Delta x^2}\right\}u_2(i,N-1)+\left\{-\frac{B_2 F}{2}\right\}u_1(i,N)$$
$$+\left\{\frac{1}{\Delta t}-\frac{1}{\Delta x^2}-\frac{B_2 G}{2}\right\}u_2(i,N)-B_2 M \qquad (8.8.12)$$

Solution of Partial Differential Equations

Therefore, a set of $2N$ simultaneous linear equations is obtained in $2N$ unknowns, and solution consists of inverting the coefficient matrix.

When the first level is solved, there is a discontinuity in the boundary conditions at the point $(0, 1)$. This discontinuity is most easily handled by assuming the initial condition to hold at $(0, 1)$ and the Danckwerts boundary condition to hold thereafter. Then, the equations written about base point $(0, 1)$ become

$$\left\{\frac{\alpha}{D}\frac{1}{\Delta x^2} + \frac{1}{\Delta t} - \frac{B_1 F}{2} + 2\Delta x r_1 \left[\frac{\alpha}{D}\frac{1}{2\Delta x^2} + \frac{r_2}{4\Delta x}\right]\right\} u_1^{n+1}(i+1,1)$$
$$+ \left\{-\frac{B_1 G}{2}\right\} u_2^{n+1}(i+1,1) + \left\{-\frac{\alpha}{D}\frac{1}{\Delta x^2}\right\} u_1^{n+1}(i+1,2) =$$
$$\left\{-\frac{\alpha}{D}\frac{1}{\Delta x^2} + \frac{B_1 F}{2} + \frac{1}{\Delta t}\right\} u_1(i,1)$$
$$+ \left\{\frac{B_1 G}{2}\right\} u_2(i,1) + \left\{\frac{\alpha}{D}\frac{1}{\Delta x^2}\right\} u_1(i,2)$$
$$+ B_1 M + 2\Delta x r_2 \left[\frac{\alpha}{D}\frac{1}{2\Delta x^2} + \frac{r_2}{4\Delta x}\right] \tag{8.8.13}$$

and

$$\left\{\frac{B_2 F}{2}\right\} u_1^{n+1}(i+1,1) + \left\{\frac{1}{\Delta t} + \frac{1}{\Delta x^2} + \frac{B_2 G}{2}\right.$$
$$+ 2\Delta x r_2 \left[\frac{1}{2\Delta x^2} + \frac{r_2}{4\Delta x}\right]\bigg\} u_2^{n+1}(i+1,1) + \left\{-\frac{1}{\Delta x^2}\right\} u_2^{n+1}(i+1,2) =$$
$$\left\{-\frac{B_2 F}{2}\right\} u_1(i,1) + \left\{\frac{1}{\Delta t} - \frac{1}{\Delta x^2} - \frac{B_2 G}{2}\right\} u_2(i,1)$$
$$+ \left\{\frac{1}{\Delta x^2}\right\} u_2(i,2) - B_2 M + 2\Delta x r_2 \left[\frac{1}{2\Delta x^2} + \frac{r_2}{4\Delta x}\right] \tag{8.8.14}$$

Two cases are considered for simulation. The first is a reactor with intermediate conversion at a single steady state, and the parameters are

$r_1 = 30$ $r_2 = 30$
$q = 17.6$ $C_0 = 0.5$ lb-mol/ft^3
$T_0 = 510°R$ $\alpha = 1$ cm^2/s
$L = 100$ cm $\rho = 60$ lb/ft^3
$k = 1.2 \times 10^4$ $D = 1$ cm^2/sec
$c_p = 1$ BTU/lb-mol°R $(-\Delta H) = 4 \times 10^4$ BTU/lb-mol

The second is a triple steady-state case for which the only changed parameters are

$$(-\Delta H) = 4.32 \times 10^4 \qquad q = 23 \qquad k = 1.2 \times 10^6$$

It should be noted that both of these cases represent liquid phase reactions (by density, concentration, etc.); however, the Peclet numbers are equal, which is a poor assumption. Again, it is emphasized that this assumption is only made as a check on the dynamics program, and subsequent use of the program does not require this assumption.

Several simulation runs follow for the two cases. The runs are represented by their initial conditions and a few selected profiles in order to show the progress of the dynamics. The spatial domain was divided into 100 segments so that $\Delta x = 1 \times 10^{-2}$. The dimensionless time increment for these runs was $\Delta t = 1 \times 10^{-3}$. At each time interval, the linearized response converged within 5 iterations to the nonlinear response. Figure 8.23 shows the dynamics of a startup from a cold reactor with a single steady state. For this run, the system exhibits uniform asymptotic stability.

In Figure 8.24, the initial profile in temperature is linear and lies just above the steady state. The system indeed does not possess uniform asymptotic stability with respect to its steady state. The physical reason behind such nonuniform behavior is clear. Small upsets in the reactor feed are exaggerated by the effects of reaction, and the magnified upset must pass through the reactor before the system settles again.

Following the technique used by Amundson (1965), the steady–state profiles for an adiabatic tubular reactor can be approximated to a desired accuracy. It should be noted that the derivation assumes that the mechanisms for heat and mass transfer are the same. Van Heerden (1958) points out that this is probably a good assumption when dealing with flowing gases. For the steady–state simulations, the Peclet numbers for mass and heat transfer are therefore specified as being the same. This is not a necessary restriction in the dynamic simulations, although it was made in order to obtain agreement between the steady states from Amundson's technique and from the dynamic solution. The mass and energy equations may be combined as

$$\frac{d^2n}{dx^2} - Pe\frac{dn}{dx} + \frac{L^2 k}{D}(n_{lim} - n)\exp(-q/n) = 0 \tag{8.8.15}$$

with the boundary conditions

$$1 - n(0) = -\frac{1}{Pe}\frac{dn}{dx}(0) \qquad \frac{dn}{dx}(1) = 0 \tag{8.8.16}$$

The solution technique is to pick a value of $n(1)$ and then integrate backwards to $x = 0$. The value of $n(1)$ is then adjusted so that the boundary condition at $x = 0$ is satisfied. Letting

$$n_{in} = n(0) - \frac{1}{Pe}\frac{dn}{dx}(0) \tag{8.8.17}$$

the values of n_{in} obtained from each integration may be plotted versus the guess for $n(1)$, and the steady–state solution is obtained where the curve crosses the line $n_{in} = 1$.

Solution of Partial Differential Equations

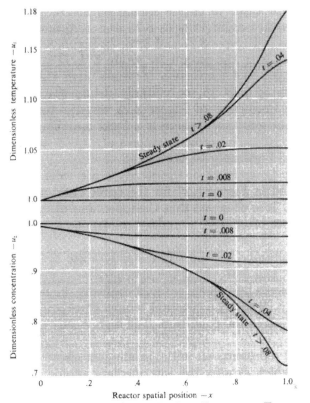

Figure 8.23: Adiabatic Tubular Reactor Dynamics.
Dimensionless Temperature and Concentration vs. Reactor
Spatial Position vs. Time Startup from Cold Reactor.

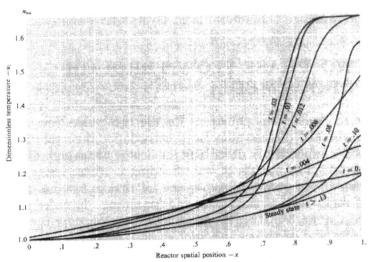

Figure 8.24: Adiabatic Tubular Reactor Dynamics.
Dimensionless Temperature vs. Reactor
Spatial Position vs. Time.

Two different types of curves should be noted, that is, those with single steady states as in Figures 8.25 and 8.26 and those with three steady states as in Figures 8.27 and 8.28. In Figure 8.25, the variation of the steady state with the parameter group L^2k/D is shown for a single steady–state case, and conversion increases with the parameter value. With the value of q at 18, only single steady–state cases are found; however, close inspection of Figure 8.25 reveals that a triple steady–state case might exist since there is a local maximum in the curve for $L^2k/D = 2 \times 10^8$. The range of values of L^2k/D which would give a triple case would certainly be very small.

The strong dependence of the steady–state value on the parameter group q is noted in Figure 8.27 for the triple case, and Figure 8.26 for the single case. It is observed that the inflections of the curves are larger for higher values of q. The boundary conditions appear to be satisfied at all times.

Three simulations for the triple steady state are presented in Figures 8.29, 8.30, and 8.31. That the intermediate conversion steady state in the triple case is never even noticed confirms the notion that it is unstable. Figure 8.29 shows a constant temperature profile close to the low steady state as an initial condition. The high steady state is attained for this run.

Figure 8.30 represents an initial profile just above the low conversion steady state which crosses the intermediate profile. From this initial condition, the system returns to the low steady state and the behavior is nonuniform.

In Figure 8.31 an initial profile just below the intermediate steady state returns to the low steady state. The high conversion steady state appears to be locally uniformly asymptotically stable, and that of low conversion possesses nonuniform local asymptotic stability.

8.9 TWO–PHASE FLOW THROUGH POROUS MEDIA

We want to develop the equations for the two–phase flow of fluids in porous media and discuss numerical methods for their solution. These problems are important in describing the flow of water and oil in petroleum reservoirs. The process of waterflooding is where water is injected into a reservoir in order to recover the oil that is residual in the void spaces of rocks such as sandstone and limestone. In order to describe this process, a mathematical model must be developed for the two–phase flow of water and oil through the porous rock material.

A material balance for the water species is

$$\phi \frac{\partial \rho_w S_w}{\partial t} = -\boldsymbol{\nabla} \cdot \rho_w \boldsymbol{v}_w \qquad (8.9.1)$$

Solution of Partial Differential Equations

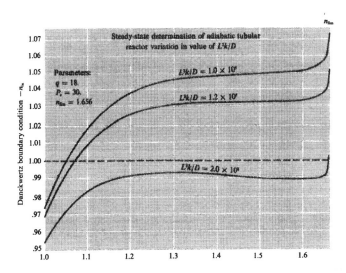

Figure 8.25: Guess of Dimensionless Temperature at Reactor Exit—$n(1)$.

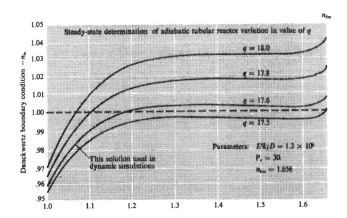

Figure 8.26: Guess of Dimensionless Temperature at Reactor Exit—$n(1)$.

394 Computational Methods for Process Simulation

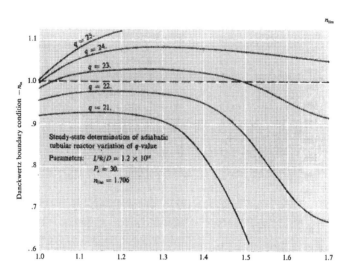

Figure 8.27: Guess of Dimensionless Temperature at Reactor Exit—$n(1)$.

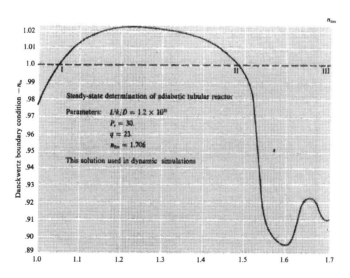

Figure 8.28: Guess of Dimensionless Temperature at Reactor Exit—$n(1)$.

Solution of Partial Differential Equations

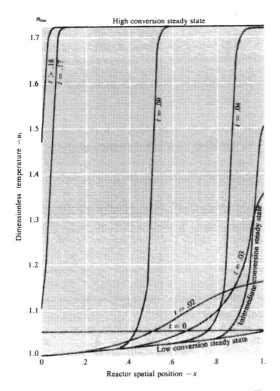

Figure 8.29: Adiabatic Tubular Reactor Dynamics. Dimensionless Temperature vs. Reactor Spatial Position vs. Time.

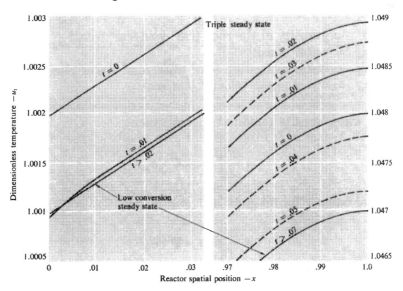

Figure 8.30: Adiabatic Tubular Reactor Dynamics. Dimensionless Temperature vs. Reactor Spatial Position vs. Time.

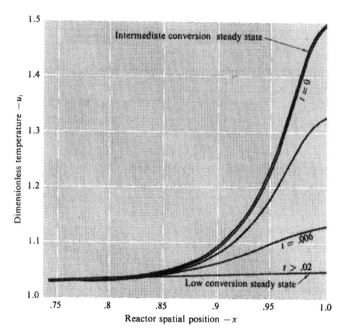

Figure 8.31: Adiabatic Tubular Reactor Dynamics. Dimensionless Temperature vs. Reactor Spatial Position vs. Time.

where ϕ = porosity of the rock (cm^3 void/cm^3 total)
ρ_w = water density
S_w = water saturation (cm^3 water/cm^3 water plus oil)
v_w = water phase velocity (cm/sec)

If the water density remains constant, then

$$\phi \frac{\partial S_w}{\partial t} = -\boldsymbol{\nabla} \cdot \boldsymbol{v}_w \qquad (8.9.2)$$

A material balance for oil gives

$$\phi \frac{\partial S_o}{\partial t} = -\boldsymbol{\nabla} \cdot \boldsymbol{v}_o \qquad (8.9.3)$$

The momentum balance is given by Darcy's law which states that (Greenkorn, 1983)

$$\boldsymbol{v}_w = -\frac{\boldsymbol{K} \, k_{rw}}{\mu_w} \boldsymbol{\nabla} p \qquad (8.9.4)$$

Solution of Partial Differential Equations

and

$$\boldsymbol{v}_o = -\frac{\boldsymbol{K}\, k_{ro}}{\mu_o} \nabla p \qquad (8.9.5)$$

where $\boldsymbol{K}$ = permeability tensor (Darcy's)
 k_{rw} = water relative permeability
 k_{ro} = oil relative permeability
 μ_w = water viscosity (cp)
 μ_o = oil viscosity (cp)
 p = system pressure

We also know that

$$S_w + S_o = 1 \qquad (8.9.6)$$

which can be used instead of equation (8.9.3).

For a one–dimensional problem, we have

$$\phi \frac{\partial S_w}{\partial t} = -\frac{\partial v_{w_z}}{\partial z} \qquad (8.9.7)$$

The velocity v_{w_z} can be expressed in terms of the fractional flow of water, f_w, as

$$v_{w_z} = \frac{Q_T}{A} f_w \qquad (8.9.8)$$

where Q_T is the total flow rate (cm^3/sec) and A is the cross–sectional area for flow (cm^2).

Therefore the material balance is for constant Q_T and A

$$\phi \frac{\partial S_w}{\partial t} = -\frac{Q_T}{A} \frac{\partial f_w}{\partial z} \qquad (8.9.9)$$

Darcy's law for one–dimensional flow is

$$\frac{Q_T}{A} f_w = -\frac{K\, k_{rw}}{\mu_w} \frac{\partial p}{\partial z} \qquad (8.9.10)$$

$$\frac{Q_T}{A} f_o = -\frac{K\, K_{ro}}{\mu_o} \frac{\partial p}{\partial z} \qquad (8.9.11)$$

We also know that

$$f_w + f_o = 1 \qquad (8.9.12)$$

Therefore, we can form the ratio

$$\frac{f_w}{f_w + f_o} = f_w = \frac{\frac{k_{rw}}{\mu_w}}{\frac{k_{rw}}{\mu_w} + \frac{k_{ro}}{\mu_o}} \qquad (8.9.13)$$

or

$$f_w = \frac{1}{1 + \frac{k_{ro}\mu_w}{k_{rw}\mu_o}} \qquad (8.9.14)$$

The relative permeabilities are functions of the water saturation and typical behavior is shown in Figure 8.32. For this case the oil phase relative permeability goes to zero at $S_w = 0.62$ so that 38 percent of the oil remains trapped in the reservoir. This means that there is a residual oil saturation of 38 percent. Also, the relative water permeability goes to zero at $S_w = 0.48$. This means that there is an irreducible water saturation of 48 percent. The residual oil saturation and irreducible water saturation vary from reservoir to reservoir.

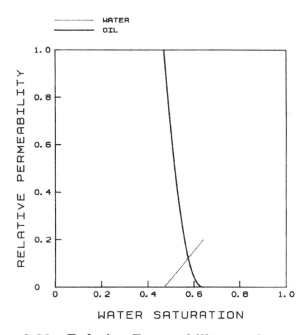

Figure 8.32: Relative Permeability vs. Saturation.

Now we want to discuss the numerical solution of equation (8.9.9). The boundary condition for waterflooding, i.e. the injection of water, is

$$f_w = 1.0 \qquad \text{at } z = 0$$

We can make the describing equation dimensionless by introducing scaled dimensionless variables

$$z^* = \frac{z}{L} \tag{8.9.15}$$

$$t^* = \frac{t\, Q_T}{AL\phi} \tag{8.9.16}$$

This gives

$$\frac{\partial S_W}{\partial t^*} = -\frac{\partial f_w}{\partial z^*} \tag{8.9.17}$$

Solution of Partial Differential Equations

The variable t^* is commonly called the number of pore volumes of fluid injected.

Equation (8.9.17) is a difficult equation for numerical methods. Taranchuk (1974) has investigated many numerical schemes and has shown that an explicit corner scheme is the best for this equation. The scheme is illustrated in Figure 8.33.

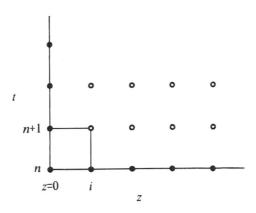

Figure 8.33: Corner Scheme.

It uses a four point grid with three known points and one unknown point. We use a forward difference scheme for time and a backward difference scheme for distance.

$$\frac{S_{w_i}^{n+1} - S_{w_i}^n}{\Delta t^*} + \frac{f_{w_i}^n - f_{w_{i-1}}^n}{\Delta z^*} = 0 \qquad (8.9.18)$$

This is a conditionally stable monotone scheme of first-order accuracy. The stability condition is

$$C = \max_i \left| \frac{\Delta t^*}{\Delta z^*} \left(\frac{\partial f_w}{\partial S_w} \right)_{i,n} \right| \leq 1 \qquad (8.9.19)$$

where C is the Courant number. The stability condition for an explicit scheme for hyperbolic equations has the general form

$$C = \left| \frac{\Delta t}{\Delta z} v \right| \leq 1 \qquad (8.9.20)$$

where v is a characteristic velocity. An approximate form to equation (8.9.19) which is easy to compute is

$$C = \max_i \left| \frac{\Delta t^*}{\Delta z^*} \left(\frac{f_{w_i}^n - f_{w_{i-1}}^n}{S_{w_i}^n - S_{w_{i-1}}^n} \right) \right| \leq 1 \qquad (8.9.21)$$

Some typical numerical results are given in Figure 8.34. Curve 1 is the analytic solution and curves 2, 3, and 4 are numerical solutions with $\Delta z^* = 1/10$ and $C = 0.5$, $\Delta z^* = 1/20$ and $C = 0.5$, and $\Delta z^* = 1/40$ and $C = 1.0$, respectively. In these cases we can see the effect of numerical dispersion in the vicinity of the shock. As the Courant number C approaches 1.0, the numerical dispersion is decreased. The corner scheme is monotone since there are no oscillations in the vicinity of the shock.

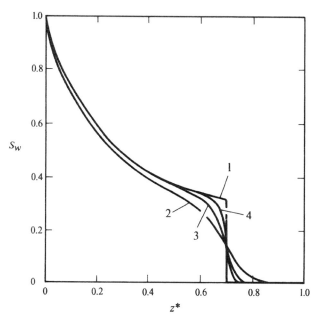

Figure 8.34: Typical Numerical Results.

8.10 TWO–DIMENSIONAL FLOW THROUGH POROUS MEDIA

We now want to develop the equations that describe the two–dimensional flow of oil, water, and gas through porous media. The material balances for water and oil are

$$\phi \frac{\partial}{\partial t} \rho_w S_w = -\nabla \cdot \rho_w v_w - q_w \qquad (8.10.1)$$

$$\phi \frac{\partial}{\partial t} \rho_o S_o = -\nabla \cdot \rho_o v_o - q_o \qquad (8.10.2)$$

where q_i = well production rate when $q_i > 0$ or a well injection rate when $q_i < 0$ (kg/m^3 s).

Solution of Partial Differential Equations

We can normalize these equations by dividing by the phase densities at standard temperature and pressure conditions

$$\phi \frac{\partial}{\partial t}\left(\frac{\rho_w}{\rho_w^s}\right) S_w = -\nabla \cdot \left(\frac{\rho_w}{\rho_w^s}\right) \boldsymbol{v}_w - \frac{q_w}{\rho_w^s} \qquad (8.10.3)$$

$$\phi \frac{\partial}{\partial t}\left(\frac{\rho_o}{\rho_o^s}\right) S_o = -\nabla \cdot \left(\frac{\rho_o}{\rho_o^s}\right) \boldsymbol{v}_o - \frac{q_o}{\rho_o^s} \qquad (8.10.4)$$

The ratios ρ_o/ρ_o^s and ρ_w/ρ_w^s can be given in terms of the formation volume factors B_o and B_w where

$$\frac{\rho_o}{\rho_o^s} = \frac{1}{B_o} \qquad \text{and} \qquad \frac{\rho_w}{\rho_w^s} = \frac{1}{B_w} \qquad (8.10.5)$$

where B_i = formation volume factor for phase i (m^3/m^3 std).
The material balances are therefore

$$\phi \frac{\partial}{\partial t}\left(\frac{S_w}{B_w}\right) = -\nabla \cdot \left(\frac{\boldsymbol{v}_w}{B_w}\right) - \frac{q_w}{\rho_w^s} \qquad (8.10.6)$$

$$\phi \frac{\partial}{\partial t}\left(\frac{S_o}{B_o}\right) = -\nabla \cdot \left(\frac{\boldsymbol{v}_o}{B_o}\right) - \frac{q_o}{\rho_o^s} \qquad (8.10.7)$$

For the water and oil phases, we have assumed that water and oil are only found in their respective phases. This is not the case for gas. There is a solubility of gas in both the water and oil phases given by the solubility coefficients R_{gw} and R_{go}. These are defined at standard conditions as

$$R_{ij} = \text{solubility of phase } i \text{ in phase } j \text{ at standard conditions} \\ (m^3 \text{ std}/m^3 \text{ std}) \qquad (8.10.8)$$

The gas phase material balance is

$$\phi \frac{\partial}{\partial t}\left(\rho_g S_g + R_{go}\frac{\rho_g^s}{\rho_o^s}S_o + R_{gw}\frac{\rho_g^s}{\rho_w^s}S_w\right)$$
$$= -\nabla \cdot \left[\rho_g \boldsymbol{v}_g + R_{go}\frac{\rho_g^s}{\rho_o^s}\boldsymbol{v}_o\rho_o + R_{gw}\frac{\rho_g^s}{\rho_o^s}\boldsymbol{v}_w\rho_w\right] - q_g \qquad (8.10.9)$$

We can normalize this equation by dividing by the gas phase density at standard conditions. This gives

$$\phi \frac{\partial}{\partial t}\left(\frac{S_g}{B_g} + \frac{R_{go}S_o}{B_o} + \frac{R_{gw}S_w}{B_w}\right) = -\nabla \cdot \left[\frac{\boldsymbol{v}_g}{B_g} + \frac{R_{go}\boldsymbol{v}_o}{B_o} + \frac{R_{gw}\boldsymbol{v}_w}{B_w}\right] - \frac{q_g}{\rho_g^s}$$
$$(8.10.10)$$

Now we introduce Darcy's law

$$v_i = -K \frac{k_{ri}}{\mu_i} \nabla p \qquad (8.10.11)$$

The material balances become

$$\phi \frac{\partial}{\partial t} \left(\frac{S_o}{B_o} \right) = \nabla \cdot \left[\frac{K \, k_{ro}}{\mu_o B_o} \nabla p \right] - \frac{q_o}{\rho_o^s} \qquad (8.10.12)$$

$$\phi \frac{\partial}{\partial t} \left(\frac{S_w}{B_w} \right) = \nabla \cdot \left[\frac{K \, k_{rw}}{\mu_w B_w} \nabla p \right] - \frac{q_w}{\rho_w^s} \qquad (8.10.13)$$

$$\phi \frac{\partial}{\partial t} \left(\frac{S_g}{B_g} + \frac{R_{go} S_o}{B_o} + \frac{R_{gw} S_w}{B_w} \right) = \nabla \cdot \left[K \left(\frac{k_{rg}}{\mu_g B_g} + \frac{R_{go} k_{ro}}{\mu_o B_o} \right. \right.$$

$$\left. \left. + \frac{R_{gw} k_{rw}}{\mu_w B_w} \right) \nabla p \right] - \frac{q_g}{\rho_g^s} \qquad (8.10.14)$$

We also know that the sum of the phase saturations is unity

$$S_o + S_w + S_g = 1 \qquad (8.10.15)$$

Equations (8.10.12) through (8.10.15) can be combined to form the pressure equation. This involves expanding the time derivative products and ratios using the chain rule, summing equations (8.10.12)–(8.10.14), and using the saturation identity of equation (8.10.15). The details are left as an exercise for the interested reader. The pressure equation is

$$\phi C_t \frac{\partial p}{\partial t} = (B_o - R_{go} B_g) \left[\nabla \cdot \left(\frac{K \, k_{ro}}{\mu_o B_o} \nabla p \right) - \frac{q_o}{\rho_o^s} \right]$$

$$+ (B_w - R_{gw} B_g) \left[\nabla \cdot \left(\frac{K \, k_{rw}}{\mu_w B_w} \nabla p \right) - \frac{q_w}{\rho_w^s} \right]$$

$$+ B_g \left[\nabla \cdot K \left(\frac{k_{rg}}{\mu_g B_g} + \frac{R_{go} k_{ro}}{\mu_o B_o} + \frac{R_{gw} k_{rw}}{\mu_w B_w} \right) \nabla p - \frac{q_g}{\rho_g^s} \right] \qquad (8.10.16)$$

where C_t is the total compressibility given by

$$C_t = C_o S_o + C_w S_w + C_g S_g \qquad (8.10.17)$$

The oil, water, and gas compressibilities are defined by

$$C_o = -\frac{1}{B_o} \frac{\partial B_o}{\partial p} + \frac{B_g}{B_o} \frac{\partial R_{go}}{\partial p} \qquad (8.10.18)$$

$$C_w = -\frac{1}{B_w} \frac{\partial B_w}{\partial p} + \frac{B_g}{B_w} \frac{\partial R_{gw}}{\partial p} \qquad (8.10.19)$$

Solution of Partial Differential Equations

$$C_g = -\frac{1}{B_g}\frac{\partial B_g}{\partial p} \qquad (8.10.20)$$

Only four of the five equations (8.10.12)–(8.10.16) are independent. The oil material balance (equation (8.10.12)), the water material balance (equation (8.10.13)), the saturation identity of equation (8.10.15), and the pressure equation (8.10.16) are normally used.

Let us consider the simulation of a five–spot geometry for oil production. The well configuration of a five–spot pattern is shown in Figure 8.35. Symmetry allows one quadrant to be isolated and modeled as shown in Figure 8.36. If injection and production wells are located at the points $y = z = 0$ and $y = z = w$, respectively, then for any system variable, x, the boundary conditions are

$$\frac{\partial x}{\partial y} = 0 \quad \text{at } y = 0 \text{ and } y = w \text{ for } 0 \leq z \leq w \text{ and } t \geq 0 \qquad (8.10.21)$$

$$\frac{\partial x}{\partial z} = 0 \quad \text{at } z = 0 \text{ and } z = w \text{ for } 0 \leq y \leq w \text{ and } t \geq 0 \qquad (8.10.22)$$

The no–flow boundary conditions of equations (8.10.21) and (8.10.22) arise because of symmetry between repeated five–spot patterns.

The saturation and pressure equations of the simulation model are not amenable to analytical solution and must be solved numerically. We use the IMPES finite–difference method to solve this black–oil model. The IMPES method develops a set of algebraic equations that are implicit in pressure and explicit in the saturations.

In order to finite–difference the model partial differential equations, we need values of the state variables at discrete distances, $y_1, y_2, ...y_N$, and $z_1, z_2, ...z_N$, and at discrete times, $0, \Delta t, 2\Delta t, ..., (n_f - 1)\Delta t$. Here N is the number of grid blocks along the quadrant boundary, Δt is the time step, and $n_f = t_f/\Delta t$. The reservoir quadrant is therefore replaced by a system of grid blocks shown in Figure 8.37. The integer i is used as the index in the y direction, and the integer j as the index in the z direction. In addition, the index n is used to denote time. Hence we use the following notation to identify a process variable

$$x(y_i, z_j, n\Delta t) = x_{ij}^n \qquad (8.10.23)$$

The point (y_i, z_j) is considered to be at the center of block (i, j). Such block–centered grid systems are common in reservoir simulators since each block can be associated with an average pressure and average phase saturations. Also, injection and production wells can be placed in appropriate blocks. We will assume that one–quarter of an injection well lies in node (1,1) and one–quarter of a producer lies in node (N,N).

Finite–difference approximations convert the partial differential equations of the modified black–oil model into a set of algebraic equations

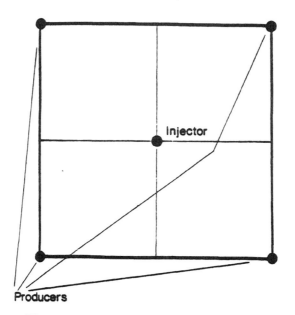

Figure 8.35: Five–Spot Pattern.

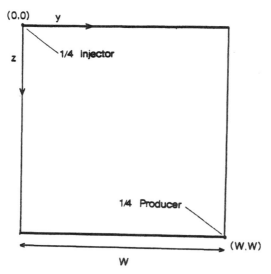

Figure 8.36: Quadrant of a Five–Spot Pattern.

Solution of Partial Differential Equations

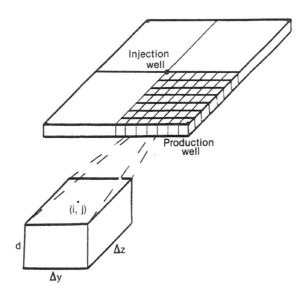

Figure 8.37: Finite–Difference Grid System.

(Aziz and Settari, 1979; Peaceman, 1977). First, each equation is multiplied through by the bulk volume, V_B, defined as

$$V_B = d \, \Delta y \, \Delta z \qquad (8.10.24)$$

where d is the reservoir thickness.

A five–point, second–order correct operator is defined for the pressure partial for each phase q as

$$\nabla (A_q \, \nabla p) = \Delta (A_q \, \Delta p) = \Delta_y (A_{qy} \, \Delta_y p) + \Delta_z (A_{qz} \, \Delta_z p) \qquad (8.10.25)$$

where

$$\Delta_y (A_{qy} \, \Delta_y p) = A_{q_{i-1/2,j}}(p_{i-1,j} - p_{i,j}) + A_{q_{i+1/2,j}}(p_{i+1,j} - p_{i,j}) \qquad (8.10.26)$$

$$\Delta_z (A_{qz} \, \Delta_z p) = A_{q_{i,j-1/2}}(p_{i,j-1} - p_{i,j}) + A_{q_{i,j_1/2}}(p_{i,j+1} - p_{i,j}) \qquad (8.10.27)$$

and the finite–differenced phase transmissibilities are

$$A_{q_{i-1/2,j}} = \frac{K \, d \, \Delta z}{\Delta y} \left(\frac{k_{rq}}{\mu_q \, B_q} \right)_{i-1/2,j} \qquad (8.10.28)$$

$$A_{q_{i+1/2,j}} = \frac{K \, d \, \Delta z}{\Delta y} \left(\frac{k_{rq}}{\mu_q \, B_q} \right)_{i+1/2,j} \qquad (8.10.29)$$

$$A_{q_{i,j-1/2}} = \frac{K\, d\, \Delta y}{\Delta z}\left(\frac{k_{rq}}{\mu_q B_q}\right)_{i,j-1/2} \qquad (8.10.30)$$

$$A_{q_{i,j+1/2}} = \frac{K\, d\, \Delta y}{\Delta z}\left(\frac{k_{rq}}{\mu_q B_q}\right)_{i,j+1/2} \qquad (8.10.31)$$

Since the quantity $k_{rg}/\mu_q B_q$ is not evaluated at a grid point, some suitable average value must be used. Typically, we use the arithmetic mean value of the phase viscosity and formation volume factor, and the value of the upstream node for the relative permeability. Such a convention is called upstream weighting. Transmissibilities are a measure of the ability of a fluid to flow in porous media. At the grid system boundaries in a five–spot pattern, no–flow constraints can be created by setting the transmissibilities to zero:

$$A_{q_{1/2,j}} = A_{q_{N+1,j}} = A_{q_{i,1/2}} = A_{q_{i,N+1/2}} = 0 \qquad (8.10.32)$$

The time derivatives can be approximated by a forward difference formula.

From the above relations, the difference equations for a black–oil simulator are

oil

$$\frac{V_p}{\Delta t}\left[\left(\frac{S_o}{B_o}\right)^{n+1} - \left(\frac{S_o}{B_o}\right)^n\right]_{i,j} = \left[\Delta\left(A_o^n \Delta p^{n+1}\right) - Q_o^n\right]_{i,j} \qquad (8.10.33)$$

water

$$\frac{V_p}{\Delta t}\left[\left(\frac{S_w}{B_w}\right)^{n+1} - \left(\frac{S_w}{B_w}\right)^n\right]_{i,j} = \left[\Delta\left(A_w^n \Delta p^{n+1}\right) - Q_w^n\right]_{i,j} \qquad (8.10.34)$$

pressure

$$\frac{V_p}{\Delta t}C_{t_{i,j}}^n\left(p^{n+1} - p^n\right)_{i,j} =$$
$$(B_o - B_g R_{go})_{i,j}^n\left[\Delta\left(A_o^n \Delta p^{n+1}\right) - Q_o^n\right]_{i,j}$$
$$+ (B_w - B_g R_{gw})_{i,j}^n\left[\Delta\left(A_w^n \Delta p^{n+1}\right) - Q_w^n\right]_{i,j}$$
$$+ B_g^n\left[\Delta\left(A_g^n \Delta p^{n+1}\right) + \Delta\left(R_{go}^n A_o^n \Delta p^{n+1}\right)\right]$$
$$+ \left[\Delta\left(R_{gw}^n A_w^n \Delta p^{n+1}\right) - Q_g\right]_{i,j} \qquad (8.10.35)$$

where

$$V_p = \phi V_B \qquad (8.10.36)$$

$$Q_{o_{i,j}}^n = q_{o_{i,j}}^n V_B/\rho_o^s \qquad (8.10.37)$$

$$Q_{w_{i,j}}^n = q_{w_{i,j}}^n V_B/\rho_w^s \qquad (8.10.38)$$

$$Q_{g_{i,j}}^n = q_{g_{i,j}}^n V_B/\rho_g^s \qquad (8.10.39)$$

Terms at the current time (evaluated at time level n) can be computed using existing data. Terms at the new time level (evaluated at time level $n+1$) are the unknown variables. The unknown saturations therefore appear explicitly in equations (8.10.33) and (8.10.34). The unknown pressure, however, appears implicitly in equation (8.10.35). Efficient matrix methods have been developed for solving the IMPES set of algebraic equations (Peaceman, 1977; Aziz and Settari, 1979). The method uses matrix methods to simultaneously solve for the new time level pressures from equation (8.10.35) and then using these values, the new time level saturations are computed explicitly from the saturation equations (8.10.33) and (8.10.34).

The IMPES method is subject to the following stability condition:

$$\Delta t \left(\frac{u_y}{\Delta y} + \frac{u_z}{\Delta z} \right) \leq 1 \qquad (8.10.40)$$

where u_y and u_z are the y and z components of the Darcy simplified velocity.

Finally, we need to describe the production rates in terms of the wellbore pressure. This is done by means of a "performance index" analysis (Muskat, 1949). The production rates are

$$Q_{o_{N,N}} = \text{PI} \left(\frac{k_{ro}}{\mu_o B_o} \right)_{N,N} (p_{N,N} - p_{wf}) \qquad (8.10.41)$$

$$Q_{w_{N,N}} = \text{PI} \left(\frac{k_{rw}}{\mu_w B_w} \right)_{N,N} (p_{N,N} - p_{wf}) \qquad (8.10.42)$$

$$Q_{g_{N,N}} = \text{PI} \left(\frac{k_{rg}}{\mu_g B_g} \right)_{N,N} (p_{N,N} - p_{wf}) + (R_{go}Q_o)_{N,N} + (R_{gw}Q_w)_{N,N} \qquad (8.10.43)$$

where PI is the performance index and for a five-spot pattern is

$$\text{PI} = \frac{\pi K h}{4 \left[\ln \left(\frac{D}{r_w} \right) - 0.619 \right]} \quad (\text{m}^3) \qquad (8.10.44)$$

where r_w = well radius, m
h = well depth, m
D = distance between wells, m

These are the numerical principles which are used to create efficient and accurate two-dimensional finite-differenced simulations of oil recovery problems. Such simulations are routinely used in the oil industry today to aid in the development of oil fields and efficient production of oil reservoirs.

8.11 WEIGHTED RESIDUALS

The method of weighted residuals is a way of reducing the number of independent variables or the problem domain dimension. The basic idea of the method is to approximate the solution of the problem over a domain by a functional form called a trial function. The trial function's form is specified but it has adjustable constants. The trial function is chosen so as to give a good solution to the original differential equation. An excellent treatment of the method is given in the book by Finlayson (1972). As an example of how the method works, let us consider the heat conduction equation

$$\frac{d^2T}{dx^2} = 0 \qquad (8.11.1)$$

with boundary conditions

$$T = T_0 \quad \text{at} \quad x = 0, \; x = 1 \qquad (8.11.2)$$

We will assume a linear combination of trial solutions

$$T = T_0 + \sum_{i=1}^{N} C_i T_i(x) \qquad (8.11.3)$$

The functions $T_i(x)$ are chosen in order that the boundary conditions $T = T_0$ are satisfied at $x = 0$ and $x = 1$. This means that $T_i = 0$ at $x = 0$ and $x = 1$. The trial solution of equation (8.11.3) therefore satisfies the boundary conditions for all values of the constants C_i.

The trial function (equation (8.11.3)) is substituted into the original differential equation (equation (8.11.1)) to form the residual,

$$R(C_i, x) = \frac{d^2T_0}{dx^2} + \sum_{i=1}^{N} C_i \frac{d^2T_i}{dx^2} \qquad (8.11.4)$$

If the trial functions were exact, then the residual $R(C_i, x)$ would be zero. A straightforward scheme to determine the residual would be to set the integral of the residual to zero

$$\int R(C_i, x) dx = 0 \qquad (8.11.5)$$

However, this scheme only generates one equation for the N unknown constants, C_i. This criterion can be modified by introducing weighting functions w_i. Setting the integral of each weighted residual to zero gives N independent equations,

$$(w_j, R) = \int w_j R \, dx = 0 \qquad (8.11.6)$$

Solution of Partial Differential Equations

The weighting factors w_j can be chosen in many ways. Two common methods are the collocation method and the Galerkin method.

For the *collocation method*, the weighting factors w_j are chosen to be the displaced Dirac delta functions

$$w_j = \delta(x - x_j) \tag{8.11.7}$$

where

$$\begin{aligned}\delta(0) &= 1 \\ \delta(\neq 0) &= 0\end{aligned} \tag{8.11.8}$$

Therefore

$$\int w_j R \, dx = R|_{x_j} = 0 \tag{8.11.9}$$

Thus, the residual, R, is zero at N specified collocation points, x_j. As N increases, the residual becomes zero at more points and hopefully approaches zero everywhere.

For the *Galerkin method*, the weighting functions are chosen to be the trial functions, $w_j = T_i$. Therefore, the weighted residual is

$$(w_j, R) = \int T_i R \, dx = 0 \tag{8.11.10}$$

8.11.1 One–Dimensional Heat Conduction

Finlayson (1972) illustrates the method by considering one–dimensional heat conduction in a slab where the thermal conductivity is a linear function of temperature. The energy balance for this system is

$$\frac{d}{dz}\left(k(T)\frac{dT}{dz}\right) = 0 \tag{8.11.11}$$

with boundary conditions

$$\begin{aligned}T(0) &= T_0 \\ T(d) &= T_1\end{aligned} \tag{8.11.12}$$

The linear temperature dependency of the thermal conductivity is given by

$$k(T) = k_0 + \alpha(T - T_0) \tag{8.11.13}$$

We can express the problem in dimensionless form by introducing

$$\theta = \left(\frac{T - T_0}{T_1 - T_0}\right) \quad \text{and} \quad x = \frac{z}{d} \tag{8.11.14}$$

The energy balance becomes

$$\frac{d}{dx}\left[(1 + a\theta)\frac{d\theta}{dx}\right] = 0 \tag{8.11.15}$$

with boundary conditions
$$\theta(0) = 0 \quad \text{and} \quad \theta(1) = 1 \tag{8.11.16}$$

Equation (8.11.15) can also be written as

$$(1 + a\theta) \frac{d^2\theta}{dx^2} + a \left(\frac{d\theta}{dx}\right)^2 = 0 \tag{8.11.17}$$

We choose a trial solution for θ an $(N + 1)^{th}$ order polynomial in the independent variable x

$$\theta_N = \sum_{i=0}^{N+1} C_i x^i \tag{8.11.18}$$

This polynomial function can be made to satisfy the boundary conditions by choosing
$$C_0 = 0 \tag{8.11.19}$$

and
$$\sum_{i=1}^{N+1} C_i = 1 \tag{8.11.20}$$

The trial solution is then

$$\theta_N = C_0 x^0 + C_1 x^1 + \sum_{j=1}^{N} C_{j+1} x^{j+1} \tag{8.11.21}$$

From equation (8.11.20) we have

$$C_1 = 1 - \sum_{j=1}^{N} C_{j+1} \tag{8.11.22}$$

Using equation (8.11.22) and the boundary condition (8.11.19), we have

$$\theta_N = x + \sum_{j=1}^{N} C_{j+1} \left(x^{j+1} - x\right) \tag{8.11.23}$$

or

$$\theta_N = x + \sum_{j=1}^{N} A_j \left(x^{j+1} - x\right) \tag{8.11.24}$$

The trial solution is then substituted into the original differential equation to generate the residual

$$R(x, \theta_N) = (1 + a\theta_N) \frac{d^2\theta_N}{dx^2} + a \left(\frac{d\theta_N}{dx}\right)^2 \tag{8.11.25}$$

The weighted residual of equation (8.11.6) is

$$(w_j, R) = \int_0^1 w_j R(x, \theta_N) dx = 0 \quad j = 1, ..., N \quad (8.11.26)$$

If we choose the order $N = 1$, then the trial solution is

$$\theta_1 = x + A_1(x^2 - x) \quad (8.11.27)$$

In order to generate the residual, we also need to evaluate

$$\frac{d\theta_1}{dx} = 1 + A_1(2x - 1) \quad (8.11.28)$$

and

$$\frac{d^2\theta_1}{dx^2} = 2A_1 \quad (8.11.29)$$

Then from equation (8.11.25), the residual is

$$R(x, \theta_1) = (1 + a\theta_1)(2A_1) + a(1 + A_1(2x - 1))^2 \quad (8.11.30)$$

If we use the collocation method for the weighting factor, w_1, then we need to specify the collocation point in the x domain. A reasonable choice for the collocation point is the midpoint $x = 1/2$. The weighted residual at $x = 1/2$ is

$$\left(1 + \frac{1}{2}a\left(1 - \frac{1}{2}A_1\right)\right) 2A_1 + a = 0 \quad (8.11.31)$$

If the model parameter a is unity, then the coefficient $A_1 = -0.317$. Therefore, the MWR solution is

$$\theta_1 = x - 0.317(x^2 - x) \quad (8.11.32)$$

Applying the Galerkin method, we use

$$w_1 = x(1 - x) \quad (8.11.33)$$

The weighted residual is

$$\int_0^1 x(1 - x) R(x, \theta_1) dx = 0 \quad (8.11.34)$$

or

$$\int_0^1 x(1 - x)\left\{\left(1 + x + A_1(x^2 - x)\right) 2A_1 + (1 + A_1(2x - 1))^2\right\} dx = 0 \quad (8.11.35)$$

This gives $A_1 = -0.326$ and the approximate MWR solution is

$$\theta_1 = x - 0.326(x^2 - x) \quad (8.11.36)$$

For this problem, we can solve the original differential equation analytically. The exact solution is

$$\theta = -1 + (1 + 3x)^{1/2} \quad (8.11.37)$$

The exact solution is compared to the approximate MWR solutions in Table 8.5. Each of the approximate methods works quite well for this example.

Table 8.5: Comparison of Exact and Approximate Heat Conduction Temperatures (Source: Finlayson, 1972)

x	Exact	Collocation	Galerkin
0.1	0.140	0.129 (−7.9%)	0.129 (−7.9%)
0.25	0.323	0.309 (−4.3%)	0.311 (−3.7%)
0.5	0.581	0.579 (−0.34%)	0.582 (+0.17%)
0.75	0.803	0.809 (+0.74%)	0.811 (+0.99%)
0.9	0.924	0.929 (+0.54%)	0.929 (+0.54%)

8.11.2 Two–Dimensional Heat Conduction

The method of weighted residuals can be used to reduce the dimensionality of a problem. To illustrate this, let us consider a two–dimensional heat conduction problem:

$$\frac{\partial}{\partial x}\left(k\frac{\partial \theta}{\partial x}\right) + \frac{\partial}{\partial y}\left(k\frac{\partial \theta}{\partial y}\right) = 0 \qquad (8.11.38)$$

with the boundary conditions

$$\theta(x, y = 0) = x(1 - x)$$
$$\theta(x = 0, y) = \theta(x = 1, y) = \theta(x, y = \infty) = 0 \qquad (8.11.39)$$

We will get an approximate solution over the x domain and then solve the resulting ordinary differential equation in y. The problem is symmetric about $x = 1/2$ and vanishes at $x = 0$ and $x = 1$. We want to choose a trial function that has these properties. As discussed by Finlayson (1972), the quadratic form

$$x(1 - x) \qquad (8.11.40)$$

satisfies these conditions. We therefore use the trial function

$$\theta_N = \sum_{i=1}^{N} x^i (1 - x)^i C_i(y) \qquad (8.11.41)$$

For $N = 1$ and k constant, we have the residual

$$R(\theta_1) = \frac{\partial^2 \theta_1}{\partial x^2} + \frac{\partial^2 \theta_1}{\partial y^2} \qquad (8.11.42)$$

or

$$R(\theta_1) = -2C_1(y) + x(1 - x)\frac{d^2 C_1(y)}{dy} \qquad (8.11.43)$$

If we use the collocation method and choose the collocation point as the midpoint $x = 1/2$, then the weighted residual of equation (8.11.9) is

$$R(\theta_1, 1/2) = -2C_1 + 1/2(1 - 1/2)\frac{d^2C_1}{dy^2} = 0 \qquad (8.11.44)$$

or

$$\frac{d^2C_1}{dy^2} - 8C_1 = 0 \qquad (8.11.45)$$

This ordinary differential equation has the analytical solution

$$C_1 = e^{-\sqrt{8}\, y} \qquad (8.11.46)$$

and the approximate solution for the temperature is

$$\theta_1 = x(1-x)e^{-\sqrt{8}\, y} \qquad (8.11.47)$$

As discussed by Finlayson (1972), a good way to check the accuracy of the approximate solution is to compare the average heat flux at the boundary using the MWR solution method to an analytical solution. The average heat flux is given by

$$\bar{N} = -\int_0^1 \frac{\partial \theta}{\partial y}\bigg|_{y=0} dx \qquad (8.11.48)$$

The exact average heat flux is 0.542 while the approximate MWR value is 0.471. This represents a 13 percent error, which is quite acceptable considering that this is just a first–order approximation ($N = 1$).

8.11.3 Finite Elements

Finite element methods for solving partial differential equations use weighted residual concepts. The idea behind the finite element method is to break the spatial domain up into a number of simple geometric elements such as triangles or quadrilaterals. The weighted residual concept is then used to approximate the solution function over each finite element domain. Care needs to be taken to ensure continuity of the dependent variables and their first partials in moving from element to element. Partial differential equations are therefore transformed into sets of ordinary differential equations in time. The method is particularly suited for solving problems involving irregular geometries and steep gradients. Good references on finite element methods are Finlayson (1972) and Reddy (1984).

8.12 ORTHOGONAL COLLOCATION

When using the method of weighted residuals, the most important factor is the choice of the trial function. Important guidelines come from the boundary conditions and the symmetry of the problem. A further extension of the method is choosing collocation points as roots of orthogonal polynomials. This is known as the method of orthogonal collocation.

The basic property of any orthogonal polynomial is

$$\int_a^b w(x) P_n(x) P_m(x) dx = 0 \quad \text{when } n \neq m \quad (8.12.1)$$

Here $P_n(x)$ is a polynomial in x. If property (8.12.1) is satisfied, then the polynomial $P_n(x)$ is orthogonal over the domain $[a, b]$.

There are two polynomial forms that are often used to solve engineering problems. These are the shifted Legendre polynomials and the Jacobi polynomials. The shifted Legendre polynomials are used for problems without symmetry and the Jacobi polynomials for problems with symmetry.

8.12.1 Shifted Legendre Polynomials

We consider the spatial domain $x \in [0, 1]$ and that the problem has no special symmetry properties over that domain. In this case we choose a trial solution

$$y(x) = b + cx + x(1-x) \sum_{i=1}^{N} a_i P_{i-1}(x) \quad (8.12.2)$$

and we choose the polynomial $P_n(x)$ to be orthogonal so that

$$\int_0^1 w(x) P_n(x) P_m(x) dx = 0 \quad n \neq m \quad (8.12.3)$$

If the weighting function $w(x)$ is unity, then the polynomials $P_n(x)$ are called shifted Legendre polynomials. The trial functional form of equation (8.12.2) is chosen to handle second-order problems since there are $N + 2$ unknown constants $(b, c, a_1, ...a_N)$. The constants b and c are chosen to satisfy the two specified boundary conditions of a second-order differential equation, and the coefficients a_i are chosen to minimize the weighted residuals. The roots x_i for the shifted Legendre polynomials are given in Table 8.6 and they specify the collocation points to be used in the method of weighted residuals. A more complete set of roots can be found in Stroud and Secrest (1966). Villadsen (1970) gives a computer algorithm for their calculation.

For the collocation method, we will force the differential equation to be satisfied at the collocation points. In order to do so, we need to

Solution of Partial Differential Equations

Table 8.6: Roots (x_i) for Shifted Legendre Polynomials (Source: Finlayson, 1972)

$N = 1$	0.5000	$N = 5$	0.0469
			0.2308
$N = 2$	0.2113		0.5000
	0.7887		0.7692
			0.9531
$N = 3$	0.1127		
	0.5000	$N = 6$	0.0338
	0.8873		0.1694
			0.3807
$N = 4$	0.0694		0.6193
	0.3300		0.8306
	0.6700		0.9662
	0.9306		

evaluate the function and its derivatives at the collocation points. The trial function of equation (8.12.2) can also be written as

$$y(x) = \sum_{i=0}^{N+1} d_i x^{(i-1)} \qquad (8.12.4)$$

At a collocation point x_j we have

$$y(x_j) = \sum_{i=0}^{N+1} d_i x_j^{(i-1)} \qquad j = 1, ..., N \qquad (8.12.5)$$

The set of collocation points (N in number) and the boundary points (2 in number) can be expressed in terms of an unknown vector of coefficients **d** as

$$\mathbf{y} = \mathbf{Q}\,\mathbf{d} \qquad (8.12.6)$$

where $\mathbf{Q}$ is a known matrix and is given in terms of the known collocation point x_j raised to an appropriate collocation order, i,

$$Q_{ij} = x_j^{i-1} \qquad (8.12.7)$$

The vectors $\mathbf{y}$ and $\mathbf{d}$ are of order ($N + 2$) and the matrix $\mathbf{Q}$ of order ($N + 2, N + 2$).

The first derivative of the trial function at a collocation point is given by

$$\frac{dy_j}{dx} = \sum_{i=0}^{N+1} (i-1) x_j^{(i-2)} d_i \qquad (8.12.8)$$

The set of relations of equation (8.12.8) at the collocation and boundary points can be expressed as

$$\frac{d\boldsymbol{y}}{dx} = \boldsymbol{C}\,\boldsymbol{d} \qquad (8.12.9)$$

where $C_{ij} = (i-1)x_j^{(i-2)}$.

Using equation (8.12.6) for $\boldsymbol{d}$ gives

$$\frac{d\boldsymbol{y}}{dx} = \boldsymbol{C}\,\boldsymbol{Q}^{-1}\boldsymbol{y} \qquad (8.12.10)$$

or

$$\frac{d\boldsymbol{y}}{dx} = \boldsymbol{A}\,\boldsymbol{y} \qquad (8.12.11)$$

where $\boldsymbol{A} = \boldsymbol{C}\,\boldsymbol{Q}^{-1}$.

The second derivative of the trial function at a collocation point is

$$\frac{d^2 y_j}{dx^2} = \sum_{i=0}^{N+1} (i-1)(i-2)x_j^{(i-3)} d_i \qquad (8.12.12)$$

The set of second derivative relations at the collocation and boundary points can therefore be expressed as

$$\frac{d^2 \boldsymbol{y}}{dx^2} = \boldsymbol{D}\,\boldsymbol{d} \qquad (8.12.13)$$

where $D_{ij} = (i-1)(i-2)x_j^{(i-3)}$ or

$$\frac{d^2 \boldsymbol{y}}{dx^2} = \boldsymbol{B}\,\boldsymbol{y} \qquad (8.12.14)$$

where $\boldsymbol{B} = \boldsymbol{D}\,\boldsymbol{Q}^{-1}$.

For $N = 1$ and $N = 2$, the $\boldsymbol{A}$ and $\boldsymbol{B}$ matrices are given in Table 8.7.

8.12.2 Heat Conduction in an Insulated Bar

We consider the unsteady heat conduction in an insulated bar. The thermal energy balance can be transformed into the dimensionless form

$$\frac{\partial T}{\partial t} = \frac{\partial^2 T}{\partial y^2} \qquad (8.12.15)$$

with the boundary conditions

$$T(0,t) = T(1,t) = 0 \qquad (8.12.16)$$

and initial conditions

$$T(y,0) = 1 \qquad (8.12.17)$$

Table 8.7: First and Second Derivative Matrices Using Shifted Legendre Polynomials (Source: Finlayson, 1972)

$$N = 1 \quad A = \begin{pmatrix} -3 & 4 & -1 \\ -1 & 0 & 1 \\ 1 & -4 & 3 \end{pmatrix} \quad B = \begin{pmatrix} 4 & -8 & 4 \\ 4 & -8 & 4 \\ 4 & -8 & 4 \end{pmatrix}$$

$$N = 2 \quad A = \begin{pmatrix} -7 & 8.196 & -2.196 & 1 \\ -2.732 & 1.732 & 1.732 & -0.7321 \\ 0.7321 & -1.732 & -1.732 & 2.732 \\ -1 & 2.196 & -8.196 & 7 \end{pmatrix}$$

$$B = \begin{pmatrix} 24 & -37.18 & 25.18 & -12 \\ 16.39 & -24 & 12 & -4.392 \\ -4.392 & 12 & -24 & 16.39 \\ -12 & 25.18 & -37.18 & 24 \end{pmatrix}$$

Since we have not used the symmetry boundary condition that $dT/dy = 0$ at $y = 1/2$, we will solve the problem by performing an orthogonal collocation in the spatial domain using the shifted Legendre polynomials as the basis function. This will reduce the problem to a set of initial-value ordinary differential equations that can be solved using IMSL ordinary differential equation routines. As discussed by Cooper et al. (1986), seven internal collocation points accurately describe the solution to the partial differential equations. Therefore, we use equation (8.12.14) to approximate the second spatial derivative. This reduces the original partial differential equation of (8.12.15) to

$$\frac{dT}{dt} = BT \qquad (8.12.18)$$

where T is a vector of dimension 9 (the two boundary temperatures which are known, and seven internal temperatures at the internal collocation points). The seven internal collocation points are

$$y^T = (0.0254, 0.1292, 0.2971, 0.5000, 0.7029, 0.8708, 0.9746) \qquad (8.12.19)$$

Since the boundary conditions are always satisfied, $dT(1)/dt = dT(9)/dt = 0$, and $T(1) = T(9) = 0$, we only need to solve for seven differential equations

$$\frac{d\tilde{T}}{dt} = \tilde{B}\tilde{T} \qquad (8.12.20)$$

where $\tilde{T}$ is the vector of seven internal collocation points.

This set of linear differential equations is easily solved using the MATLAB routines ode45 or ode113. The solution of the problem is shown at two different times in Figure 8.38.

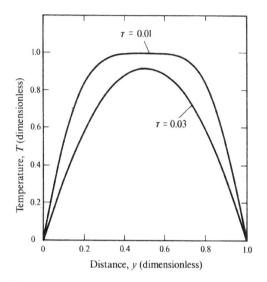

Figure 8.38: Temperature Profiles.

8.12.3 Jacobi Polynomials

For problems with symmetry, we construct a different set of orthogonal polynomials. We consider the problem defined over the spatial domain [0,1] and that the problem solution is symmetric about $x = 0$. That is to say that one of the boundary conditions is $dy/dx = 0$ at $x = 0$. We use a trial solution that is in terms of x^2 and assume a Dirichet boundary condition at $x = 1$. The trial function is

$$y(x) = y(x=1) + (1-x^2)\sum_{i=1}^{N} a_i P_{i-1}(x^2) \qquad (8.12.21)$$

We construct the polynomial $P_n(x^2)$ so that it satisfies the orthogonality condition

$$\int_0^1 w(x^2) P_n(x^2) P_m(x^2) x^{a-1} dx = 0 \qquad n \neq m \qquad (8.12.22)$$

The orthogonal polynomials described by equation (8.12.22) are Jacobi polynomials (Carnahan et al., 1969). The value of the parameter

Solution of Partial Differential Equations

a specifies the geometry for the problem. If $a = 1$, then we have cartesian coordinates, $a = 2$ is for cylindrical coordinates, and $a = 3$ is for spherical coordinates.

There are two common values for the weighting $w(x^2)$. These are $w = 1$ and $w = 1 - x^2$. Roots for these Jacobi polynomials are presented in Table 8.8 for all three geometries, both values of w and various orders of N. The roots presented in Table 8.8 specify specific collocation points to be used in the method of weighted residuals. The collocation method satisfies the describing differential equation at the collocation points. In order to implement this, we need to evaluate the function and its derivatives at the collocation points. The trial function of equation (8.12.21) can also be written as

$$y(x) = \sum_{i=1}^{N+1} d_i x^{2i-2} \qquad (8.12.23)$$

Table 8.8: Roots of Jacobi Polynomials (Source: Finlayson, 1972)

	planar geometry $a=1$		cylind. geometry $a=2$		spherical geometry $a=3$	
	$w=1$	$w=1-x^2$	$w=1$	$w=1-x^2$	$w=1$	$w=1-x^2$
N=1	0.57735	0.44721	0.70710	0.57735	0.77460	0.65465
N=2	0.33998	0.28523	0.45970	0.39377	0.53847	0.46885
	0.86114	0.76506	0.88807	0.80309	0.90618	0.83022
N=3	0.23862	0.20930	0.33571	0.29764	0.40585	0.36312
	0.66121	0.59170	0.70711	0.63990	0.74153	0.67719
	0.93247	0.87174	0.94197	0.88750	0.94911	0.89976
N=4	0.18343	0.16528	0.26350	0.23896	0.32425	0.29576
	0.52553	0.47792	0.57446	0.52616	0.61337	0.56524
	0.79667	0.73877	0.81853	0.76393	0.83603	0.78448
	0.96029	0.91953	0.96466	0.92749	0.96816	0.93400
N=5	0.14887	0.13655	0.21659	0.19952	0.26954	0.24929
	0.43340	0.39953	0.48038	0.44499	0.51910	0.48291
	0.67941	0.63288	0.70711	0.66179	0.73015	0.68169
	0.86506	0.81930	0.87706	0.83395	0.88706	0.84635
	0.97391	0.94490	0.97626	0.94946	0.97823	0.95331
N=6	0.12523	0.11633	0.18375	0.17122	0.23046	0.21535
	0.36783	0.34272	0.41158	0.38481	0.44849	0.42064
	0.58732	0.55064	0.61700	0.58050	0.64235	0.60625
	0.76990	0.72887	0.78696	0.74744	0.80156	0.76352
	0.90412	0.86780	0.91138	0.87706	0.91760	0.88508
	0.98156	0.95993	0.98297	0.96278	0.98418	0.96525

At a collocation point x_j, we have

$$y(x_j) = \sum_{i=1}^{N+1} d_i x_j^{2i-2} \qquad j = 1, N \qquad (8.12.24)$$

The set of collocation points (N) and the boundary point at $x = 1$ can be expressed as

$$\boldsymbol{y} = \boldsymbol{Q}\,\boldsymbol{d} \qquad (8.12.25)$$

where

$$Q_{ij} = x_j^{2i-2} \qquad (8.12.26)$$

The vectors $\boldsymbol{y}$ and $\boldsymbol{d}$ are of order $(N+1)$ and the matrix $\boldsymbol{Q}$ of order $(N+1, N+1)$.

The first derivative of the trial function at a collocation point is given by

$$\frac{dy_j}{dx} = \sum_{i=1}^{N+1} \left(\frac{dx^{2i-2}}{dx}\right)_j d_i \qquad (8.12.27)$$

The set of relations is

$$\frac{d\boldsymbol{y}}{dx} = \boldsymbol{C}\,\boldsymbol{d} \qquad (8.12.28)$$

or using equation (8.12.6), we have

$$\frac{d\boldsymbol{y}}{dx} = \boldsymbol{A}\,\boldsymbol{y} \qquad (8.12.29)$$

where

$$\boldsymbol{A} = \boldsymbol{C}\,\boldsymbol{Q}^{-1} \qquad (8.12.30)$$

The set of second derivatives at the collocation points and the boundary $x = 1$ can be expressed by the linear relation

$$\frac{d^2\boldsymbol{y}}{dx^2} = \boldsymbol{B}\,\boldsymbol{y} \qquad (8.12.31)$$

For $N = 1$ and $N = 2$, the matrices $\boldsymbol{A}$ and $\boldsymbol{B}$ are given in Table 8.9 for all three geometries and both $w = 1$ and $w = 1 - x^2$.

8.12.4 Diffusion in Spherical Coordinates

Let us consider the solution of an unsteady diffusion problem in spherical coordinates

$$\frac{\partial C}{\partial t} = \frac{1}{x^2} \frac{\partial}{\partial x}\left(x^2 \frac{\partial C}{\partial x}\right) \qquad (8.12.32)$$

with the initial condition

$$C = 0 \quad \text{at } t = 0 \qquad (8.12.33)$$

Solution of Partial Differential Equations

Table 8.9: Matrices of First and Second Derivatives for Orthogonal Collocation Using Jacobi Polynomials (Source: Finlayson, 1972)

Planar Geometry ($a = 1$)

$$N = 1 \quad A = \begin{pmatrix} -1.118 & 1.118 \\ -2.500 & 2.500 \end{pmatrix} \quad B = \begin{pmatrix} -2.5 & 2.5 \\ -2.5 & 2.5 \end{pmatrix}$$

$$N = 2 \quad A = \begin{pmatrix} -1.753 & 2.508 & -0.7547 \\ -1.371 & -0.6535 & 2.024 \\ 1.972 & -8.791 & 7 \end{pmatrix}$$

$$B = \begin{pmatrix} -4.740 & 5.677 & -0.9373 \\ 8.323 & -23.26 & 14.94 \\ 19.07 & -47.07 & 28 \end{pmatrix}$$

Cylindrical Geometry ($a = 2$)

$$N = 1 \quad A = \begin{pmatrix} -1.732 & 1.732 \\ -3 & 3 \end{pmatrix} \quad B = \begin{pmatrix} -6 & 6 \\ -6 & 6 \end{pmatrix}$$

$$N = 2 \quad A = \begin{pmatrix} -2.540 & 3.826 & -1.286 \\ -1.378 & -1.245 & 2.623 \\ 1.715 & -9.715 & 8 \end{pmatrix}$$

$$B = \begin{pmatrix} -9.902 & 12.30 & -2.397 \\ 9.034 & -32.76 & 23.73 \\ 22.76 & -65.42 & 42.67 \end{pmatrix}$$

Spherical Geometry ($a = 3$)

$$N = 1 \quad A = \begin{pmatrix} -2.291 & 2.291 \\ -3.5 & 3.5 \end{pmatrix} \quad B = \begin{pmatrix} -10.5 & 10.5 \\ -10.5 & 10.5 \end{pmatrix}$$

$$N = 2 \quad A = \begin{pmatrix} -3.199 & 5.015 & -1.816 \\ -1.409 & -1.807 & 3.215 \\ 1.697 & -10.70 & 9 \end{pmatrix}$$

$$B = \begin{pmatrix} -15.67 & 20.03 & -4.365 \\ 9.965 & -44.33 & 34.36 \\ 26.93 & -86.93 & 60 \end{pmatrix}$$

and boundary conditions

$$C = 1 \quad \text{at} \quad x = 1 \tag{8.12.34}$$

$$\frac{\partial C}{\partial x} = 0 \quad \text{at} \quad x = 0 \tag{8.12.35}$$

Because of the last boundary condition, we have symmetry at $x = 0$ and the Jacobi polynomials will be used with $a = 3$ because of the spherical geometry. We will use the weighting of $w = 1$.

The basic diffusion equation of (8.12.32) can be written as

$$\frac{\partial C}{\partial t} = \nabla^2 C \tag{8.12.36}$$

The set of second derivatives at the collocation points and the $x = 1$ boundary are given by (8.12.31) as

$$\nabla^2 \boldsymbol{y} = \boldsymbol{B}\, \boldsymbol{y} \tag{8.12.37}$$

where $\boldsymbol{B}$ is given in Table 8.9 with $a = 3$ and $w = 1$.

The orthogonal collocation solution is at all collocation points and the boundary $x = 1$ is

$$\frac{d\boldsymbol{C}}{dt} = \boldsymbol{B}\, \boldsymbol{C} \tag{8.12.38}$$

This is a set of $N+1$ differential equations but since the condition at $x = 1$ is constant, we have $dC_{N+1}/dt = 0$ and (8.12.36) can be reduced to the N differential equations

$$\frac{d\boldsymbol{C}}{dt} = \boldsymbol{B}\, \boldsymbol{C} + \boldsymbol{f} \tag{8.12.39}$$

where

$$f_j = B_{j,N+1}\, C_{N+1} \quad \text{with} \quad C_{N+1} = 1 \tag{8.12.40}$$

The linear set of ordinary differential equations is easily solved using standard MATLAB ordinary differential equation subroutines.

As a convenient way to evaluate the accuracy of the solution, we will focus on the mass flux at the surface $x = 1$. The mass flux is given by

$$J = \nabla C \tag{8.12.41}$$

and from equation (8.12.29) the mass flux at the collocation points and the boundary $x = 1$ is

$$\boldsymbol{J} = \boldsymbol{A}\, \boldsymbol{C} \tag{8.12.42}$$

The flux at the boundary $x = 1$ is

$$J_{N+1} = \sum_{i=1}^{N+1} A_{N+1,i}\, C_i \tag{8.12.43}$$

The results for $N = 1, 2$, and 6 are given in Figure 8.39. As can be seen, it takes approximately six collocation points in order to obtain an accurate solution.

Solution of Partial Differential Equations

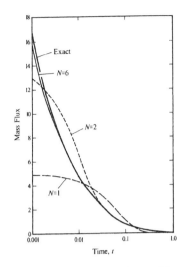

Figure 8.39: Comparison Results.

8.12.5 Summary

Experience has shown that low-order weighted residual approximations work well when the solution varies smoothly over the entire domain of the trial function. When there are rapid changes over a varying portion of the domain, then the method requires high-order approximations. In that case, the finite-difference approach is more effective. Convective-diffusion problems are particularly difficult for the weighted residual method, whereas pure conduction problems can be treated very efficiently by this approach.

PROBLEMS

8.1. The Kiil artificial kidney is a flat-plate dialyzer used to remove undesirable toxic materials such as urea from the bloodstream. In this flat-plate design, blood flows in a membrane envelope and the dialysate (cleaning solution) flows between the membrane and grooved polypropylene boards. The advantages of this dialyzer are that the priming volume is low, blood loss is low, and resistance to blood flow is low (thus blood pumps are usually not required). Figure 8.40 shows the pertinent concentration and velocity profiles for this artificial kidney design.

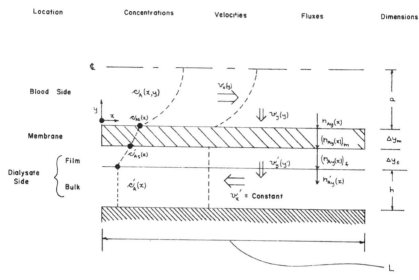

Figure 8.40: Artificial Kidney Schematic.

Derive the material balance equation for both the blood and dialysate side with their appropriate boundary conditions. The velocity profiles for the blood side are

$$v_x(y) = \frac{3}{2}\bar{v}_x \left(\frac{y}{a}\right)\left(2 - \frac{y}{a}\right) \qquad v_y(y) = -V_w\left(1 - \frac{y}{a}\right)$$

The velocity profiles on the dialysate side are

$$v'_x = 6\bar{v}'_x \left(\frac{y'}{a'}\right)\left(1 - \frac{y'}{a'}\right) \qquad v'_y = V_w\left(1 - \frac{y'}{a'}\right)$$

The membrane is characterized by parameters which reflect its resistance to mass transfer:

$$(n_{Ay})_m = P(C_{AS} - C'_{AS}) + \frac{T_R}{2}V_w(C_{AS} + C'_{AS})$$

where P is the permeability and T_R the transmission coefficient.
By continuity of flux at the membrane, we have

$$n_{Ay}(x)\,|_{y=0} = n_{Ay}(x)_m = n'_{Ay}(L - x)\,|_{y'=0}$$

Via finite–difference techniques, solve the resulting mathematical model for the Kiil dialyzer. For your simulation results, use the following dimensionless variables:

$$C = \frac{C_A}{C_{in}} \qquad X = \frac{x}{L} \qquad Y = \frac{y}{a} \qquad Y' = \frac{y'}{a'}$$

$$V_x = \frac{v_x}{\langle v_x \rangle} \qquad V_y = \frac{v_y}{V_w} \qquad V'_x = \frac{v'_x}{\langle v'_x \rangle} \qquad V'_y = \frac{v'_y}{V_w}$$

The Peclet numbers are

$$Pe_m = \frac{a^2 v_x}{L \mathcal{D}_{AB}} \qquad Pe'_m = \frac{(a')^2 v'_x}{L \mathcal{D}_{AB}} \qquad Pe_y = \frac{a V_w}{\mathcal{D}_{AB}} \qquad Pe'_y = \frac{a' V_W}{\mathcal{D}_{AB}}$$

Typical values are

$$Pe_m = 0.5 \qquad Pe_y = 0.01 \qquad \frac{a^2}{L^2} = 10^{-8}$$

$$Pe'_m = 5 \qquad Pe'_y = 0.07 \qquad \frac{(a')^2}{L^2} = 5 \times 10^{-6}$$

8.2. Dynamic Adsorption of Surfactants

In this problem, the mass transport of a surfactant during miscible displacement in a saturated medium is evaluated. Besides convection and dispersion, the additional mechanism which must be considered is adsorption. Most studies have assumed that the adsorption mechanism occurs rapidly compared with the mechanisms of convection and dispersion. Thus, an equilibrium isotherm usually of the Langmuir type is assumed.

$$\Gamma = \frac{C_b}{M + N C_b}$$

Γ = amount of surfactant adsorbed per surface area (g mol/cm^2)

C_b = bulk surfactant concentration (g mol/cm^3)

Trogus et al. (1977) have proposed a kinetic model

$$\frac{d\Gamma}{dt} = k_1 C_b (Q_a - \Gamma) = k_2 \Gamma$$

where k_1 = adsorption rate constant
k_2 = desorption rate constant
Q_a = total adsorbent capacity of the solid

At equilibrium

$$N = \frac{1}{Q_a} \qquad M = \frac{1}{K_e Q_a}$$

where $K_e = K_1/k_2$ = equilibrium constant.

Show that a one-dimensional convective–dispersion model with dynamic adsorption becomes (Ramirez et al., 1980)

$$\frac{\partial C_D}{\partial \tau_{cD}} + \frac{\partial C_D}{\partial z_D} = \frac{1}{N_{Pe_s}} \frac{\partial^2 C_D}{\partial z_D^2} - N_{La} \frac{d\Gamma_D}{d\tau_{cD}}$$

$$\tau_{cD}^* \frac{d\Gamma_D}{d\tau_{cD}} = C_D(1 - \Gamma_D) - E\Gamma_D$$

where

N_{Pe_s}	$= \dfrac{v_w L}{K}$	K	$=$	dispersion coefficient
τ_{cD}^*	$= \dfrac{v_w}{k_1 C_{0s} L}$	C_{0s}	$=$	inlet surfactant concentration
E	$= \dfrac{1}{K_e C_{0s}}$	A_{wr}	$=$	rock area/unit volume
N_{La}	$= \dfrac{A_{wr} Q_a}{\phi C_{0s}}$	ϕ	$=$	porosity
C_D	$= C_b/C_{0s}$	τ_{cD}	$=$	$t\, v_w/L$
τ_{cD}^*	$= \dfrac{v_w}{k_1 C_{0s} L}$	z_D	$=$	z/L
Γ_D	$= \Gamma/\Gamma_{max}$			

The boundary conditions are

$$C_D = 1 \quad \text{at } z_D = 0$$

$$\frac{dC_D}{dz_D} = 0 \quad \text{at } z_D = 1$$

and the initial condition at $\tau_{cD} = 0 \qquad C_D = 0,\ C_{s0} = 0,\ \Gamma_D = 0$

Solve the model using the Crank–Nicolson finite–difference scheme after quasilinearization.

Typical Parameters

$$\tau_{cD}^* = 0.12, \quad N_{La} = 1.53, \quad N_{Pe} = 256, \quad E = 0.178$$

and

$$\tau_{cD}^* = 0.041, \quad N_{La} = 1.61, \quad N_{Pe} = 308, \quad E = 0.174$$

8.3. Two–Phase Flow and Surfactant Flooding

Develop a finite–difference solution to the following one–dimensional surfactant flooding problem (Fathi and Ramirez, 1984). You should use the corner scheme for the saturation equation and the Crank–Nicolson scheme for the concentration equation with quasilinearization.

Show that the saturation equation becomes

$$\frac{\partial S_w}{\partial t} = -\frac{\partial f_w}{\partial z}$$

$$f_w = \frac{1}{1 + \frac{k_{ro}\mu_w}{\mu_o k_{rw}}}$$

where S_w = water saturation
t = $\theta Q_T/AL\theta$
θ = time
Q_T = total volumetric flow rate
A = cross–sectional area
ϕ = porosity
L = length of core
z = length/core length

and the surfactant balance is

$$\left[S_w + \frac{(1-S_w)}{K_s}\right]\frac{\partial C}{\partial t} + \left[f_w + \frac{(1-f_w)}{K_s}\right]\frac{\partial C}{\partial z} = \frac{1}{N_{Pe}}\frac{\partial^2 C}{\partial z^2} - \frac{A_{wr}}{\phi C_o}\frac{\partial \Gamma}{\partial t}$$

where K_s = surfactant partition coefficient
C = C_m/C_o
C_m = surfactant concentration
C_o = characteristic surfactant concentration
N_{Pe} = $v_s L/K$
A_{wr} = specific surface area of core

with $\Gamma = \dfrac{C_o C}{M + N\, C_o C}$ M and N are constants.

The initial conditions are

$$C = 0, \quad S_w = 1 - S_{OR\phi} \quad \text{at } t = 0$$

where $S_{OR\phi}$ is the initial residual oil saturation The boundary conditions are

$$f_w = 1 \quad \text{at } z = 0, \quad \text{or} \quad S_w = 1 - S_{OR} \quad \text{at } z = 0$$

$$\frac{\partial C}{\partial z} = N_{Pe}(C - C_{slug}) \quad \text{at } z = 0$$

$$\frac{\partial C}{\partial z} = 0 \quad \text{at } z = 1$$

Consider the following representative data for surfactant flooding. The interfacial tension as a function of surfactant concentration is given as three straight line segments on a semilog plot. See Figure 8.41.

The residual oil saturation is expressed as a function of the capillary number

$$N_{cap} = \frac{v_w \mu_a}{\sigma_{wo}}$$

A specific correlation is shown in Figure 8.42, where

$$S_{OR} = 0.280 + 0.187 \ln N_{cap} + 0.0365 (\ln N_{cap})^2$$
$$+ \; 0.00176 \, (\ln N_{cap})^3 \qquad 5 \times 10^{-5} \leq N_{cap} \leq 0.02$$

The relative permeabilities are

$$k_{rw} = (S_w - S_{OR})^2$$

$$k_{ro} = \exp(4.5 S_{OR})(1.0 - S_{OR} - S_w)^2$$

if $S_w \geq 1 - S_{OR}$ then $k_{ro} = 0$
if $S_w \leq S_{will}$ then $k_{rw} = 0$
Also,

ϕ	$= 0.225$
A	$= 10 \text{ cm}^2$
L	$= 100 \text{ cm}$
A_{wr}	$= 15000 \text{ cm}^{-1}$
μ_o	$= 3$
C_o	$= 4000 \text{ ppm}$
K_s	$= 10$
MW_{surf}	$= 400$
N_{Pe}	$= 50$
v_w	$= 4.445 \times 10^{-3} \text{ cm/s}$

Solution of Partial Differential Equations

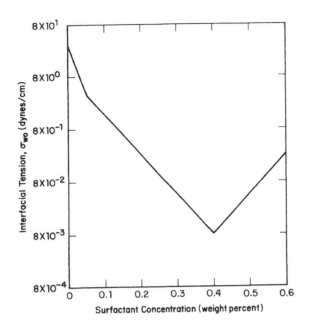

Figure 8.41: Hypothetical Interfacial Tension Behavior.

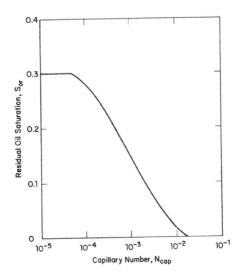

Figure 8.42: Hypothetical Behavior of Oil Residual Saturation vs. the Capillary Number.

REFERENCES

Amundson, N. R., *Can. J. Ch.E.* **43**, 99 (1965).

Aziz, K. and Settari, A., *Petroleum Engineering Simulation*, Applied Science Publishers, London (1979).

Carnahan, B., Luther, H. A., and Wilkes, J. O., *Applied Numerical Methods*, Wiley, New York (1969).

Clough, D. E. and Ramirez, W. F., "Stability of Tubular Reactors," *Simulation* **16**, No. 5, 207–216 (May 1971).

Cooper, D. J., Ramirez, W. F., and Clough, D. E., "Comparison of Linear Distributed Filters to Lumped Approximations," *AIChE J.* **32**, No. 2, 186–194 (1986).

Fathi, Z. and Ramirez, W. F., "Optimal Injection Policies for Enhanced Oil Recovery: Part 2—Surfactant Flooding," *Soc. Pet. Engr. Journal* **30**, No. 4, 333–342 (June 1984)..

Finlayson, B. A., *The Method of Weighted Residuals and Variational Principles*, Academic Press, New York (1972).

Friedman, F. and Ramirez, W. F., "A Single Phase Model of Mechanisms Effecting Miscible Oil Recovery," *Chem. Engr. Sci.* **32**, 687 (1977).

Greenkorn, R. A., *Flow Phenomena in Porous Media*, Dekker, New York (1983).

McCracken, E. A., Leefe, C. L., and Weaver, R. C. C., *Chemical Engineering Progress Symposium Series 101* **66**, 37 (1970).

Muskat, M., *Physical Principles of Oil Production*, McGraw–Hill, New York (1949).

Peaceman, D. W., *Fundamentals of Numerical Reservoir Simulation*, Elsevier, New York (1977).

Ramirez, W. F., Shuler, P. J., and Friedman, F., "Convection, Dispersion, and Adsorption of Surfactants in Porous Media," *Soc. of Pet. Engrs. J.* **20**, No. 6, 430–438 (December 1980).

Reddy, J. N., *An Introduction to the Finite Element Method*, McGraw–Hill (1984).

Slattery, J. C., *Momentum, Energy, and Mass Transfer in Continua*, McGraw–Hill, New York (1972).

Stroud, A. and Secrest, D., *Gaussian Quadrature Formulas*, Prentice–Hall, Englewood Cliffs, N.J. (1966).

Taranchuk, V. B., "Numerical Method for Calculating Pressure and Saturation in One–Dimensional Displacement of Oil by Water," *Chislenny Methody Mechaniki Sploshnoy Sredy. U.S.S.R., Novasibirsk* **5**, No. 3, 88–95 (1974).

Trogus, F., Sophany, T., Schechter, R. S., and Wade, W. H., "Static and Dynamic Adsorption of Anions and Nonionic Surfactants," *SPE Journal* **17**, (Oct. 1977), 337–344.

Villadsen, J., *Selected Approximation Methods for Chemical Engineering Problems*, Ph.D. Thesis, Copenhagen (1970).

Van Heerden, C., *Chem. Engr. Sci.*, 133 (1958).

Von Rosenberg, D. U., *Methods for Numerical Solution of Partial Differential Equations*, Elsevier, New York (1969).

NOMENCLATURE

A	area (cm^2)
$\hat{A}$	Helmholtz free energy per unit mass (cal/g)
A	Helmholtz free energy (cal)
$\boldsymbol{A}$	linear matrix
A_q	transmissibility of phase q
$\boldsymbol{b}$	arbitrary vector
B_i	formation volume factor for phase i
C	Courant number
c_i	phase compressibility
C_i	molar concentration (g mol/cm^3)
C_p	heat capacity at constant pressure
C_t	total compressibility
C_v	heat capacity at constant volume
C_v	value coefficient
D	diameter (cm)
$\tilde{D}$	dispersion coefficient (cm^2/sec)
$\mathcal{D}$	diffusivity (cm^2/sec)
D/Dt	substantial derivative
E	total energy (cal)
E	activation energy (cal/°C)
$\hat{E}$	total energy per unit mass (cal/g)
E_v	irreversible energy losses (cal/sec)
f	Fanning friction factor
f	fugacity
$\boldsymbol{f}$	force (dynes)
$\boldsymbol{f}$	arbitrary nonlinear function
f_i	fractional flow of species i
F	volumetric flow rate (cm^3/sec)
$\tilde{F}$	partial molar Gibbs free energy
g	acceleration of gravity (cm/sec^2)
G	Gibbs free energy (cal)
$\hat{G}$	Gibbs free energy per unit mass (cal/g)
h	heat transfer coefficient (cal/cm^2 °C)
h_f	friction head loss (cm^2/sec^2)
h_L	head loss due to friction (cm^2/sec^2)
H	enthalpy (cal)
H	hold-up (g or g mol)
H	heat of reaction (cal/g mol)
$\hat{H}$	enthalpy per unit mass (cal/g)
$I_i(x)$	Bessel function of x of order i
$\boldsymbol{j}$	diffusional flux (g/cm^2 sec)
$\boldsymbol{J}^*$	diffusional flux (g mol/cm^2 sec)
k	rate constant

k	thermal conductivity
$\tilde{k}$	thermal dispersion coefficient
k_0	frequency factor
k_{ri}	relative permeability of species i
K	permeability (cm^2)
K_i	Michaelis constant (g mol/m^3)
K_i'	inhibition constant (g mol/m^3)
L	liquid flow rate (g mol/sec)
L	length (cm)
m	mass (g)
M	total mass (g)
MW	molecular weight
$\mathbf{n}$	normal vector
$\mathbf{n}_i$	mass flux (g/cm^2 sec)
N_i	number of moles (mol)
$\mathbf{N}_i$	molar flux (g mol/cm^2 sec)
NTU	number of transfer units
p	pressure (dynes/cm^2)
P_e	Peclet number
P_i	partial pressure (dynes/cm^2)
PI	performance index for a well
q	mass flow rate of coolant (cm/sec)
q	rate of heat (cal/sec)
q_i	injection rate of species i (g/sec)
Q	heat transfer rate (cal/sec)
Q_{rev}	heat transfer rate of reversible process (cal/sec)
r	rate of generation per unit volume (g/cm^3 sec)
r	radial direction (cm)
$\mathbf{r}$	resultant force (dynes)
R	gas constant
R	radius (cm)
R	molar rate of generation per unit volume (g mol/cm^3 sec)
Re	Reynolds number
R_{ij}	solubility of phase i in phase j at standard conditions
R_{ji}	stoichiometric yield of j per mole of i
S	cross-sectional area (cm^2)
S_i	saturation of species i
t	time (sec)
T	temperature (°C)
T_c	coolant temperature (°C)
u	general state variable
U	overall heat-transfer coefficient (cal/cm^2 °C)
$\hat{U}$	internal energy per unit mass (cal/g)
v	velocity (cm/sec)
v	vapor flow rate (g mol/sec)
$\langle v \rangle$	average velocity (cm/sec)

Nomenclature

V	volume (cm^3)
V	vapor flow rate (g mol/sec)
$\hat{V}$	volume/mass (cm^3/g)
V_i	maximum reaction velocity (hr^{-1})
w	mass flow rate (g/sec)
$\dot{W}$	rate of work (Jules/sec)
$\hat{W}$	work per unit mass (J/g)
$\dot{W}_n$	viscous work rate (J/sec)
$\dot{W}_s$	rate of mechanical work (J/sec)
$\dot{W}_{rev}$	rate of work of reversible process (J/sec)
x	liquid phase mole fraction
x	distance coordinate (cm)
$\boldsymbol{x}$	process or state variables
X	biomass concentration (g mol/m^3)
y	vapor phase mole fraction
y	distance coordinate (cm)
z	distance coordinate (cm)
z_i	compressibility factor for species i

Greek Letters

α	angle (radians)
ϕ	porosity
γ_i	activity coefficient of component i
ϵ	porosity of packed bed
ϵ	error
ϵ	effectiveness factor
Δ	difference operator
ΔH	heat of reaction (J/mol)
ΔH_F	heat of fermentation (J/mol)
η_p	pump efficiency
η_t	turbine efficiency
∇	backward difference operator
∇	gradient operator
λ	heat of vaporization (cal/g)
μ	viscosity (cp)
$\tilde{\mu}$	viscous dispersion coefficient
μ_i	specific growth rate of species i (hr^{-1})
ω	mass fraction
ϕ	spherical angle (radians)
ρ	density (g/cm^3)
τ	residence time (sec)
$\boldsymbol{\tau}$	shear stress tensor
θ	angle (radians)

Superscript

$*$	dimensionless variable

Appendix A
ANALYTICAL SOLUTIONS TO ORDINARY DIFFERENTIAL EQUATIONS

A.1 FIRST–ORDER EQUATIONS

First–order equations have the form

$$P(t,x)dt + Q(t,x)dx = 0 \qquad (A-1)$$

They arise in solving single macroscopic balance equations. Several special forms will be considered.

A.1.1 Separation of Variables

When the variables are separable, then the general first–order equation (A-1) can be simplified to

$$P(t)dt + Q(x)dx = 0 \qquad (A-2)$$

The general solution to this equation is

$$\int P(t)dt + \int Q(x)dx = C \qquad (A-3)$$

where C is a constant whose value is specified by the boundary data of the problem.

EXAMPLE A.1.1
Find the solution to

$$(1-x^2)dt = x(1-t)dx \qquad (A-4)$$

This can be placed in the general separable form of equation (A-3) as

$$\left(\frac{1}{1-t}\right)dt - \left(\frac{x}{1-x^2}\right)dx = 0 \qquad (A-5)$$

Performing the simple integrations gives

$$-2\ \ln(1-t) + \ln(1-x^2) = C \qquad (A-6)$$

This equation can be rearranged as

$$(1-t)^2 = \lambda(1-x^2) \qquad (A-7)$$

where $\lambda = e^{-C}$

A.1.2 Exact Differential Equations

If the general first-order differential equation of (A-1) corresponds to

$$dF = \frac{\partial F}{\partial t} dt + \frac{\partial F}{\partial x} dx = 0 \qquad (A-8)$$

then the solution to $dF = 0$ is

$$F(t, x) = C \qquad (A-9)$$

where C is a constant.

Since

$$\frac{\partial F}{\partial t} = P(t, x) \qquad (A-10)$$

and

$$\frac{\partial F}{\partial x} = Q(t, x) \qquad (A-11)$$

then because

$$\frac{\partial^2 F}{\partial t \partial x} = \frac{\partial^2 F}{\partial x \partial t} \qquad (A-12)$$

we must have as a necessary condition for the existence of an exact differential equation

$$\frac{\partial P}{\partial x} = \frac{\partial Q}{\partial t} \qquad (A-13)$$

When condition (A-13) holds then the solution to the differential equation is found by integrating equation (A-10) with respect to t, holding x constant, to give

$$F = \int P(t, x) \, dt + f(x) \qquad (A-14)$$

where $f(x)$ is determined so as to satisfy equation (A-11).

Since

$$\frac{\partial F}{\partial x} = \frac{\partial}{\partial x} \left(\int P(t, x) \, dt + f(x) \right) = Q(t, x) \qquad (A-15)$$

Solving (A-15) for $f(x)$ gives the general formula,

$$f(x) = \int \left(Q(t, x) - \frac{\partial}{\partial x} \int P(t, x) \, dt \right) dx \qquad (A-16)$$

Appendix A: Analytical Solutions

EXAMPLE A.1.2
Find the solution to

$$(2tx + 1)dt + (t^2 + 4x)dx = 0 \qquad (A-17)$$

This is an exact equation since the test of equation (A–13) holds,

$$\frac{\partial P}{\partial x} = \frac{\partial Q}{\partial t} = 2t \qquad (A-18)$$

To find the solution we first compute the constant F from equation (A–14) as

$$F = \int (2tx + 1)\, dt + f(x)$$

or

$$F = t^2 x + t + f(x) \qquad (A-19)$$

The function $f(x)$ is obtained from equation (A–16),

$$f(x) = \int \left[(t^2 + 4x) - \frac{\partial}{\partial x}(t^2 x + t) \right] dx$$

or

$$f(x) = \int 4x\, dx$$

$$f(x) = 2x^2 \qquad (A-20)$$

Therefore, the solution is

$$F = \text{const.} = t^2 x + t + 2x^2 \qquad (A-21)$$

A.1.3 Homogeneous Equations

For the general first–order equation (A–1) if $P(t, x)$ and $Q(t, x)$ have the property that substitution of λt for t and λx for x converts them into the expressions

$$\lambda^n\, P(t, x) \quad \text{and} \quad \lambda^n\, Q(t, x)$$

then equation (A–1) is homogeneous and can be reduced to a separable form by the transformation

$$\frac{x}{t} = u \qquad (A-22)$$

or

$$\frac{t}{x} = v \qquad (A-23)$$

EXAMPLE A.1.3
Solve the first–order equation

$$x^2\, dt + (t^2 - tx)dx = 0 \qquad (A-24)$$

Substituting λt for t and λx for x gives

$$\lambda^2 x^2 \, dt + \lambda^2(t^2 - tx)dx = 0 \qquad (A-25)$$

which means that the equation is homogeneous and we seek a separation of variables solution using $u = x/t$ or $v = t/x$.

Rearranging equation (A–24) gives

$$\frac{dx}{dt} = \frac{\left(\frac{x}{t}\right)^2}{\left(\frac{x}{t}\right) - 1} \qquad (A-26)$$

Therefore, we make the substitution

$$u = \frac{x}{t} \quad \text{or} \quad x = ut \qquad (A-27)$$

Using the chain rule and direct substitution, equation (A–26) becomes,

$$\frac{dx}{dt} = u + t\frac{du}{dt} = u^2/(u-1) \qquad (A-28)$$

or

$$\frac{dt}{t} + \frac{1-u}{u} du = 0 \qquad (A-29)$$

which integrates directly to

$$\ln t + \ln u - u = C \qquad (A-30)$$

or

$$\ln t + \ln\left(\frac{x}{t}\right) - \left(\frac{x}{t}\right) = C \qquad (A-31)$$

A.1.4 Linear First–Order Equations

A first–order equation is linear in x if it can be placed in the form

$$\frac{dx}{dt} + R(t)x = M(t) \qquad (A-32)$$

We will seek an integrating factor $\mu = f(t)$ which will transform the equation into an exact equation. If we let

$$\ln \mu = \int R(t) \, dt \qquad (A-33)$$

then the solution to equation (A–32) is

$$\mu x = \int \mu M(t) \, dt + C \qquad (A-34)$$

We can show this by letting

$$x = \eta v$$

Appendix A: Analytical Solutions

where η and v are functions to be determined later.
Equation (A–32) becomes

$$\eta \frac{dv}{dt} + v \frac{d\eta}{dt} + R(t)\eta v = M(t)$$

or

$$v \left[\frac{d\eta}{dt} + R(t)\eta \right] + \eta \frac{dv}{dt} = M(t)$$

We force the term in parenthesis to zero by letting

$$\frac{d\eta}{dt} + R(t)\eta = 0$$

This is a separable differential equation with its solution given by

$$\ln \eta = -\int R(t)dt$$

Now the solution to

$$\eta \frac{dv}{dt} = M(t)$$

is

$$v = \int \frac{1}{\eta} M(t)\, dt + C$$

or

$$x/\eta = \int \frac{1}{\eta} M(t)\, dt + C$$

If $\mu = \frac{1}{\eta}$, then

$$\ln \mu = \int R(t)dt$$

and

$$\mu x = \int \mu M(t)dt + C$$

which is equation (A–34), the solution to the linear differential equation with μ defined by equation (A–33).

EXAMPLE A.1.4
Solve the linear first–order equation

$$t^2 dx + (2tx - t + 1)dt = 0 \qquad (A-35)$$

This equation can be placed in the linear form of equation (A–32) as

$$\frac{dx}{dt} + \left(\frac{2}{t}\right) x = \frac{t-1}{t^2} \qquad (A-36)$$

The integrating factor μ is defined by

$$\ln \mu = \int \frac{2}{t} \, dt = 2\ln t \qquad (A-37)$$

or

$$\mu = t^2 \qquad (A-38)$$

The solution of the differential equation is given by equation (A–34) to be

$$t^2 x = \int t^2 \, \frac{(t-1)}{t^2} \, dt + C \qquad (A-39)$$

or

$$x = \frac{1}{2} - \frac{1}{t} + \frac{C}{t^2} \qquad (A-40)$$

A.2 N^{th} ORDER LINEAR DIFFERENTIAL EQUATIONS WITH CONSTANT COEFFICIENTS

In order to consider a general n^{th} order differential equation with constant coefficients, we introduce the linear differential operator, D, which is defined such that

$$Dx = \frac{dx}{dt}, \qquad (A-41)$$

$$D^2 x = \frac{d^2 x}{dt^2}, \qquad (A-42)$$

and

$$\frac{1}{D} x = \int x \, dt + C \qquad (A-43)$$

Expressions for higher–order derivatives follow by analogy. This differential operator is linear because the Distributive Law holds:

$$D(\alpha x_1 + \beta x_2) = \alpha D(x_1) + \beta D(x_2) \qquad (A-44)$$

where α and β are constants.

Of interest is to compute $D \, e^{\lambda t} f(t)$ where $f(t)$ is an arbitrary function of t. Using the chain rule, we obtain

$$D \, e^{\lambda t} f(t) = D\left(e^{\lambda t} f(t)\right) = \lambda e^{\lambda t} f(t) + e^{\lambda t} f'(t)$$

or

$$D \, e^{\lambda t} f(t) = e^{\lambda t} (\lambda + D) f(t) \qquad (A-45)$$

In general, we have the Commuting Law for moving $e^{\lambda t}$ outside of the D operator

$$D^n \, e^{\lambda t} f(t) = e^{\lambda t} \, (D + \lambda)^n f(t) \qquad (A-46)$$

A.2.1 Homogeneous Solutions

A general n^{th} order homogeneous differential equation with constant coefficients has the form

$$(D - \lambda_1)(D - \lambda_2) \cdots (D - \lambda_n) \, x(t) = 0 \qquad (A-47)$$

The solution of equation (A–47) depends upon the nature of the roots λ_i.

Case 1 Distinct λ_i's

When the roots λ_i are distinct, we seek the solution to

$$(D - \lambda_i) \, x(t) = 0 \qquad (A-48)$$

If we let

$$x(t) = e^{\lambda_i t} \, z(t) \qquad (A-49)$$

then

$$(D - \lambda_i) e^{\lambda_i t} z(t) = 0 \qquad (A-50)$$

and from the Commuting Law (A–46) we have

$$e^{\lambda_i t} \, (D - \lambda_i + \lambda_i) \, z(t) = 0 \qquad (A-51)$$

or

$$Dz(t) = 0 \qquad (A-52)$$

which has the solution

$$z(t) = A_i \qquad \text{where } A_i = \text{constant} \qquad (A-53)$$

Therefore, the solution of equation (A–48) is

$$x(t) = A_i e^{\lambda_i t} \qquad (A-54)$$

and the solution to

$$(D - \lambda_i)(D - \lambda_2) \cdots (D - \lambda_n) \, x(t) = 0 \qquad (A-55)$$

from the principle of superposition is

$$x = A_1 e^{\lambda_1 t} + A_2 e^{\lambda_2 t} + \cdots + A_n e^{\lambda_n t} \qquad (A-56)$$

Case 2 Multiple Roots

If we have a multiple root such as

$$(D - \lambda_i)^2 \, x(t) = 0 \qquad (A-57)$$

we again will let
$$x(t) = e^{\lambda_i t} z(t) \qquad (A-49)$$
but now equation (A-57) becomes
$$(D - \lambda_i)^2 \, e^{\lambda_i t} z(t) = 0 \qquad (A-58)$$
which from the Commuting Law is
$$e^{\lambda_i t} \, D^2 z(t) = 0 \qquad (A-59)$$
which has the solution
$$z(t) = A_i t + B_i \qquad (A-60)$$
Therefore the solution to equation (A-57) is
$$x(t) = A_i t e^{\lambda_i t} + B_i e^{\lambda_i t} \qquad (A-61)$$

Case 3 Complex Roots

Complex roots always occur in conjugate pairs so that we have a solution
$$x(t) = A e^{(a+bi)t} + B e^{(a-bi)t} \qquad (A-62)$$
However, we want real and not complex solutions. With the aid of Euler's formula (Sokolnikoff and Redheffer, 1958), it can be shown that equation (A-62) becomes
$$x(t) = A e^{at} \cos bt + B e^{at} \sin bt \qquad (A-63)$$

A.2.2 General Solutions

A general n^{th} order differential equation is of the form
$$(D - \lambda_1)(D - \lambda_2) \cdots (D - \lambda_n) \, x(t) = R(t) \qquad (A-64)$$
and its solution is the sum of the homogeneous solution and a particular integral,
$$x(t) = x_h(t) + x_p(t) \qquad (A-65)$$
Several types of particular integrals will be considered.

Case 1 Polynomial Particular Integrals

When equation (A-64) is of the form
$$(D - \lambda_1)(D - \lambda_2) \cdots (D - \lambda_n) \, x(t) = P(t) \qquad (A-66)$$

Appendix A: Analytical Solutions

where $P(t)$ is a polynomial, then the method of undetermined coefficients can be used. We assume that the particular solution $x_p(t)$ is also a polynomial of equal order and equate terms of the same order in the independent variable to determine the coefficients of the polynomial solution form.

EXAMPLE A.2.1
Find the general solution to

$$\frac{d^2x}{dt^2} + 9x = 2t^2 + 4t + 7 \qquad (A-67)$$

The homogeneous solution is given by

$$(D^2 + 9)\, x_h = 0 \qquad (A-68)$$

which has the solution

$$x_h = A\cos 3t + B\sin 3t \qquad (A-69)$$

For a particular solution we assume the polynomial

$$x_p = a_0 t^2 + a_1 t + a_2 \qquad (A-70)$$

Substituting this solution into (A–67) gives

$$2a_0 + 9(a_0 t^2 + a_1 t + a_2) = 2t^2 + 4t + 7 \qquad (A-71)$$

Equating coefficients of t^2, t, and the constant term t^0 gives

$$9a_0 = 2 \quad \text{or} \quad a_0 = 2/9$$
$$9a_1 = 4 \quad \text{or} \quad a_1 = 4/9$$
$$2a_0 + 9a_2 = 7 \quad \text{or} \quad a_2 = 59/81$$

Therefore the general solution to the equation is

$$x = A\cos 3t + B\sin 3t + \frac{2}{9}t^2 + \frac{4}{9}t + \frac{59}{81} \qquad (A-72)$$

Case 2 Particular Integrals when e^t Appears.

When the term e^t appears in the right–hand side of equation (A–64), then we first want to eliminate the e^t term using the Commuting Law and then follow the method of undetermined coefficients on the remaining polynomial.

EXAMPLE A.2.2
Solve for the particular integral of

$$(D-3)(D-2)\, x(t) = 4t^2 e^t \qquad (A-73)$$

First we cancel the e^t by assuming

$$x_p(t) = e^t z(t) \qquad (A-74)$$

Therefore, we have

$$(D-3)(D-2)\, e^t z(t) = e^t 4t^2 \qquad (A-75)$$

or using the Commuting Law gives

$$e^t(D-2)(D-1)\, z(t) = e^t 4t^2 \qquad (A-76)$$

or

$$(D-2)(D-1)\, z(t) = 4t^2 \qquad (A-77)$$

Now we assume

$$z(t) = a_0 t^2 + a_1 t + a_2 \qquad (A-78)$$

which makes equation (A-77)

$$(D^2 - 3D + 2)(a_0 t^2 + a_1 t + a_2) = 4t^2 \qquad (A-79)$$

Equating terms of equal powers of t gives

$$a_0 = 2$$

$$a_1 = 6$$

$$a_2 = 7$$

Therefore, the particular solution is

$$x_p = (7 + 6t + 2t^2)e^t \qquad (A-80)$$

and the general solution is

$$x = Ae^{3t} + Be^t + e^t \cdot (7 + 6t + 2t^2) \qquad (A-81)$$

REFERENCE

Sokolnikoff, I. S. and Redheffer, R. M., *Mathematics of Physics and Modern Engineering*, Chapter 2, McGraw–Hill, New York (1958).

Appendix B
MATLAB REFERENCE TABLES

MATLAB provides main categories of functions. Some of MATLAB's functions are built into the interpreter, while others take the form of M-files. The M-file functions, and in the case of the built-in functions, M-files containing only help text, are organized into directories, each containing the files associated with a category. The MATLAB command `help` displays an online table of these main categories.

Table B.1 – MATLAB's Main Categories of Functions	
color	Color control and lighting model functions.
datafun	Data analysis and Fourier transform functions.
demos	Demonstrations and samples.
elfun	Elementary math functions.
elmat	Elementary matrices and matrix manipulation.
funfun	Function functions – nonlinear numerical methods.
general	General purpose commands.
graphics	General purpose graphics functions.
iofun	Low-level file I/O functions.
lang	Language constructs and debugging.
matfun	Matrix functions – numerical linear algebra.
ops	Operators and special characters.
plotxy	Two dimensional graphics.
plotxyz	Three dimensional graphics.
polyfun	Polynomial and interpolation functions.
sparfun	Sparse matrix functions.
specfun	Specialized math functions.
specmat	Specialized matrices.
sounds	Sound processing functions.
strfun	Character string functions.

The `help` command will also display the toolboxes that are available to you.

The following tables give General Purpose Commands:

Table B.2 Managing Commands and Functions

demo	Run demos.
help	Online documentation.
info	Information about MATLAB and the MathWorks.
lookfor	Keyword search through the help entries.
path	Control MATLAB's search path.
type	List M-file.
what	Directory listing of M-, MAT- and MEX-files.
which	Locate functions and files.

Table B.3 Managing Variables and the Workspace

clear	Clear variables and functions from memory.
disp	Display matrix or text.
length	Length of vector.
load	Retrieve variables from disk.
pack	Consolidate workspace memory.
save	Save workspace variables to disk.
size	Size of matrix.
who	List current variables.
whos	List current variables, long form.

Table B.4 Working with Files and the Operating System

cd	Change current working directory.
delete	Delete file.
diary	Save text of MATLAB session.
dir	Directory listing.
getenv	Get environment value.
unix	Execute operating system command; return result.
!	Execute operating system command.

Table B.5 Controlling the Command Window

clc	Clear command window.
echo	Echo commands inside script files.
format	Set output format.
home	Send cursor home.
more	Control paged output in command window.

Table B.6 Starting and Quitting from MATLAB

matlabrc	Master startup M-file.
quit	Terminate MATLAB.
startup	M-file executed when MATLAB is invoked.

Appendix B: IMSL Routines

Specified Operators and Special Characters are given in the next two tables.

Table B.7	Operators and Special Characters
+	Plus.
−	Minus.
*	Matrix multiplication.
.*	Array multiplication.
^	Matrix power.
.^	Array power.
kron	Kronecker tensor product.
\	Backslash or left division.
/	Slash or right division.
./	Array division.
:	colon.
( )	Parentheses.
[]	Brackets.
.	Decimal point.
..	Parent directory.
...	Continuation.
,	Comma.
;	Semicolon.
%	Comment.
!	Exclamation point.
'	Transpose and quote.
.'	Nonconjugated transpose.
=	Assignment.
==	Equality.
<>	Relational operators.
&	Logical AND.
|	Logical OR.
~	Logical NOT.
xor	Logical EXCLUSIVE OR.

Table B.8	Logical Functions
all	True if all elements of vector are true.
any	True if any element of vector is true.
exist	Check if variables or functions exist.
find	Find indices of non-zero elements.
finite	True for finite elements.
isempty	True for empty matrix.
isieee	True for IEEE floating point arithmetic.
isinf	True for infinite elements.
ufbab	True for Not-A-Number.
issparse	True for sparse matrix.
isstr	True for text string.

The following are Language Constructs

Table B.9 MATLAB as a Programming Language

eval	Execute string with MATLAB expression.
feval	Execute function specified by string.
function	Add new function.
global	Define global variable.
nargchk	Validate number of input arguments.

Table B.10 Control Flow

break	Terminate execution of loop.
else	Used with `if`.
elseif	Used with `if`.
end	Terminate the scope of `for`, `while` and `if` statements.
error	Display message and abort function.
for	Repeat statements a specific number of times.
if	Conditionally execute statements.
return	Return to invoking function.
while	Repeat statements an indefinite number of times.

The following tables give Elementary Matrices and Matrix Manipulation

Table B.11 Elementary Matrices

eye	Identity matrix.
linspace	Linearly spaced vector.
logspace	Logarithmically spaced vector.
meshgrid	X and Y arrays for 3-D plots.
ones	Ones matrix.
rand	Uniformly distributed random numbers.
randn	Normally distributed random numbers.
zeros	Zeros matrix.
:	Regularly spaced vector.

Table B.12 Special Variables and Constants

ans	Most recent answer.
computer	Computer type.
eps	Floating point relative accuracy.
flops	Count of floating point operations.
i, j	Imaginary unit.
inf	Infinity.
NaN	Not-a-Number.
nargin	Number of function input arguments.
nargout	Number of function output arguments.
pi	3.1415926535897...
realmax	Largest floating point number.
realmin	Smallest floating point number.

Appendix B: IMSL Routines

Table B.13 Time and Dates

clock	Wall clock.
cputime	Elapsed CPU time.
date	Calendar.
etime	Elapsed time function.
tic, toc	Stopwatch timer functions.

Table B.14 Matrix Manipulation

diag	Create or extract diagonals.
fliplr	Flip matrix in the left/right direction.
flipud	Flip matrix in the up/down direction.
reshape	Change size.
rot90	Rotate matrix 90 degrees.
tril	Extract lower triangular part.
triu	Extract upper triangular part.
:	Index into matrix, rearrange matrix.

The next table gives Elementary Functions available in MATLAB

Table B.15 Elementary Math Functions

abs	Absolute value.
acos	Inverse cosine.
acosh	Inverse hyperbolic cosine.
angle	Phase angle.
asin	Inverse sine.
asinh	Inverse hyperbolic sine.
atan	Inverse tangent.
atan2	Four quadrant inverse tangent.
atanh	Inverse hyperbolic tangent.
ceil	Round towards plus infinity.
conj	Complex conjugate.
cos	Complex conjugate.
cosh	Hyperbolic cosine.
exp	Exponential.
fix	Round towards zero.
floor	Round towards minus infinity.
imag	Complex imaginary part.
log	Natural logarithm.
log10	Common logarithm.
real	Complex real part.
rem	Remainder after division.
round	Round towards nearest integer.
sign	Signum function.
sin	Sine.
sinh	Hyperbolic sine.
sqrt	Square root.
tan	Tangent.
tanh	Hyperbolic tangent.

Specialized Math Functions are given below.

Table B.16 Specialized Math Functions

bessel	Bessel function.
besselh	Hankel function.
beta	Beta function.
betainc	Incomplete beta function.
betaln	Logarithm of beta function.
ellipj	Jacobi elliptic functions.
ellipke	Complete elliptic integral.
erf	Complementary error function.
erfc	Complementary error function.
erfcx	Scaled complementary error function.
erfinv	Inverse error function.
gamma	Gamma function.
gammainc	Incomplete gamma function.
gammaln	Logarithm of gamma function.
log2	Dissect floating point numbers.
pow2	Scale floating point numbers.
rat	Rational approximation.
rats	Rational output.

Matrix Functions and Numerical Linear Algebra capabilities are given in the next four tables.

Table B.17 Matrix Analysis

cond	Matrix condition number.
det	Determinant.
norm	Matrix or vector norm.
null	Null space.
orth	Orthogonalization.
rcond	LINPACK reciprocal condition estimator.
rank	Number of linearly independent rows or columns.
rref	Reduced row echelon form.
trace	Sum of diagonal elements.

Table B.18 Linear Equations

chol	Cholesky factorization.
inv	Matrix inverse.
lscov	Least squares in the presence of known covariance.
lu	Factors from Gaussian elimination.
nnls	Non-negative least-squares.
pinv	Pseudoinverse.
qr	Orthongonal-triangular decomposition.
\ and /	Linear equation solution.

Table B.19 Eigenvalues and Singular Values

balance	Diagonal scaling to improve eigenvalue accuracy.
cdf2rdf	Complex diagonal form to real block diagonal form.
eig	Eigenvalues and eigenvectors.
hess	Hessenberg form.
poly	Characteristic polynomial.
qz	Generalized eigenvalues.
rsf2csf	Real block diagonal form to complex diagonal form.
schur	Schur decomposition.
svd	Singular value decomposition.

Table B.20 Matrix Functions

expm	Matrix exponential.
expm1	M-file implementation of expm.
expm2	Matrix exponential via Taylor series.
expm3	Matrix exponential via eigenvalues and eigenvectors.
funm	Evaluate general matrix function.
logm	Matrix logarithm.
sqrtm	Matrix square root.

Data Analysis tools available are given in the next four tables.

Table B.21 Basic Operations

cumprod	Cumulative product of elements.
cumsum	Cumulative sum of elements.
max	Largest component.
mean	Average or mean value.
median	Median value.
min	Smallest component.
prod	Product of elements.
sort	Sort in ascending order.
std	Standard deviation.
sum	Sum of elements.
trapz	Numerical integration using trapezoidal method.

Table B.22 Finite Differences

del2	Five-point discrete Laplacian.
diff	Difference function and approximate derivative.
gradient	Approximate gradient (see online help).

Table B.23 Correlation

corrcoef	Correlation coefficients.
cov	Covariance matrix.

Table B.24 Filtering and Convolution	
conv	convolution and polynomial multiplication.
conv2	Tow-dimensional convolution (see online help).
deconv	Deconvolution and polynomial division.
filter	One-dimensional digital filter (see online help).
filter2	Two-dimensional digital filter (see online help).

Polynomial and Interpolation Functions are given below.

Table B.25 Polynomials	
conv	Multiply polynomials.
deconv	Divide polynomials.
poly	Construct polynomial with specified roots.
polyder	Differentiate polynomial (see online help).
polyfit	Fit polynomial to data.
polyval	Evaluate polynomial.
polyvalm	Evaluate polynomial with matrix argument.
residue	Partial-fraction expansion (residues).
roots	Find polynomial roots.

Table B.26 Data Interpolation	
griddata	Data gridding.
interp1	1-D interpolation (1-D table lookup).
interp2	2-D interpolation (2-D table lookup).
interpft	2-D interpolation using FFT method.

Functions and Numerical Methods available are given in the next table.

Table B.27 Function Functions – Nonlinear Numerical Methods	
fmin	Minimize function of one variable.
fmins	Minimize function of several variables.
fplot	Plot function.
fzero	Find zero of function of one variable.
ode23	Solve differential equations, low order method.
ode45	Solve differential equations, high order method.
quad	Numerically evaluate integral, low order method.
quad8	Numerically evaluate integral, high order method.

Appendix B: IMSL Routines

Two Dimensional Graphic capabilities are given below.

Table B.28 Elementary X-Y Graphs	
fill	Draw filled 2-D polygons.
loglog	Log-log scale plot.
plot	Linear plot.
semilogx	Semi-log scale plot.
semilogy	Semi-log scale plot.

Table B.29 Specialized X-Y Graphs	
bar	Bar graph.
compass	Compass plot.
errorbar	Error bar plot.
feather	Feather plot.
fplot	Plot function.
hist	Histogram plot.
polar	Polar coordinate plot.
rose	Angle histogram plot.
stairs	Stairstep plot.

Table B.30 Graph Annotation	
grid	Grid lines.
gtext	Mouse placement of text.
text	Text annotation.
title	Graph title.
xlabel	X-axis label.
ylabel	Y-axis label.

Table B.31 Hardcopy and Storage	
orient	Set paper orientation.
print	Print graph or save graph to file.
printopt	Configure local printer defaults.

Index

absorption column, 54,111
activity coefficient, 229
adiabatic, 152
adjoints, 330
alcohol distillation, 59,112
annular chemical reactor, 320
Arrhenius plot, 198
Arrhenius temperature dependency, 188,204
artificial kidney, 39,297
atmospheric pollutant dispersion, 211
autocatalytic reactions, 182,209
average velocity, 14
axial dispersion, 350

backward difference operator, 142
backward integration, 308
backward pass, 51
band algorithm, 362
Band.m, 364
basic tank dynamics, 135
batch distillation, 229
batch fermentation, 198
Berea, 378
Bessel equation, 293
binary distillation, 231
binary distillation column, 171
bisec.m, 69
bisection, 68
black–oil simulator, 406
block–centered grid, 403
body force, 33
boiler dynamics, 218
boiling, 217
boil–off rate, 217
bubble columns, 350
bulk flow, 260

catalyzed fluidized beds, 164
centered–difference technique, 353
chlorination of benzene, 177
collocation, 409
commuting law, 441
component material balance, 259
complex roots, 442
compressibility, 402
compressibility factor, 228
cond, 65
condition number, 64
conservation of energy, 19,275
conservation of mass, 12,257
conservation of momentum, 32,264
conservation of species, 14,259
control volume, 12
convective heat transfer coefficient, 281
convective problems, 353
converging nozzle, 34
corner scheme, 399
countercurrent extraction, 111
countercurrent heat exchanger, 359
Courant number, 399
cp1.m, 67
cp2.m, 67
cp3.m, 78

Crank–Nicolson, 370
cstr.m, 86

Dalton's Law, 227
Danckwert's boundary conditions, 306
Darcy's Law, 291,397,402
decomposition hydrogen peroxide, 212
decomposition sulfuryl chloride, 41
design equation analysis, 87
design of a mixer–exchanger–mixer system, 101
diagonal matrix, 64
diffusion coefficient, 260
diffusion in spherical coordinates, 420
diffusional flux, 260
diffusive problems, 370
direct substitution, 73
Dirichet boundary condition, 418
dispersion, 260,266
distillation column, 255
distributive law, 440
drag force, 33
dtank32n.m, 140
dtank32o.m, 140
dynamic adsorption of surfactants, 425
dynamic programming, 322

effectiveness factor, 196,295
ejection molder, 38
enclosed tank dynamics, 152
energy balances with variable properties, 158
enthalpy, 21,22
entropy, 28
enzyme reaction, 210
eq51.m, 221

equal–molar overflow, 231
equation of continuity, 257
equation of motion, 264
equation ordering algorithm, 93
equilibrium still, 41
euclidian norm, 52
Euler algorithm, 125
evaporator, 112
ex310.m, 166
ex41.m, 182
ex41.m, 184
ex43.m, 188
ex43a.m, 188
ex432.m, 205
ex51.m, 222
ex56.m, 247
ex72.m, 314
ex74.m, 322
ex76.m, 330
ex77.m, 334
ex82.m, 356
ex83.m, 364
ex85.m, 374
exact differential, 436
exothermic CSTR, 85,209
expdata.dat, 175
explicit method, 356

Fanning friction factor, 31
fermentation reactor, 169
Fick's first law, 260
fictitious points, 387
fig2_28.matrix, 100
first order, 435
fixed bed catalytic reactor, 349
five–spot geometry, 301
flash distillation, 122
fluid catalytic cracking, 117
force vector, 32
foreign protein secretion dynamics, 211
formation volume factor, 401

Index

Fourier's Law, 280
forward pass, 51
fractional flow, 397
freezer–crystallizer, 174
friction factor, 274
friction head, 31
friction loss, 30
fsolve, 84
fugacity, 227
fun32.m, 137
fun34.m, 151
fun.f, 99, 105
fundamental property of adjoint system, 331
functional forms for design equations, 89
functionality matrix, 88,100
fzero, 69

Galerkin, 409
Gauss–Jordan, 49
Gaussian elimination, 48
generalized shooting technique, 311
general solutions, 442
Gibbs free energy, 29,227
glycerol, 215
gravity force, 33,265
gravity head, 30

half–interval, 68
head loss due to friction, 30
heat capacity, 21
heat conduction, 275
heat conduction equation, 408
heat conduction in a rod, 372
heat conduction in an insulated bar, 416
heat conduction with chemical heat source, 282
heat exchanger, 116
Helmholtz free energy, 29

help, 53
hemodialysis, 168
Heun's method, 130
hold–up, 233
homogeneous equations, 437
homogeneous solutions, 317,441
http://optimal.colorado.edu/~ramirez/chen4580.html, 56,99,105,132,151
hydraulic time constant, 233
hydrogen peroxide reactor, 189

ideal stage, 235
IMPES, 403
implicit method, 361
improved Euler method, 130
information–flow diagram, 16
integrating factor, 438
internal energy, 21
interpolating polynomial, 141
inv, 54
issparse, 66
iterative improvement, 52

jacketed vessel, 175
Jacobi polynomials, 418
Jacobian, 83,163
jet ejector, 36

Kiil artificial kidney, 423
kinetic energy, 20
Kravoskii's Theorem, 189

labmodel.m, 198
labreact.m, 198
Leibniz's formula, 159
Liapunov's Direct Method, 189
limitations of process simulation, 8
lin2_1.m, 56,57
lin2_2.m, 61
linear differential equations, 440
linear differential operator, 440

linear first-order equations, 438
linear inverse interpolation, 70
local stability, 214
lumped systems, 125
LU decomposition, 50,54,84
lua.m, 64

M-files, 53
macroscopic balances, 11
mass average velocity, 260
mass flow rate, 14
mass matrix, 164
MATLAB as a programming language, 448
MATLAB basic operations, 451
MATLAB control flow, 448
MATLAB controlling the command window, 446
MATLAB correlation, 451
MATLAB data interpolation, 452
MATLAB eigenvalues, 451
MATLAB elementary math functions, 449
MATLAB elementary matrices, 448
MATLAB filtering and convolution, 452
MATLAB finite differences, 451
MATLAB graphs, 453
MATLAB hardcopy and storage, 453
MATLAB linear equations, 450
MATLAB logical function, 447
MATLAB main categories, 445
MATLAB managing commands, 446
MATLAB managing variables, 446
MATLAB matrix analysis, 450
MATLAB matrix functions, 451
MATLAB matrix manipulation, 449
MATLAB nonlinear numerical methods, 452
MATLAB operators, 447
MATLAB polynomials, 452
MATLAB reference tables, 445
MATLAB software, 53
MATLAB special variables, 448
MATLAB specialized math functions, 450
MATLAB starting and quitting, 446
MATLAB time and dates, 449
MATLAB working with files, 446
mean value theorem, 126
mechanical energy balance, 27
mechanical work, 20
methanol synthesis, 112
Michaelis constant, 204
microscopic mechanical energy balance, 272
Milne Predictor Corrector Formula, 144
minimum difficulty strategy, 93
miscible flow in porous media, 378
mix1.matrix, 106
mixer.matrix, 105
model310.m, 166
model41.m, 182
model42.m, 184
model43.m, 188
model432.m, 205
model51.m, 221
model56.m, 239
model71.m, 308
model72.m, 314
model74.m, 322
model76.m, 330
model77.m, 334
model77a.m, 334

Index 459

molar average velocity, 260
Monod expression, 203
multicomponent boiling, 227
multicomponent distillation, 235
multicomponent feeds, 160
multiple roots, 441
multistep methods, 141

NAG, 45,53,333
natural form, 146
Newton Method, 80
Newtonian fluid, 266
Newton–Raphson, 83,313
Newton's second law, 265
Newton's Fundamental Formula, 142
Newton's law of cooling, 24,170
newtrap.m, 86
no–flow boundary conditions, 403
nonlinear algebraic equations, 67,82
no–slip, 272
norm, 64

ode23,132
ode45, 132
ode113, 132,151,156
ode15s, 132,163
ode13s, 132,163
ode45n.m, 136
odesuite.ps 163
one–dimensional heat conduction, 409
optimal solution strategy, 92
orange juice concentration, 39
order–of–magnitude analysis for chlorination of benzene, 179
order–of–magnitude scaling, 139,146
orthogonal collocation, 414
orthogonal matrix, 64

orthogonal polynomial, 414
overall heat transfer coefficient, 281
oxidation of CO, 215
oxidation of methane, 114

parabolic velocity profile, 272
parasitic differential equation, 173
particular solution, 318,444
partitioning equations, 46
Peclet number, 328
performance index, 407
phase plane, 189
phase transmissibility, 405
pinv, 65
pipe flow, 266
pipeline gas flow, 274
pivot element, 52
pivoting, 52
platform variables, 93
polyfit, 175
polyval, 175
pore volume, 399
porosity, 396
porous tubular reactor, 298
potential energy, 20
preheating of boiler, 256
pressure equation, 402
pressure force, 33,265
pressure head, 30
pseudo inverse, 65
pump efficiency, 30

qc.m, 189
quasilinearization, 322,382

Raoult's Law, 227
rank, 64
rate of work, 20
rate of work for viscous forces, 20
reactor stability, 189
reboiler, 234

regfals.m, 71
Regula Falsi, 70
relative permeability, 398
residual, 408
resultant force, 33
reversible heat, 28
reversible work, 28
Reynolds number, 32
Runge–Kutta algorithms, 129
Runge–Kutta–Gill algorithm, 131

saturation, 398
secant method, 118
second–order correct, 354
separation of variables, 435
sewage treatment, 209
shifted Legendre polynomials, 414
shooting system dynamics, 311
shooting techniques, 311
simultaneous solution algebraic equations, 47
singular value decomposition, 64
slash, 54
slurry grinder, 297
slurry reactor, 26,213
solubility coefficients, 401
solvent extraction, 113
sparse format, 364
sparse matrices, 66
specific reaction rate, 203
split boundary value problems, 305
stability, 128
stability analysis, 390
stability condition, 407
stable.m, 189
stagnant film diffusion, 263
state–variable form, 319
state–variable formulation, 374
steam heat exchanger, 355
steam power boiler, 221
stiff differential equations, 162

stirred tank modeling, 189
stirred tank with heating jacket, 160
strategy of process simulation, 6
structural analysis, 87
stress tensor, 265
styrene monomer tubular reactor, 283,300
substantial derivative, 259
superheaters, 334
superposition, 316
surface force, 33
surfactant adsorption, 300
surfactant flooding, 426
surge tank, 168
SVD, 64
symbolic toolbox, 214

tangent algorithm, 80
tangent.m, 81
tank32n.m, 137
tank32.m, 151
tank32o.m, 136
tank34.m, 151
tank35.m, 156
tanks with multicomponent feeds, 160
tapered tube geometry, 13
tearing equations, 47
temperature effects of CSTR, 184
thermal energy equation, 277
toolboxes, 53
total condenser, 235
total error propagation, 126
transmissibilities, 405
tray hydraulics, 233
trial function, 408
triangle inequality, 125
triethylamine, 122
truncation error, 127
tubular reactor, 313,382

Index 461

tubular reactor with dispersion, 306,327
turbine efficiency, 30
turbojet engine, 43
two–dimensional flow in porous media, 400
two–dimensional heat conduction, 412
two–phase flow in porous media, 392

upstream weighting, 406

valve equation, 145
vapor pressure, 227
variable group algorithm, 95
velocity head, 30
velocity vector, 32
viscous dissipation, 278
viscous work, 20
volume averaging, 262

wegstein.m, 77
Wegstein method, 75
weighted residuals, 408
weighting function, 408
well production rate, 407
work, 20
work per unit mass, 30
world wide web, 56